INTERNATIONAL SERIES OF MONOGRAPHS ON PHYSICS

## INTERNATIONAL SERIES OF MONOGRAPHS ON PHYSICS

144. T. R. Field: *Electromagnetic scattering from random media*
143. W. Götze: *Complex dynamics of glass-forming liquids – a mode-coupling theory*
142. V. M. Agranovich: *Excitations in organic solids*
141. W. T. Grandy: *Entropy and the time evolution of macroscopic systems*
140. M. Alcubierre: *Introduction to 3 + 1 numerical relativity*
139. A. L. Ivanov, S. G. Tikhodeev: *Problems of condensed matter physics – quantum coherence phenomena in electron–hole and coupled matter–light systems*
138. I. M. Vardavas, F. W. Taylor: *Radiation and climate*
137. A. F. Borghesani: *Ions and electrons in liquid helium*
136. C. Kiefer: *Quantum gravity*, Second edition
135. V. Fortov, I. Iakubov, A. Khrapak: *Physics of strongly coupled plasma*
134. G. Fredrickson: *The equilibrium theory of inhomogeneous polymers*
133. H. Suhl: *Relaxation processes in micromagnetics*
132. J. Terning: *Modern supersymmetry*
131. M. Mariño: *Chern–Simons theory, matrix models, and topological strings*
130. V. Gantmakher: *Electrons and disorder in solids*
129. W. Barford: *Electronic and optical properties of conjugated polymers*
128. R. E. Raab, O. L. de Lange: *Multipole theory in electromagnetism*
127. A. Larkin, A. Varlamov: *Theory of fluctuations in superconductors*
126. P. Goldbart, N. Goldenfeld, D. Sherrington: *Stealing the gold*
125. S. Atzeni, J. Meyer-ter-Vehn: *The physics of inertial fusion*
123. T. Fujimoto: *Plasma spectroscopy*
122. K. Fujikawa, H. Suzuki: *Path integrals and quantum anomalies*
121. T. Giamarchi: *Quantum physics in one dimension*
120. M. Warner, E. Terentjev: *Liquid crystal elastomers*
119. L. Jacak, P. Sitko, K. Wieczorek, A. Wojs: *Quantum Hall systems*
118. J. Wesson: *Tokamaks*, Third edition
117. G. E. Volovik: *The universe in a helium droplet*
116. L. Pitaevskii, S. Stringari: *Bose–Einstein condensation*
115. G. Dissertori, I. G. Knowles, M. Schmelling: *Quantum chromodynamics*
114. B. DeWitt: *The global approach to quantum field theory*
113. J. Zinn-Justin: *Quantum field theory and critical phenomena*, Fourth edition
112. R. M. Mazo: *Brownian motion – fluctuations, dynamics, and applications*
111. H. Nishimori: *Statistical physics of spin glasses and information processing – an introduction*
110. N. B. Kopnin: *Theory of nonequilibrium superconductivity*
109. A. Aharoni: *Introduction to the theory of ferromagnetism*, Second edition
108. R. Dobbs: *Helium three*
107. R. Wigmans: *Calorimetry*
106. J. Kübler: *Theory of itinerant electron magnetism*
105. Y. Kuramoto, Y. Kitaoka: *Dynamics of heavy electrons*
104. D. Bardin, G. Passarino: *The standard model in the making*
103. G. C. Branco, L. Lavoura, J. P. Silva: *CP violation*
102. T. C. Choy: *Effective medium theory*
101. H. Araki: *Mathematical theory of quantum fields*
100. L. M. Pismen: *Vortices in nonlinear fields*
99. L. Mestel: *Stellar magnetism*
98. K. H. Bennemann: *Nonlinear optics in metals*
96. M. Brambilla: *Kinetic theory of plasma waves*
94. S. Chikazumi: *Physics of ferromagnetism*
91. R. A. Bertlmann: *Anomalies in quantum field theory*
90. P. K. Gosh: *Ion traps*
88. S. L. Adler: *Quaternionic quantum mechanics and quantum fields*
87. P. S. Joshi: *Global aspects in gravitation and cosmology*
86. E. R. Pike, S. Sarkar: *The quantum theory of radiation*
83. P. G. de Gennes, J. Prost: *The physics of liquid crystals*
73. M. Doi, S. F. Edwards: *The theory of polymer dynamics*
69. S. Chandrasekhar: *The mathematical theory of black holes*
51. C. Møller: *The theory of relativity*
46. H. E. Stanley: *Introduction to phase transitions and critical phenomena*
32. A. Abragam: *Principles of nuclear magnetism*
27. P. A. M. Dirac: *Principles of quantum mechanics*
23. R. E. Peierls: *Quantum theory of solids*

# Mathematical Theory of Quantum Fields

HUZIHIRO ARAKI

*Department of Mathematics*
*Faculty of Science and Technology*
*The Science University of Tokyo*
*Japan*

Translated by
Ursula Carow-Watamura

*Department of Physics*
*Tohoku University*
*Japan*

Great Clarendon Street, Oxford OX2 6DP

Oxford University Press is a department of the University of Oxford.
It furthers the University's objective of excellence in research, scholarship,
and education by publishing worldwide in

Oxford New York

Auckland Cape Town Dar es Salaam Hong Kong Karachi
Kuala Lumpur Madrid Melbourne Mexico City Nairobi
New Delhi Shanghai Taipei Toronto

With offices in

Argentina Austria Brazil Chile Czech Republic France Greece
Guatemala Hungary Italy Japan South Korea Poland Portugal
Singapore Switzerland Thailand Turkey Ukraine Vietnam



Published in the United States
by Oxford University Press Inc., New York



Reprinted 2010



ISBN 978-0-19-956640-2

Printed in the United Kingdom by
Lightning Source UK Ltd., Milton Keynes

# PREFACE TO THE ENGLISH EDITION

The author is pleased to have this English edition of his Japanese book and would like to express his gratitude to the translator, Dr Ursula Carow-Watamura.

In addition to some minor corrections of the text, a number of references are newly added in this English edition in order to enable interested readers to find original papers or more involved formulations and/or detailed proofs of theorems, whenever they are omitted for the sake of simpler and easily readable presentation as well to keep the length of the book within the limit set by the original publisher.

These references are now collected at the end of the book under the heading 'General literature and references' (instead of appearing as footnotes as in the Japanese version). They are organized as follows.

The first group of literature (which under the heading 'General' appeared in the original Japanese edition) provides background material, and consists largely of textbooks.

The other references are either newly added or taken from the footnotes in the Japanese edition, and are grouped as ($n$) References for Chapter $n$ ($n = 1, \dots, 6$) and (7) Literature for appendices.

References, some with explanatory notes, are numbered in the order of citation in each chapter. Works in the 'General' section and in 'Literature for appendices' are not cited in the text.

References numbers will be placed between a pair of single square brackets except those for books (in contrast to journal articles and explanatory notes) which will be placed between pairs of double square brackets.

The present book presents a theory which is supposed to be valid for massive quantum field theories in $3+1$ space–time dimensions (although the fields do not appear in the theory, bounded local observables are playing a central role). From the late 1980s, interests in additional particle statistics which are possible only in lower space–time dimensions ($2+1$ and $1+1$), such as braid statistics, arose and a number of articles have been written. This topic is not discussed in the present book. However, some references (those that seem to be related to the formulation of local quantum theory in the text) are collected as Reference [8] for Chapter 6.

# PREFACE

Research on the mathematical foundation of physics, with quantum field theory as a central subject, has been developed for about 40 years. The physical meaning and its mathematical backup for basic concepts in the formulation of the theory have now been established to a fairly satisfactory extent, in a form where the technical assumptions and logical gaps have been reduced to a minimum. This book is intended to be an introduction to this subject.

As for the underlying physics, there were two profound conceptional revolutions in the twentieth century, namely quantum theory and the special theory of relativity. The physical meaning of the basic concepts of quantum theory will be explained in the general framework of a probabilistic description in comparison with the classical theory. The basic ingredients in this description are states and observables.

By incorporating the special theory of relativity into such a general framework of quantum mechanics we come to the theory of unitary representations of the group of motion of special relativity. This was formulated in the 1930s by Wigner, ahead of his time.

Another important principle which originates in the special theory of relativity is the so-called relativistic causality, which means that a physical effect cannot propagate faster than the speed of light. To formulate this principle it is necessary to specify a space–time domain in which the measurement of an observable is going to be performed. This naturally leads to the concept of local observables: namely, with each space–time domain we associate a set of observables which can be measured there. A further basic assumption we require throughout this book is the stability of the vacuum state, represented by the condition of positivity of the energy, which completes the list of basic assumptions.

We adopt the particle concept formulated as the smallest unit of relativistic quantum field theory. In our approach we analyse multiple particle scattering according to a model in which the asymptotic behaviour of local excitations, corresponding to such particles, is to approach motion in a straight line in the limit of infinite time, and they are measured by a large number of counters. In this way we can show the existence of 'in' states of incoming particles and 'out' states of outgoing particles, as well as the LSZ reduction formulae of the S-matrix.

In the formulation of this book, our basic objects are the observables, and observables localized in mutually space-like regions are assumed to commute. Thus those states such as one-particle states of a Fermion which are distinguished from the vacuum by a superselection rule do not appear in the vacuum representation, but they appear in a representation space disjoint with the vacuum representation. The key connecting the two representations is the idea of (transportable) localized excitations and this is formulated as a theory of localized endomorphisms. As a result we obtain the classification of representation spaces (called sectors) carrying localized excitations in terms of statistical parameters, a general understanding of para-Bose and para-Fermi statistics, the connection of spin and statistics for the case of finite mass multiplicity, and introduction of the lattice of operator algebras of fields connecting different sectors and satisfying normal commutation relations. In particular, it is a great success of this sector theory due to Doplicher, Haag, and Roberts that the anticommutation relations among the Fermi fields in two regions located space-like to each other are derived without being assumed.

The mathematical ingredients which give the basis for these physical contents are mainly Hilbert spaces and bounded linear operators on them, and the corresponding mathematical theory is the operator algebra theory which has recently been rapidly developed.

This book is written as an introduction, confining ourselves to the simplest case so that readers can grasp general ideas without consulting other books. Therefore, even if one can treat the general cases, we take up only the simplest case as an example to avoid complicated notation; as for the mathematical proofs, we take up only simple ones, avoiding complicated proofs or those requiring mathematical explanations of the theory of functions of many complex variables, theory of hyperfunctions, etc., if not related with the general flow of ideas, and if necessary, we explain only the results and give the references for the proofs. The operator algebras which are the main mathematical tool of this book are exceptions, but even there we also skip some topics.

Chapter 1 of this book presents the general framework of the probabilistic description and we are already able to perform a fair amount of analysis at this level. In Chapter 2 we describe the quantum theory based on this general framework. The main point in these two chapters is the concept of states and observables. In Appendix A we summarize those items of theory of Hilbert spaces which are necessary as mathematical tools. In Appendix B we also give a summary of operator algebra theory, but some of the fundamental topics about operator algebras are explained inside Chapter 2.

In Chapter 3 the special theory of relativity is discussed within the framework of quantum theory and is formulated as the theory of unitary representations of its group of motions, namely the inhomogeneous Lorentz group $\mathscr{P}_+^\uparrow$ and its universal covering group $\tilde{\mathscr{P}}_+^\uparrow$. Further, the Fock representation of free scalar fields is described. The cases for a general spin are discussed in Appendix C.

In Chapter 4, we introduce the system of operator algebras $\mathfrak{A}(D)$ ($D$ is a double cone domain) formed by the local observables and discuss its fundamental properties. We then discuss scattering theory in Chapter 5 and sector theory in Chapter 6.

The starting point for this book consists of the author's lectures given from 1961 until 1962 at the Swiss Federal Institute of Technology (ETH), Institute of Theoretical Physics in Zürich, Switzerland. Up to Chapter 5 we basically follow the lecture notes. However, at various points we include items based on later developments, such as the Geiger counter interpretation. Concerning detailed discussion about analyticity, we only give references. Chapter 6 mainly follows the series of papers by Doplicher, Haag, and Roberts.

Since the author is one of the persons involved in the development of the theory given in this book, there may be places where necessary explanation is omitted because it looks obvious to us. Therefore, comments from the reader will be much appreciated.

Tokyo
August 1992

Huzihiro Araki

# CONTENTS

# 1
# STATES AND OBSERVABLES

The states and the observables are basic concepts in the description of a physical system and their description has undergone a drastic fundamental change in the transition from the classical theory to the quantum theory.

In this chapter we present the general theory of states and observables in the framework of a general probabilistic description, which covers both the classical and quantum theories.

## 1.1 Probabilistic description

When we perform a physical measurement we usually distinguish the following four parts of the entire physical system of the laboratory:

1. the physical system being measured, i.e. the object on which the measurement is performed;
2. the measuring instruments which are the tools by which the measurement is performed;
3. the observer who performs the measurement;
4. the environment, besides the above three parts.

Usually, (3) and (4) are arranged in such a way that they do not affect the measuring process, and are not considered in the discussion of the result of the measurement. However, in discussions of a quantum mechanical measurement, (3) is often taken into account and also the influence of (4) is sometimes discussed. Here, we consider only (1) and (2).

In order to distinguish the measured objects we shall use in the following the Greek symbols $\alpha_1$, $\alpha_2$, etc. To obtain a meaningful measurement the measured object has to be prepared very carefully by an appropriate method into a certain definite initial condition. It should be understood that we distinguish by $\alpha_1$, $\alpha_2$, etc. the detailed prescriptions of this preparation procedure.

In a physical measurement, we measure certain properties of the measured object, and as a tool we are using certain measuring apparatus. Let us distinguish them by the symbols $Q_1$, $Q_2$, etc. In this case, also, the measuring apparatus has to be carefully prepared and it should be understood that we distinguish by $Q_1$, $Q_2$, etc. the detailed instructions of their preparation.

A physical measurement is performed by causing an interaction between the measured object and the measuring instrument, for example by putting them into contact with each other. After the measurement is finished, the measuring apparatus settles into a state among the various possible (finite number of) final states. For example, the needle of the meter is pointing to a certain number. These various possible results are distinguished by certain symbols or numbers denoted here as

$p, q$, etc. and one measurement is completed by recording the occurrence of a specific result among these various possibilities.

A physical measurement is a process which we usually repeat a large number of times. If the measured values always give the same result it means that we have obtained a definite value. However, there may be measurements which give a different result each time. Even in such a case we often observe that when repeating the measurement a large number of times, the measured values tend to show a definite distribution. Thus, if we measure the object $\alpha$ by the measuring apparatus $Q$ and the measured value $q$ is obtained $n_q$ times out of $N$ runs, then as $N$ becomes large, we may observe that the ratio

$$n_q/N \tag{1.1}$$

seems to approach a definite value for each $q$.

By assuming the existence of the limit of the ratio (1.1) for large $N$

$$w_\alpha^Q(q) = \lim_{N\to\infty} (n_q/N) \tag{1.2}$$

we can apply the probabilistic description for the measured object. Thus eq. (1.2) is interpreted as the probability that the physical quantity of the object $\alpha$ measured with the instrument $Q$ has the value $q$.

To make such a probabilistic interpretation possible, it is necessary that (1.2) satisfies the following mathematical properties:

$$0 \leq w_\alpha^Q(q) \leq 1, \tag{1.3}$$

$$\sum_q w_\alpha^Q(q) = 1. \tag{1.4}$$

The quantities given by (1.1) satisfy these conditions and hence if they have a definite limit, then the limit (1.2) also satisfies the same conditions.

In the following we shall develop the general theory of states and observables, considering only the case where the results of a physical measurement allow such a probabilistic interpretation. The case where the measured result has a definite value $p$ can be included in this general case, by taking the probability $w_\alpha^Q(q)$ to be 1 for the specific result $q = p$, and to be 0 for the remaining possible measured results $q \neq p$.

## 1.2 States and observables

Let us define the concept of states and observables within the above described framework.

Two measured objects $\alpha_1$ and $\alpha_2$ may be identical although they were prepared by completely different procedures. In such a case they cannot be distinguished by a measurement with any instrument, i.e. for any $Q$ and $q$ the following equation will be satisfied:

$$w^Q_{\alpha_1}(q) = w^Q_{\alpha_2}(q). \tag{1.5}$$

Conversely, if (1.5) holds for all $Q$ and $q$, then $\alpha_1$ and $\alpha_2$ cannot be distinguished by any physical measurement.

Representing the set of all measured objects $\alpha$ under consideration by the symbol $\Sigma$, let us define two objects $\alpha_1$ and $\alpha_2$ in $\Sigma$ to be equivalent if (1.5) holds for all $Q$ and $q$. This relation satisfies the three properties of an equivalence relation:

reflexive: $\alpha \sim \alpha$;
symmetric: if $\alpha_1 \sim \alpha_2$ then $\alpha_2 \sim \alpha_1$;
transitive: if $\alpha_1 \sim \alpha_2$ and $\alpha_2 \sim \alpha_3$ then $\alpha_1 \sim \alpha_3$.

Thus it defines equivalence classes in $\Sigma$. Each such equivalence class is called a *state*. In the following we represent a measured object and its equivalence class by the same letter $\alpha$, and we represent the set of all states again by $\Sigma$. It follows from the definition of the equivalence class that $w^Q_\alpha(q)$ is determined by the equivalence class of $\alpha$, i.e. it has the same value for all elements $\alpha$ in the same equivalence class. Therefore we can regard $w^Q_\alpha(q)$ as the probability determined by the state $\alpha$ which is defined as an equivalence class.

Next, let us discuss the measuring instruments. Taking into account the possibility that the quantities measured by two different instruments $Q_1$ and $Q_2$ are the same, we consider the following equivalence relation for the measuring instruments: If for all states $\alpha$ and measured results $q$

$$w^{Q_1}_\alpha(q) = w^{Q_2}_\alpha(q) \tag{1.6}$$

holds, then $Q_1$ and $Q_2$ are identified in the sense that they are instruments which measure the same quantity. This defines the equivalence relation among the measuring instruments.

In the following we denote by $\mathscr{A}$ the set of all measuring instruments. Among the measuring instruments in $\mathscr{A}$ we introduce the above equivalence relation and call this equivalence class an *observable*. Furthermore, we represent a measuring instrument and the observable of the corresponding equivalence class by the same letter $Q$, and the set of all observables is denoted by $\mathscr{A}$. Due to the definition (1.6) of an equivalence class, the probability $w^Q_\alpha(q)$ is determined by the equivalence class of $Q$, and therefore the probability $w^Q_\alpha(q)$ is a quantity determined by the state $\alpha$ and the observable $Q$, both of which are defined as equivalence classes.

## 1.3 Functions of observables

First let us make several remarks about the quantity $q$ which distinguishes the results of the measurement. To label the measured results we can in principle use elements of an arbitrary set. However, we usually use real numbers since they are convenient in the description of physical laws. For example, to distinguish locations in space it would be natural to use the geometrical points in the space. However, to describe physical laws we usually use coordinates relative to fixed coordinate axes. For example, if we consider only the $x$ coordinate, the object corresponding to $q$ is a real number. Therefore, in the following we assume that the label $q$ of a measured result is a real number. We call $q$ the *measured value*.

In each measurement we can only distinguish a finite number of values. We have neither the ability to distinguish an infinite number of results nor the ability to record them all. In this sense it is sufficient for the general theory to discuss only those observables which have a finite number of different observed values. Therefore, with this fact in mind we introduced the probability in eq. (1.4) by simply considering a discrete sum and not quantities like the probability density or integral.

Of course, when developing the theory, the introduction of observables with an infinite number of observed values or a continuous range of observed values may become natural, and when measuring a certain physical system we may realize the existence of such an observable behind the scene. We want to stress that we do not intend to exclude such observables with a continuous range of observed values. Rather we would like to include freely all such observables after we have sufficiently developed the theory. In this chapter, however, we want to discuss the relations between theory and phenomena within a range where we can still confirm the relations by experiments, and therefore we want to use for the basic ingredients of the theory only those observables which have a finite number of observed values as the *primitive observables*. The mathematical relations between such primitive observables and observables with an infinite discrete or continuous range of observed values will be briefly discussed in the last part of this section after we have explained the concept of simultaneous measurability.

Even when using the same instrument to perform a measurement, there is still a large freedom in assigning a real number to the result. There are various possibilities to distinguish two results: for example, we can use 0 and 1, or 1 and 2, or $\pm 1$. Except for the case where a specific labelling is required by the theory, any labelling is allowed and all labellings are equivalent. Now, choose and fix one labelling. Then an observable $Q$ is accordingly defined. A reassignment of the label may then be considered as a certain function.

Given a real valued function $f$ defined over the real numbers, the *function* $f(Q)$ of the observable $Q$ is defined as an observable with the same measuring apparatus as $Q$, the measured value of which is $f(q)$ if the measured value of $Q$ is $q$. In brief, the result of measurement is labelled by $f(q)$ instead of $q$. In terms of the probability, if for all states $\alpha$ and for all values $q'$

$$w_\alpha^{Q'}(q') = \sum_{q:\, f(q)=q'} w_\alpha^{Q}(q) \tag{1.7}$$

holds, then the observable $Q'$ is defined as a function $f(Q)$ of $Q$:

$$Q' = f(Q). \tag{1.8}$$

If the function $f$ gives a one-to-one correspondence, then the sum of eq. (1.7) has only one term with a unique value $q$ for a given $q'$, satisfying $f(q) = q'$, and this case corresponds to a change of the labelling explained above. If the function $f$ is not one-to-one in the range of observed values of $Q$, then we can say that $f(Q)$ is a coarser measurement than $Q$ itself, i.e. there are measured values which are distinguished by $Q$, and which are not distinguished by $f(Q)$.

A special case is the constant function $f(q) = c$. This means that for the observed value of any observable $Q, f(Q)$ is always fixed as $c$ and we do not need to measure it. Usually it is called a *trivial observable*. Especially when $c = 1$ we simply use the symbol $\mathbf{1}$ to represent it and for a general constant $c$ we often write the symbol $c\mathbf{1}$.

When a finite number of observables $Q_1, \ldots, Q_n$ are functions of a single observable $Q$:

$$Q_i = f_i(Q) \tag{1.9}$$

then $Q_1, \ldots, Q_n$ are said to be *simultaneously measurable*, i.e. by measuring $Q$ we know $Q_1, \ldots, Q_n$ at the same time.

As a typical example let us consider two space–time domains $D_1$ and $D_2$ which are space-like separated and two observables $Q_1$ and $Q_2$ which can be measured entirely in $D_1$ and $D_2$, respectively. Since it follows from the theory of relativity that no physical effect can propagate faster than the speed of light, the measurements of $Q_1$ and $Q_2$ cannot interfere with each other and thus can be performed simultaneously. By considering the two observed values together as a single measured result, and introducing in this sense an observable $Q$ which describes these two values simultaneously, the observables $Q_1$ and $Q_2$ initially considered both become functions of $Q$ and $Q_1$ and $Q_2$ are simultaneously measurable in the sense defined here.

When a finite set of observables $Q_1, \ldots, Q_n$ is a function of one observable, then as in eq. (1.9) the probability of $Q_1, \ldots, Q_n$ taking observed values $q_1, \ldots, q_n$ for a state $\alpha$ can be represented as

$$w_\alpha^Q \begin{pmatrix} Q_1, \ldots, Q_n \\ q_1, \ldots, q_n \end{pmatrix} = \sum_q \{ w_\alpha^Q(q); f_i(q) = q_i, \; i = 1, \ldots, n \}. \tag{1.10}$$

Here, the r.h.s. means that we sum $w_\alpha^Q(q)$ over all $q$ which satisfy $f_1(q) = q_1, \ldots, f_n(q) = q_n$.

Consider the set $\mathscr{C}$ of observables (not necessarily finite). If any finite subset of observables in $\mathscr{C}$ is simultaneously measurable, we call $\mathscr{C}$ a *simultaneously measurable system*: moreover, we call it a *full set of simultaneously measurable observables* if the following two conditions hold:

1. Any finite number of observables in $\mathscr{C}$ can be represented as a function of an observable $Q$ in $\mathscr{C}$ in the form of eq. (1.9).
2. Any function of any observable in $\mathscr{C}$ is in $\mathscr{C}$.

Then, for any finite number of observables in $\mathscr{C}$, we can consider their joint probability given by (1.10). However, if the joint probability depends on $Q$ the situation becomes harder to deal with. If the following condition is satisfied we call the set a *compatible system*:

3. If the observables $Q_1, \ldots, Q_n$ in $\mathscr{C}$ are functions of an observable $Q$ in $\mathscr{C}$ and also functions of another observable $Q'$ in $\mathscr{C}$, then the following equation holds for any state $\alpha$:

$$w_\alpha^Q \begin{pmatrix} Q_1, \ldots, Q_n \\ q_1, \ldots, q_n \end{pmatrix} = w_\alpha^{Q'} \begin{pmatrix} Q_1, \ldots, Q_n \\ q_1, \ldots, q_n \end{pmatrix}. \tag{1.11}$$

In this case the joint probability defined by eq. (1.10) does not depend on the observable $Q \in \mathscr{C}$.

Usually, as in the example described above, we specify a certain $Q$ by making a simultaneous measurement of $Q_1, \ldots, Q_n$, and by this specification the joint probability is defined. Hence, a natural choice of $Q$ is nearly decided. Still, if there are many equivalent measuring apparatuses for each $Q_i$, we cannot theoretically exclude the possibility that the joint probability changes according to the choice of the measuring apparatus for each $Q_i$.

As for the physical systems treated in this book, including classical systems as well as quantum systems, the joint probability in eq. (1.10) does not depend on $Q$. In the following we continue to analyse these cases.

In the full set $\mathscr{C}$ of simultaneously measurable, compatible observables, functional calculus can be performed freely as follows. Given the observables $Q_1, \ldots, Q_n$ and $Q$ in $\mathscr{C}$ satisfying the relation (1.9), we can define the function $f(Q_1, \ldots, Q_n)$ of the observables $Q_1, \ldots, Q_n$ for a real function $f(q_1, \ldots, q_n)$ in $n$ variables by

$$f(Q_1, \ldots, Q_n) = F(Q), \tag{1.12}$$

$$F(q) \equiv f(f_1(q), \ldots, f_n(q)), \tag{1.13}$$

as the element $F(Q)$ in $\mathscr{C}$ and the following standard relations hold:

$$g(f_1(Q_1, \ldots, Q_n), \ldots, f_k(Q_1, \ldots, Q_n)) = h(Q_1, \ldots, Q_n), \tag{1.14}$$

$$h(q_1, \ldots, q_n) \equiv g(y_1, \ldots, y_k), \qquad y_j \equiv f_j(q_1, \ldots, q_n), \quad j = 1, \ldots, k. \tag{1.15}$$

Note that algebraic calculus like summation and multiplication can be freely performed and the product is commutative.

Let us denote the set of all Borel measurable simple functions (i.e. functions which take only a finite number of values) by $\mathscr{F}$. The set of all simple functions in only one observable $Q$

$$\mathscr{C} = \{f(Q);\ f \in \mathscr{F}\} \tag{1.16}$$

is a full set of simultaneously measurable, compatible observables. Conversely, if there exists an observable $Q_f$ for each function $f$ in $\mathscr{F}$ satisfying the following two conditions:

1. $\mathscr{C} = \{Q_f : f \in \mathscr{F}\}$ is a full set of simultaneously measurable, compatible observables,
2. to each state $\alpha$, there corresponds a measure $\mu_\alpha$ over the real numbers satisfying

$$w_\alpha^{Q_f}(q) = \mu_\alpha(f^{-1}(q)) \tag{1.17}$$

for all $f$ in $\mathscr{F}$, where $f^{-1}(q)$ denotes the set of all real numbers $x$ satisfying $f(x) = q$,

then we may imagine that there exists an extended observable $Q$ (not necessarily an observable with a finite number of values) behind the scene, satisfying

$$Q_f = f(Q)$$

and especially $\mu_\alpha(B)$ can be interpreted as the probability that the measured value of the extended observable $Q$ with respect to the state $\alpha$ is in the Borel set $B$. In this way we can also consider observables with an infinite number of values as well as observables which have a continuous spectrum.

## 1.4 The expectation value of an observable

We define the expectation value $\alpha(Q)$ of an observable $Q$ in the state $\alpha$ by

$$\alpha(Q) = \sum_q q w_\alpha^Q(q). \tag{1.18}$$

Conversely, the probability $w_\alpha^Q(q)$ can be represented by the expectation value of a function of $Q$ as

$$w_\alpha^Q(q) = \alpha(\chi_B(Q)). \tag{1.19}$$

Here, $B$ can be any Borel set which contains $q$ and does not contain other measured values of the observable $Q$ (the number of measured values being assumed to be finite). For example, the set $B = \{q\}$ containing only one point $q$ will do. If we take the point of view that along with the observable $Q$ we always include functions of $Q$

in our observables as well, then the set of expectation values $\alpha(Q)$ for all observables $Q$ contains the equivalent information as contained in the set of probabilities $w_\alpha^Q(q)$. In particular, if

$$\alpha_1(Q) = \alpha_2(Q) \tag{1.20}$$

holds for all $Q$, then $\alpha_1 = \alpha_2$. Therefore we can proceed by regarding $\alpha(Q)$ as coordinates of $\alpha$ (we may call $\alpha(Q)$ the $Q$ coordinate, analogously to the terminology we use for the $x$ coordinate). Of course we may as well consider $w_\alpha^Q(q)$ as coordinates of $\alpha$ ($(Q, q)$ coordinates). However, we will see in the course of the considerations in Chapter 2 that it is more convenient and easier to use $\alpha(Q)$.

For the trivial observables it is clear that

$$\alpha(c\mathbf{1}) = c$$

holds for all $\alpha$. Among functions of observables, the equation

$$\alpha(f(Q)) = c\alpha(Q) + d, \qquad (f(Q) = cQ + d)$$

holds only for the linear function $f(q) = cq + d$. Furthermore, if the $Q_1, \ldots, Q_n$ are simultaneously observable, then for their linear function the following equation holds as well:

$$\alpha(c_1Q_1 + \cdots + c_nQ_n) = c_1\alpha(Q_1) + \cdots + c_n\alpha(Q_n). \tag{1.21}$$

These relations can be proved by a simple calculation using the definition of an expectation value (1.18) and the definition of a function of observables (1.17) and (1.12). In particular, for the proof of (1.21) the compatibility is not necessary.

## 1.5 Mixture and pure states

First let us use an example to explain the concept of a mixture of states. We are given a particle beam produced by an accelerator (we denote this state by $\alpha$), and let us assume that we have performed various measurements concerning its properties. As a result suppose that we find the following relation between the measured results for this beam and the known results about a proton beam $\alpha_1$ and a $\pi^+$ meson beam in a state $\alpha_2$:

$$w_\alpha^Q(q) = \lambda w_{\alpha_1}^Q(q) + (1 - \lambda) w_{\alpha_2}^Q(q), \tag{1.22}$$

for all $Q$ and $q$ which are measured. Here $\lambda$ is a real number between 0 and 1, independent of $Q$ and $q$. We will then judge that the beam in question is a mixture of a proton in the state $\alpha_1$ and a $\pi^+$ meson in the state $\alpha_2$, with the ratio $\lambda$ to $(1 - \lambda)$.

In general, if eq. (1.22) holds for all $Q$ and $q$, we say that the state $\alpha$ is a *mixture* (or mixture state) of the states $\alpha_1$ and $\alpha_2$, and write

$$\alpha = \lambda\alpha_1 + (1-\lambda)\alpha_2. \tag{1.23}$$

More generally, for states $\alpha, \alpha_1, \ldots, \alpha_n$, if

$$w_\alpha^Q(q) = \sum_{k=1}^{n} \lambda_k w_{\alpha_k}^Q(q) \qquad \left(\lambda_k \geq 0, \ \sum_{k=1}^{n} \lambda_k = 1\right) \tag{1.24}$$

holds for all observables $Q$ and all observed values $q$, $\alpha$ is said to be a mixture of the states $\alpha_1, \ldots, \alpha_n$, and we write

$$\alpha = \lambda_1\alpha_1 + \cdots + \lambda_n\alpha_n. \tag{1.25}$$

The general form (1.25) of a mixture can be reproduced by repeated use of the mixture (1.23) of two states. For example

$$\begin{aligned} \lambda_1\alpha_1 + \lambda_2\alpha_2 + \lambda_3\alpha_3 &= \lambda_1\alpha_1 + (1-\lambda_1)\beta, \\ \beta \equiv \lambda_2'\alpha_2 + (1-\lambda_2')\alpha_3, \quad \lambda_2' &\equiv (1-\lambda_1)^{-1}\lambda_2, \end{aligned} \tag{1.26}$$

where $(1-\lambda_1)(1-\lambda_2') = \lambda_3$ follows from the condition $\lambda_1 + \lambda_2 + \lambda_3 = 1$.

The states $\alpha_1$ and $\alpha_2$ can be concretely given by specifying the preparation procedure of each state. Let us first choose 1 or 2 with probability $\lambda$ and $(1-\lambda)$, respectively, by using an appropriate method such as throwing dice. Then, according to this choice we prepare $\alpha_1$ or $\alpha_2$. Preparing a state in this way, we will be able to produce exactly the mixture given by eq. (1.23). In this sense we assume that the set $\Sigma$ of all states is closed under the operation (1.23) of taking a mixture. Thus, we assume that for any two elements $\alpha_1$ and $\alpha_2$ in $\Sigma$ and for any real number of $\lambda$, $0 \leq \lambda \leq 1$, a state $\alpha$ satisfying (1.23) is in the set $\Sigma$.

Naturally it can be proved in the same way as (1.26) that if $\alpha_1, \ldots, \alpha_n$ are in $\Sigma$ then the state $\alpha$ of eq. (1.25) is in $\Sigma$.

Next, let us take the point of view that (1.23) and (1.25) are decompositions of a state $\alpha$. This means that we consider the decomposition of a mixed object into purer elements. We expect that ultimately we will arrive at objects which are no longer decomposable, and we call these indecomposable objects pure states. Of course, for any state $\alpha$, eq. (1.23) holds trivially in the following cases:

1. $\lambda = 1$ (necessarily $\alpha_1 = \alpha$);
2. $\lambda = 0$ (necessarily $\alpha_2 = \alpha$);
3. $\alpha_1 = \alpha_2$ ($\lambda$ arbitrary, necessarily $\alpha = \alpha_1 = \alpha_2$).

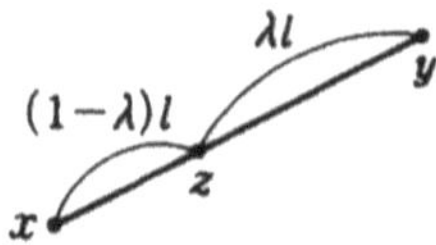


FIG 1.1 A segment connecting two points.

In these three cases, it is called a *trivial mixture*. If a state $\alpha$ cannot be written as a mixture (1.23) except for trivial mixtures, then we say that the state $\alpha$ is a *pure state*. In other words, if $\alpha$ is not a mixture of the states $\alpha_1$ and $\alpha_2$ both different from $\alpha$, then $\alpha$ is a pure state.

Next let us explain the geometric interpretation of the concept of a mixture.

In a linear space $L$, we can write any point $z$ lying on the straight line segment joining two points $x, y \in L$, as

$$z = \lambda x + (1 - \lambda)y \qquad (0 \leq \lambda \leq 1). \tag{1.27}$$

Taking the length of the segment $xy$ to be $l$, then the length of $xz$ is $(1 - \lambda)l$ and the length of $zy$ is $\lambda l$, their ratio being $(1 - \lambda) : \lambda$ (Fig. 1.1). Representing points $x$ of $L$ by their coordinates $(x_i)$, we can write

$$z_j = \lambda x_j + (1 - \lambda)y_j. \tag{1.28}$$

If we consider $w_\alpha^Q(q)$ as the coordinates of $\alpha$, and regard the pair $Q, q$ as the suffix $j$ which distinguishes the coordinate axes, eqs. (1.22) and (1.28) coincide after identification of $z$, $x$, and $y$ with $\alpha$, $\alpha_1$, and $\alpha_2$, respectively. If for a subset $S$ of $L$ the segment between any two points in $S$ is again in $S$, then $S$ is called a *convex set*. Since for the set of all states $\Sigma$ we have made the assumption that any mixture of states in $\Sigma$ is again in $\Sigma$, we can consider $\Sigma$ to be a convex set.

Given points $x^1, \ldots, x^n$ of a linear set $E$, and a set of real positive numbers the sum of which is one,

$$\lambda_1 \geq 0, \ldots, \lambda_n \geq 0, \qquad \lambda_1 + \cdots + \lambda_n = 1,$$

the linear combination

$$x = \lambda_1 x^1 + \cdots + \lambda_n x^n \tag{1.29}$$

is called a convex combination. The $\alpha$ of eq. (1.25) is a convex combination of $\alpha_1, \ldots, \alpha_n$. For the case $n = 2$ this convex combination is given by eq. (1.27). As we remarked for mixtures, the convex combinations (1.29) can be constructed by repeating the formation of the convex combination (1.27) of two points. Therefore, the convex combination of points $x^1, \ldots, x^n$ in a convex set $S$ again belongs to $S$. For given points $x^1, \ldots, x^n$, the full set of their convex combinations given by all choices of the real positive numbers $\lambda_1, \ldots, \lambda_n$ satisfying $\sum \lambda_i = 1$ becomes a polyhedron, the vertices of which are given by some or all of the points $x^1, \ldots, x^n$.

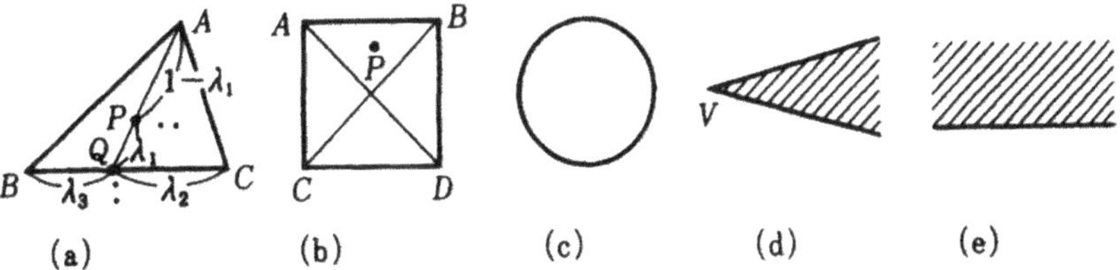


FIG 1.2 Convex sets and extremal points: (a) the triangle; (b) the square; (c) the disc; (d) the wedge; (e) the half-plane.

(If $x^k$ is a convex combination of other points $x^j$, then $x^k$ itself does not become a vertex.)

Expressing (1.29) in terms of components, we obtain

$$x_j = \lambda_1(x^1)_j + \cdots + \lambda_n(x^n)_j. \tag{1.30}$$

Equation (1.24) is of exactly the same form.

If a point $x$ in a convex set $S$ satisfies the following condition, $x$ is called an *extremal point* of $S$: If any segment $yz$ in $S$ contains $x$ then either $y = x$ or $z = x$. The extremal points of the set of all states $\Sigma$ are then exactly the pure states.

In Fig. 1.2 we have given several examples of convex sets and their extremal points. In each case the extremal points lie on the boundaries of the figure. Thus, when we consider the open sets where the boundaries are removed from the figures, there are no extremal points. In (d), the wedge extending to infinity on a plane has only one extremal point at the vertex; in the case of the upper half plane given in (e) there are no extremal points. (If the angle of a wedge exceeds 180° it is no longer a convex set.) For the disc given in (c) all points on the circle are extremal points, and in the cases of the polygons given in (a) and (b) the vertices are the extremal points.

There is a theorem which guarantees sufficiently many extremal points.

**The Krein–Mil'man extremal point theorem** [1]
Any compact convex set in a locally convex space coincides with the closed convex hull of its extremal points.

Here, the convex hull of the extremal points is the set of all convex combinations of the extremal points, and the closed convex hull is the closure of the convex hull, i.e. all limit points are added.

In the examples of Fig. 1.2, the necessary and sufficient condition for compactness is that the set is closed (including all boundary points) and bounded. Therefore (a), (b), and (c) are compact.

For the triangle $ABC$ of Fig. 1.2(a), any point $P$ in this triangle can be uniquely represented as a convex combination of the vertices $A$, $B$, $C$: let the intersection point of the extension of $AP$ with $BC$ be $Q$. Then, by $QP : PA = \lambda_1 : (1 - \lambda_1)$ we can fix $\lambda_1$, and by $BQ : QC = \lambda_3 : \lambda_2$ and $\lambda_2 + \lambda_3 = 1 - \lambda_1$ we can fix $\lambda_2$ and $\lambda_3$.

Contrary to this, for the square $ABCD$ in Fig. 1.2(b) the point $P$ is contained in the triangle $ABC$ as well as in the triangle $ABD$. So we can write it as a convex combination of the three points $A, B, C$ and also as that of the three points $A, B, D$. Thus the representation as a convex combination of extremal points is not unique. Similarly, it is not unique for the case of the disc in (c).

If the convex combination representation by extremal points is unique, the convex set is called a *simplex*.

In the case of infinite dimensions, convex combinations alone are not sufficient, but it becomes necessary to take a certain limit. In a comparatively "good" case we can construct an integral representation by using a measure on the set of extremal points.

The expectation value for the convex combination (1.25) of states can be given by the following equation:

$$\alpha(Q) = \lambda_1\alpha_1(Q) + \cdots + \lambda_n\alpha_n(Q). \tag{1.31}$$

Conversely, since we can obtain eq. (1.25) from eq. (1.31), by adopting the point of view that a function of observables is again an observable, we can as well take eq. (1.31) as the definition of a mixture. In this case we consider $\alpha(Q)$ as the coordinate of the state $\alpha$, where $Q$ plays the role of the suffix $j$.

We want to give some remarks about the relation to the description of classical physics for the case that there are sufficiently many pure states.

An observable $Q$ is called a *classical observable* if it always takes a definite value in any pure state $\alpha$, i.e. if it satisfies

$$w_\alpha^Q(q) = \begin{cases} 0 & (q \neq q_\alpha) \\ 1 & (q = q_\alpha) \end{cases} \qquad (q_\alpha = \alpha(Q)). \tag{1.32}$$

As far as classical observables are concerned, the responsibility for the situation that a measured value is not definite lies with the state. If we make a measurement in a pure state, we obtain a definite value, while in a mixture we obtain the value for each pure state with a frequency proportional to its coefficient in the mixture.

In particular, if all observables are classical observables, we have a *probabilistic* (statistical) *description of classical physics*. Namely, pure states give all observables definite values and their convex combination, being the probabilistic mixture of pure states, gives a statistical distribution of measured values of observables. For example, the states in classical statistical mechanics can be understood as a limit of such mixture states, and under an appropriate condition they can be described by probability measures on the set of pure states. In such a situation the set of all pure states is called *phase space*, the observables are described by functions on the phase space and a general state is described by a probability measure on the phase space. The expectation value can be described by an integral over the phase space $\Omega = \{\omega\}$

$$\alpha(Q) = \int Q(\omega)\,\mathrm{d}\alpha(\omega), \tag{1.33}$$

where the function $Q(\omega)$ represents the observable $Q$ and the probability measure $\mathrm{d}\alpha(\omega)$ represents the state $\alpha$. In particular, for a pure state $\omega_0$ the measure $\mathrm{d}\omega_0(\omega)$ is a point measure with support at the points $\omega_0$.

## 1.6 The physical topology of the states

Physical measurements are accompanied by errors. In the probabilistic description, for example, suppose we measure the value $q$ of the observable $Q$, and among $N$ runs ($N$ being a finite number) the value $q$ is obtained $n_q$ times, then the measured value of the probability $w_\alpha^Q(q)$ is given by $n_q/N$. However, $N$ is always finite. Besides this kind of statistical error, we can also think of various errors caused by the measurement itself. Taking also the fact into account that we can measure only a finite number of observables, the information about a state $\alpha$ obtained by measurement may be represented, after some simplifications, in the form

$$|w_\alpha^{Q_j}(q_j) - w_j| < \varepsilon_j \qquad (j = 1, \ldots, N). \tag{1.34}$$

In this equation $w_j$ is the value of the probability obtained by the measurement, $\varepsilon_j$ represents the error, and the number of the pairs $Q_j, q_j$ is finite ($N < \infty$).

Defining the system of neighbourhoods by the collection of all sets of states $\alpha$ satisfying (1.34) with $N < \infty$, $Q_j, q_j, w_j, \varepsilon_j > 0$ varying over all possibilities, we introduce a topology into the set $\Sigma$ of all states. We call this topology the *physical topology* of $\Sigma$.

Suppose that a point $x$ in a linear space $L$ is described by a set $\{x_j\}_{j\in J}$ of real coordinates $x_j$ labelled by the suffix $j$ belonging to a certain set $J$, and suppose all sets $\{x_j\}$ are allowed. Adopting as the system of neighbourhoods of a point $x \in L$ a family of subsets

$$\{y \in L;\ |y_{j_k} - x_{j_k}| < \varepsilon_k \quad (k = 1, \ldots, n)\} \tag{1.35}$$

indexed by a natural number $n$, $n$-tuple of suffices $j_1, \ldots, j_n$ and $n$ positive numbers $\varepsilon_1, \ldots, \varepsilon_n$, we obtain a linear topological space $L$ called the *topological product* of the real numbers $\mathbf{R}$, with its topology called the *direct product topology*.

In particular, the subset of $L$,

$$L_{(1)} = \{x \in L;\ 0 \le x_j \le 1 \text{ for all } j \in J\},$$

is the topological product of the interval $[0, 1]$, and is known to be compact in the direct product topology (*Tikhonov's product theorem* [2]).

If we take as $J$ the set of all pairs of observables $Q$ and measured values $q$, then the set $\Sigma$ of all states can be considered as a subset of $L_{(1)}$. As we explained, we can use $w_\alpha^Q(q)$ as coordinates of the state $\alpha$. In this case the physical topology defined above coincides with the direct product topology of $L$ restricted to $\Sigma$.

Now let us consider the closure $\bar{\Sigma}$ of $\Sigma$ in $L$. We can also consider $\bar{\Sigma}$ as the completion of $\Sigma$ in the physical topology. A point $\alpha$ of $\bar{\Sigma}$ is the limit point of a net $\alpha_\nu$

in $\bar{\Sigma}$. For any observable $Q$ and its (finite number of) measured values $q_1, \ldots, q_n$

$$w_\beta^Q(q_j) \geq 0 \quad (j = 1, \ldots, n), \qquad \sum_{j=1}^{n} w_\beta^Q(q_j) = 1, \tag{1.36}$$

$$w_\beta^Q(q) = 0 \quad (q \neq q_1, \ldots, q_n), \tag{1.37}$$

hold for each $\beta = \alpha_\nu \in \Sigma$, and hence also for their limit $\alpha$ in the direct product topology. (See (1.3) and (1.4).) Also for a function $f(Q)$

$$w_\beta^{f(Q)}(q') = \sum_{j: f(q_j) = q'} w_\beta^Q(q_j) \tag{1.38}$$

holds for each $\beta = \alpha_\nu \in \Sigma$, and since the r.h.s. is a sum of a finite number of coordinates, it also holds for their limit $\alpha$ in the direct product topology. Therefore, points in $\bar{\Sigma}$ fulfil the various relations which we introduced for states.

Since $\Sigma$ is a convex set in $L$ its closure $\bar{\Sigma}$ is also a convex set. Moreover, since $\bar{\Sigma}$ is a closed subset of the compact set $L_{(1)}$, it is itself compact. Since $\bar{\Sigma}$ satisfies the condition of the Krein–Mil'man theorem cited in the previous section, there are sufficiently many extremal points of $\bar{\Sigma}$ and their mixtures are dense in $\bar{\Sigma}$.

The concept of states has been introduced in order to describe the results of a measurement of a physical system. The elements of $\bar{\Sigma}$ have been additionally introduced for convenience as described above. For example, the extremal points of $\bar{\Sigma}$ can be considered as pure states in an idealized description. Even if a pure state is an extremal point of the set $\Sigma$, remember, for instance, that we can prepare a pure state only approximately due to effects like the environmental disturbance listed in point (4) of Section 1.1. So it is not unnatural to introduce them as idealized objects into our description. The discussion of the next section can be viewed as a further justification of this point.

In the description using the expectation value, the physical topology of the set $\Sigma$ of all states is defined by taking as the neighbourhood system of $\alpha$ the family of all sets

$$\{\beta \in \Sigma, |\beta(Q_j) - \alpha(Q_j)| < \varepsilon_j \qquad (j = 1, \ldots, n)\} \tag{1.39}$$

with all possible values for a natural number $n$, observables $Q_1, \ldots, Q_n$ and positive numbers $\varepsilon_1, \ldots, \varepsilon_n$. Under our standing assumption in this chapter that the number of measured values of any observable is finite, and any function of observables is again an observable, the topology introduced in this way is equivalent to the topology introduced above in terms of $w_\alpha^Q(q)$.

### 1.7 Equivalent theories

A physical theory starts from a description of the information about a physical system obtained by measurements. While the aim is to understand the laws governing mutual relations among various measured results, the description basically provides the foundation for them.

In the general theory described in the preceding sections the information obtained from measurements on a state of a physical system is described by a point in $\Sigma$ which is a subset of $L_{(1)}$. If we choose a different set for $\Sigma$, we will obtain (the starting point of) a different theory.

As long as we are able to describe the information obtained from physical measurements, different choices of $\Sigma$ will be equally qualified. A measurement about a state $\alpha$ provides information as given in (1.34). It means that $\alpha$ lies in a certain neighbourhood of the physical topology.

Given two subsets $\Sigma_1$ and $\Sigma_2$ of $L_{(1)}$, suppose that in any neighbourhood of any point in $\Sigma_1$ there is always a point of $\Sigma_2$. This means that if there is a point $P_\alpha^1$ in $\Sigma_1$ representing a measured result about a state $\alpha$ (namely satisfying (1.34)), then we can always find a point $P_\alpha^2$ of $\Sigma_2$ in that sufficiently small neighbourhood which satisfies (1.34) and hence represents the same measured result, because all points in a sufficiently small neighbourhood of $P_\alpha^1$ satisfy (1.34). Therefore, a state which we can describe by a point in $\Sigma_1$, can also be described by a point in $\Sigma_2$. Conversely, if we require that this property holds under all circumstances, then in any neighbourhood of any point in $\Sigma_1$ there always exists a point of $\Sigma_2$. (This can be immediately concluded from the form of (1.34). We may rephrase eq. (1.34) as saying that we measure neighbourhoods (in the sense of the physical topology).) If the property under discussion holds, we say that $\Sigma_2$ *physically contains* $\Sigma_1$. The necessary and sufficient condition for this is

$$\bar{\Sigma}_1 \subset \bar{\Sigma}_2. \tag{1.40}$$

Further, for the case that $\Sigma_2$ physically contains $\Sigma_1$, and $\Sigma_1$ physically contains $\Sigma_2$, we say that $\Sigma_1$ and $\Sigma_2$ are *physically equivalent*. This means that measured results which can be described by $\Sigma_1$ and those which can be described by $\Sigma_2$ are identical. The necessary and sufficient condition for this is

$$\bar{\Sigma}_1 = \bar{\Sigma}_2, \tag{1.41}$$

i.e. the closures in $L$ coincide.

Physically equivalent sets cannot be distinguished by any kind of experiment. Therefore, we can choose among them the one which is convenient from the theoretical point of view.

Since $\Sigma$ and its closure $\bar{\Sigma}$ are equivalent from this point of view, we can use $\bar{\Sigma}$, which is obtained by adding the idealized states, instead of $\Sigma$. In practice we may also make use of an appropriate set between $\Sigma$ and $\bar{\Sigma}$.

When describing a physical result there is some arbitrariness in using points of $\Sigma$ or $\bar{\Sigma}$. This choice can also be made according to theoretical convenience. Let us give a physical example. We consider an equilibrium state of a gas filling an infinite space. Let $\Sigma$ be the set of states of a finite number of particles (we may choose this number as large as we like). In fact, actual observed systems in our measurement, strictly speaking, consist of a finite number of particles. (The influence of the boundary walls is usually neglected.) However, for a measurement of finitely many observables where the measured values are obtained within a finite time period, accompanied by a non-zero error, we can also use the states of a gas homogeneously distributed in a finite but sufficiently large space, which cannot be distinguished from the equilibrium states of a gas homogeneously distributed in the whole space. The gas near the boundary soon starts to disperse to the outside, but if the boundary is far enough, it takes a long time until its effect on the measured area exceeds the range of the error, and we can finish the measurement without disturbance.

Let us consider a similar example from elementary particle theory. We assume that the charge conservation law holds, i.e. in a physical process the total charge is invariant. Starting from a state where a certain number of elementary particles are distributed in the vacuum, we can confine our considerations to the states of fixed total charge due to the charge conservation law. (Even in the case of an infinite number of particles, the same situation prevails automatically by considering only charge conserving observables.) Then the question is whether it is possible under this circumstance to describe a state with an extra particle with unit charge added. By the approximation using the physical topology this is possible in the following way. Assume that we measure a finite number of observables in a bounded space–time region, and let us describe the measured result approximately. We add a particle with charge 1 in the space–time region where we are making the measurement, and we add another particle with charge $-1$ in a very remote place, for example we put it behind the moon. Then the total sum of charges is unchanged and the corresponding state can be found among the states allowed by the initial condition. However, since the effect of the far remote particle is within the range of the error, we can approximately describe the aimed measured result without changing the total charge. If the moon is considered to be too near, we may also put the second particle beyond the galaxy. This kind of argument was favoured by Haag and is called the *behind-the-moon argument*.

## 1.8 Symmetries

For the probabilistic description given in Section 1.1, it is necessary to determine the probabilistic distribution of the observed values, by repeating a measurement a large number of times. However, if we take the space–time coordinates into consideration, we cannot really repeat exactly the same measurement. Due to the shift of the space–time coordinates, by passage of time and also by the displacement of the spatial coordinates (for example caused by the rotation of the Earth) we are obliged to repeat measurements which we believe to be equivalent except for a shift of space–time location and up to a certain range of errors. The equivalence in this case will be in the sense of the symmetry described in the following.

Let us suppose that we have specified a transformation of states and observables of a physical system. Concretely, this is a modification in the instruction of the preparation procedure for the measured object as well as for the measuring apparatus; for example, we may shift the whole time schedule by one day. Let us denote this transformation at the level of measured objects as well as at the level of measuring apparatus (i.e. the level before considering the equivalence class given in Section 1.2) by a letter $s$ as

$$\alpha \in \Sigma \to s\alpha \in \Sigma, \quad Q \in \mathscr{A} \to sQ \in \mathscr{A}. \tag{1.42}$$

There are two assumptions for $s$:

1. $s$ is a *bijection* on $\Sigma$ as well as on $\mathscr{A}$ (i.e. a one-to-one correspondence onto).
2. For all measured objects $\alpha$, measuring instruments $Q$ and observed values $q$, the following equation holds:

$$w_{\alpha}^{Q}(q) = w_{s\alpha}^{sQ}(q). \tag{1.43}$$

Due to eq. (1.43) in condition (2) the transformation $s$ can be considered as a map between equivalence classes introduced in Section 1.2. Therefore it defines a bijection on the set $\Sigma$ of all states as well as on the set $\mathscr{A}$ of all observables, and it satisfies (1.43) also for all states $\alpha$ and all observables $Q$, i.e. it conserves the probability. As a result, for any function $f$,

$$w_{\beta}^{sf(Q)}(q') = w_{\beta}^{f(sQ)}(q') \tag{1.44}$$

holds for all states $\beta$ (where $\beta = s\alpha$ for $\alpha = s^{-1}\beta$) and measured values $q'$. Therefore,

$$sf(Q) = f(sQ) \tag{1.45}$$

also holds. For the expectation value, the following equation is satisfied:

$$s\alpha(sQ) = \alpha(Q). \tag{1.46}$$

Conversely, if (1.45) and (1.46) hold for all states $\alpha$, observables $Q$ and functions $f$, then (1.43) is satisfied.

We call such a transformation $s$ a *symmetry in the active sense*. For example a homogeneous shift of time is considered as such a symmetry in many cases, and we can make use of it to implement repeated measurements which are needed in Section 1.2. In this case, the convergence of the probability distribution to a definite limit when the number of trials is increased can be taken as a proof of the assumed symmetry. (Conversely, if the probability distribution does not converge to a definite limit we must suspect that for some reason equivalence is broken.)

In a physical theory we sometimes consider another type of symmetry called a *symmetry in the passive sense*. For example, when we shift the origin of the time, then we assign to the same measured objects a different real number as a time.

The invariance of the formalism of the theory with respect to such a kind of variation in its description is a symmetry in the passive sense. This example of the time is intimately related with the symmetry in the active sense. However, gauge invariance for example is about a change of the description which cannot be observed, and there is no corresponding symmetry in the active sense.

2

# QUANTUM THEORY

In this chapter we explain how we can describe quantum theory within the general framework given in Chapter 1. We do not discuss various theories concerning the question of how quantum theory can (or cannot) be deduced from experimental facts or from a verifiable axiomatic system. Instead we start by introducing the assumptions of quantum theory. The main content is the algebraic viewpoint of quantum theory based on the theory of operator algebras, the GNS construction which supplies the fundamental relation between the description of the states by Hilbert space vectors and the description of the states in this general framework, the Wigner theorem which is basic for the mathematical description of quantum mechanical symmetries, and its generalizations. The basic definitions and theorems about Hilbert spaces and operator algebras are collected in the appendix.

## 2.1 The quantum mechanical description

The usual description of quantum mechanics uses a Hilbert space $\mathcal{H}$ and linear operators defined on it. (For Hilbert spaces and linear operators we refer to Appendix A.) An observable $Q$ is represented by a linear self-adjoint operator $Q_{\text{op}}$. To a function of an operator $f(Q)$ we assign

$$f(Q)_{\text{op}} = f(Q_{\text{op}}), \tag{2.1}$$

where the r.h.s. is a function of a self-adjoint linear operator $Q_{\text{op}}$ (See Appendix A, Section A.7). For the moment we consider a state $\alpha$ which can be represented as an expectation value with respect to a unit vector $\Psi_\alpha$ in $\mathcal{H}$

$$\alpha(Q) = (\Psi_\alpha, Q_{\text{op}}\Psi_\alpha). \tag{2.2}$$

(A unit vector means a vector $\Psi_\alpha$ satisfying $\|\Psi_\alpha\| = 1$.) Such a state is also called a vector state.

When measuring an observable $Q$, the probability that the measured value is an element in a Borel set $B$ is given by

$$w_\alpha^Q(B) = (\Psi_\alpha, E_Q(B)\Psi_\alpha). \tag{2.3}$$

$E_Q(B)$ is the spectral projection operator (See Appendix A, Section A.7) of $Q_{\text{op}}$ and can be expressed in terms of the defining function $\chi_B$ of $B$ as

$$E_Q(B) = \chi_B(Q_{\text{op}}). \tag{2.4}$$

(The defining function is defined by $\chi_B(\lambda) = 1$ for $\lambda \in B$ and $\chi_B(\lambda) = 0$ for $\lambda \notin B$.) From the condition $\|\Psi_\alpha\| = 1$ we obtain $w_\alpha^Q(\mathbf{R}) = (\Psi_\alpha, \Psi_\alpha) = \|\Psi_\alpha\|^2 = 1$, thus $w_\alpha^Q$ is a probability measure.

As a special case, the operators corresponding to observables with a finite discrete spectrum considered in Chapter 1 are of the form

$$Q_{\mathrm{op}} = \sum_q q E_Q(q) \quad \text{(finite sum)}. \tag{2.5}$$

$E_Q(q)$ is the projection operator to the eigenspace belonging to an eigenvalue $q$ of $Q_{\mathrm{op}}$. Abstractly, the projection operators $E_Q(q)$ can also be characterized as a finite number of linear operators $E(q)$ satisfying

$$E(q)^2 = E(q) = E(q)^*, \quad \sum_q E(q) = \mathbf{1}. \tag{2.6}$$

Next let us discuss how to characterize a general state $\varphi$. Since $\varphi(E(q))$ should have the meaning of a probability $w_\alpha^Q(q)$ for any family of operators $E(q)$ characterized by (2.6), the following properties should be required:

$$0 \le \varphi(E(q)), \quad \sum_q \varphi(E(q)) = 1. \tag{2.7}$$

Applying (2.4) to (2.5), $E_Q(B)$ is equal to the sum of $E_Q(q)$ for all $q \in B$ and, using the probability interpretation, we can require

$$\varphi\left(\sum_q E(q)\right) = \sum_q \varphi(E(q)) \quad \text{(finite sum)}. \tag{2.8}$$

(Since by the interpretation of (2.4), $\varphi(E_Q(B))$ represents the probability that the measured value of $Q$ is in $B$, it must be equal to $\sum w_\alpha^Q(q)$, summation being taken over all $q \in B$.) We note that the orthogonality relation

$$E(q)E(q') = E(q')E(q) = 0 \tag{2.9}$$

between $E(q)$ and $E(q')$ for $q \neq q'$ follows from (2.6) and we require the additivity (2.8) for the projection operators satisfying this orthogonality.

For an observable corresponding to a projection operator $E$ (characterized by $E^2 = E = E^*$ with eigenvalues 1 and 0 and eigenspaces $E\mathscr{H}$ and $(1 - E)\mathscr{H}$, respectively), its measured value is a choice between two values, 1 and 0. We call such an observable a *question* and interpret the observed values as the answers YES and NO, respectively. In quantum mechanics we usually assume that the set of all questions corresponds exactly to the set of all projection operators $\mathscr{P}(\mathscr{H})$ on a Hilbert space $\mathscr{H}$. In this case the requirements about a state collected in (2.7) and (2.8) above can be understood as the defining property for a finitely additive measure on $\mathscr{P}(\mathscr{H})$:

**Definition 2.1**
Let $\varphi$ be a function over $\mathscr{P}(\mathscr{H})$ taking values in $[0, 1]$. Such a $\varphi$ is called a *finitely additive measure* over $\mathscr{P}(\mathscr{H})$ if the following condition is satisfied (finite additivity). If a finite number of projection operators $E_1, \ldots, E_n$ in $\mathscr{P}(\mathscr{H})$ are mutually orthogonal ($E_iE_j = 0$ for $i \neq j$), then

$$\varphi(E_1 + \cdots + E_n) = \varphi(E_1) + \cdots + \varphi(E_n). \tag{2.10}$$

While the above characterization is for the expectation value of a question, a characterization of the expectation value of a general observable is as follows. Let us denote the set of all bounded linear operators on $\mathscr{H}$ by $\mathscr{B}(\mathscr{H})$.

**Definition 2.2**
A complex valued function over $\mathscr{B}(\mathscr{H})$ is called a *state* over $\mathscr{B}(\mathscr{H})$ if the following conditions are satisfied:

1. *linearity*: for $A_1, A_2 \in \mathscr{B}(\mathscr{H})$ and $c_1, c_2 \in \mathbf{C}$
$$\varphi(c_1A_1 + c_2A_2) = c_1\varphi(A_1) + c_2\varphi(A_2);$$
2. *positivity*: for $A \in \mathscr{B}(\mathscr{H})$
$$\varphi(A^*A) \geq 0;$$
3. *normalization*: $\varphi(\mathbf{1}) = 1$.

Here $\mathbf{1}$ represents the identity operator (for all $\Psi \in \mathscr{H}$, $\mathbf{1}\Psi = \Psi$) and $\mathbf{C}$ represents the complex numbers. Due to the condition of positivity, $A^*A$ satisfies

$$(\Psi, A^*A\Psi) = \|A\Psi\|^2 \geq 0,$$

and hence is a positive operator. Conversely, any positive operator $B$ can be written as $B = A^*A$ for $A = B^{1/2}$ because its spectrum lies in the interval $[0, \infty)$ and we can define its positive square root

$$B^{1/2} = f(B),\ f(\lambda) = \sqrt{\lambda}, \quad (\lambda \geq 0).$$

Thus, positivity is the condition that any positive operator has a positive expectation value.

Definitions 2.1 and 2.2 are equivalent definitions in the sense which we shall explain below, and therefore Definition 2.2 represents the necessary condition for a quantum mechanical state.

First, let us assume that a state $\varphi$ as defined in 2.2 is given. Since the projection operator $E$ satisfies $E = E^*E$, the positivity of $\varphi$ implies $\varphi(E) \geq 0$. Since $\mathbf{1} - E$

is also a projection operator we get

$$0 \leq \varphi(\mathbf{1} - E) = \varphi(\mathbf{1}) - \varphi(E) = 1 - \varphi(E),$$

where we have used the linearity and the normalization condition. As a consequence, the value $\varphi(E)$ lies in the interval $[0, 1]$. Again by repeated application of the linearity we can show that eq. (2.10) also holds. Therefore, if a state over $\mathscr{B}(\mathscr{H})$ as defined in Definition 2.2 is restricted to $\mathscr{P}(\mathscr{H})$ it becomes a finitely additive measure over $\mathscr{P}(\mathscr{H})$ as defined in Definition 2.1.

Next, suppose that a finitely additive measure $\varphi$ over $\mathscr{P}(\mathscr{H})$ as defined in Definition 2.1 is given. In this case the question is whether there exists a state $\bar{\varphi}$ over $\mathscr{B}(\mathscr{H})$ with its restriction to $\mathscr{P}(\mathscr{H})$ coinciding with the given $\varphi$, and whether such a state $\bar{\varphi}$ is unique. We call a state $\bar{\varphi}$ over $\mathscr{B}(\mathscr{H})$, which coincides with $\varphi$ when restricted to $\mathscr{P}(\mathscr{H})$, an *extension of* $\varphi$.

**Theorem 2.3** (Gleason's theorem [1])
If the dimension of $\mathscr{H}$ is 3 or larger, any state $\varphi$ over $\mathscr{P}(\mathscr{H})$ has an extension $\bar{\varphi}$ to $\mathscr{B}(\mathscr{H})$ and this extension $\bar{\varphi}$ is uniquely determined by $\varphi$.

The proof of this theorem is difficult and we do not give it in this book. Instead we shall explain below various forms of the theorem.

The restriction of the vector state given in eq. (2.2) to $\mathscr{P}(\mathscr{H})$ has the following property of *complete additivity* in addition to the property in Definition 2.1.

**Definition 2.4**
A state $\varphi$ over $\mathscr{P}(\mathscr{H})$ is called *completely additive* if it satisfies the following property. If a finite or infinite number of elements $E_\nu$ in $\mathscr{P}(\mathscr{H})$ are mutually orthogonal, then

$$\varphi\left(\sum_\nu E_\nu\right) = \sum_\nu \varphi(E_\nu). \tag{2.11}$$

If the $E_\nu$ are mutually orthogonal, their sum strongly converges to a projection operator. The subspace $(\sum E_\nu)\mathscr{H}$ is the closure of all finite sums of vectors in $E_\nu\mathscr{H}$ (for various $\nu$).

The above property (2.11) when required only for any countable set of $E_\nu$ is called *$\sigma$-additivity*. If $\mathscr{H}$ is separable, then any family of non-zero, orthogonal projection operators is always countable and thus complete additivity and $\sigma$-additivity are identical.

If $\varphi$ is $\sigma$-additive, then $w^Q_\alpha$ given in eq. (2.3) becomes a probability measure and, by the extension $\bar{\varphi}$ of $\varphi$ defined on $\mathscr{P}(\mathscr{H})$ to $\mathscr{B}(\mathscr{H})$, the expectation value of

$A = A^* \in \mathscr{B}(\mathscr{H})$ can be expressed as

$$\bar{\varphi}(A) = \int \lambda\varphi(E_A(\mathrm{d}\lambda)) \tag{2.12}$$

in terms of the spectral projection operators $E_A(\cdot)$ of $A$.

For the states defined over $\mathscr{B}(\mathscr{H})$ the property corresponding to the complete additivity described above is the following.

**Definition 2.5**
A state $\varphi$ over $\mathscr{B}(\mathscr{H})$ is called *normal* if it satisfies the following property. For any bounded increasing net of positive operators $A_\nu$ over $\mathscr{B}(\mathscr{H})$,

$$\varphi(\sup A_\nu) = \sup \varphi(A_\nu). \tag{2.13}$$

Let us explain the terminology used here. Given two linear operators $A, A'$, we write $A \geq A'$ or $A' \leq A$ if the operator $A - A'$ is positive. An ordered set $I$ is called *cofinal* if for any given pair $\nu, \nu' \in I$ there exists a $\mu \in I$ satisfying $\nu \leq \mu$ and $\nu' \leq \mu$. A family of linear operators $A_\nu$ ($\nu \in I$) indexed by the elements of a cofinal ordered set is called an *increasing net* if $\nu \leq \mu$ implies $A_\nu \leq A_\mu$. If $\|A_\nu\|$ is bounded, we call it a *bounded increasing net* and the strong limit $A = \lim A_\nu$ exists. Among the upper bounds of $A_\nu$ (i.e. the linear operators $B$ satisfying $B \geq A_\nu$ for all $\nu$), this $A$ is the smallest ($B \geq A$) and therefore it is also denoted as $A = \sup A_\nu$. (Usually we use this symbol for the case $A_\nu^* = A_\nu$.) By the positivity and linearity of $\varphi$, $\varphi(A_\nu) \geq \varphi(A_\mu)$ if $\nu \geq \mu$. Hence the definition of a normal state is that $\varphi(A_\nu)$ has a limit and

$$\varphi(\lim A_\nu) = \lim \varphi(A_\nu) \tag{2.14}$$

is satisfied. In (2.13) $A_\nu$ is restricted to positive operators. However, if (2.13) holds for positive operators, then it is possible to prove (2.14) also for the case of general linear operators $A_\nu$ by applying (2.13) to the bounded increasing net of positive operators $A_\nu - A_{\nu_0}$, $\nu \in I$, $\nu \geq \nu_0$, for a fixed $\nu_0 \in I$.

**Theorem 2.6**
If a state $\varphi$ over $\mathscr{P}(\mathscr{H})$ is completely additive, then its extension $\bar{\varphi}$ over $\mathscr{B}(\mathscr{H})$ is normal, and if a state over $\mathscr{B}(\mathscr{H})$ is normal then its restriction to $\mathscr{P}(\mathscr{H})$ is completely additive.

If $\mathscr{H}$ is separable then in the above definition of normal states it is sufficient to require (2.13) for any bounded increasing sequence $A_n$ ($n = 1, 2, \ldots$) instead of the net $A_\nu$, and this corresponds to $\sigma$-additivity on $\mathscr{P}(\mathscr{H})$.

A normal state over $\mathscr{B}(\mathscr{H})$ can be concretely described as follows. Consider a positive operator $\rho$ with trace 1:

$$\operatorname{Tr}\rho \equiv \sum_{\nu} (e_\nu, \rho e_\nu) = 1$$

where $\{e_\nu\}$ is an orthonormal basis in $\mathscr{H}$. (For the trace, we refer to Appendix A, Section A.8). For any $A \in \mathscr{B}(\mathscr{H})$

$$\operatorname{Tr}(\rho A) \equiv \sum_{\nu} (e_\nu, \rho A e_\nu) = \sum_{\nu} (\rho e_\nu, A e_\nu) \tag{2.15}$$

is absolutely convergent, and its value is independent of the choice of the orthonormal basis $\{e_\nu\}$.

**Theorem 2.7** [2]
The functional

$$\rho(A) = \operatorname{Tr}(\rho A)$$

over $\mathscr{B}(\mathscr{H})$ defined in (2.15) by the positive operator $\rho$ with trace 1 is a normal state over $\mathscr{B}(\mathscr{H})$. Conversely, any normal state $\varphi$ over $\mathscr{B}(\mathscr{H})$ is of the form of eq. (2.15) and $\rho$ is uniquely determined by $\varphi$.

We call $\rho$ the *density matrix* of $\varphi$.

For any unit vector $\Psi$ in $\mathscr{H}$ it is easy to prove that the state

$$\psi(A) = (\Psi, A\Psi) \tag{2.16}$$

is a normal state. Defining $E_\Psi$ by

$$E_\Psi \Phi = (\Psi, \Phi)\Psi, \tag{2.17}$$

$E_\Psi$ is a projection operator and $E_\Psi \mathscr{H} = C\Psi$ is a one dimensional subspace determined by $\Psi$. If we take as the orthonormal basis the one which includes $\Psi$, then

$$(E_\Psi e_\nu, A e_\nu) = \begin{cases} 0 & (\text{for } e_\nu \neq \Psi) \\ (\Psi, A\Psi) & (\text{for } e_\nu = \Psi) \end{cases}$$

and we obtain

$$\operatorname{Tr}(E_\Psi A) = (\Psi, A\Psi) = \psi(A).$$

Thus $E_\Psi$ is the density matrix of $\psi$.

For a general density matrix $\rho$, denoting the orthonormal basis given by the eigenvectors of $\rho$ by $\{e_\nu\}$ and the corresponding eigenvalues by $\lambda_\nu$ ($\rho e_\nu = \lambda_\nu e_\nu$), we have

$$\lambda_\nu \geq 0 \text{ (positivity of } \rho), \qquad \sum_\nu \lambda_\nu (= \mathrm{Tr}\rho) = 1, \tag{2.18a}$$

and from eq. (2.14) we obtain

$$\mathrm{Tr}(\rho A) = \sum_\nu \lambda_\nu (e_\nu, Ae_\nu) = \sum_\nu \lambda_\nu \psi_\nu (A_\nu) \tag{2.18b}$$

where we denote $\psi_\nu(A) = (e_\nu, Ae_\nu)$. This means that a normal state is a mixture of a countable number of vector states.

If a directed family of states $\varphi_\nu$ over $\mathscr{B}(\mathscr{H})$ has the limit

$$\varphi(A) = \lim \varphi_\nu(A)$$

for any $A \in \mathscr{B}(\mathscr{H})$, then $\varphi$ is also a state over $\mathscr{B}(\mathscr{H})$ and $\varphi_\nu$ *weakly converges* to this $\varphi$. The topology which is characterized by this weak convergence is identical with the physical topology introduced in Section 1.6.

By Theorem 2.21 in Section 2.3 the set of all normal states on $\mathscr{B}(\mathscr{H})$ is dense in the set of all states over $\mathscr{B}(\mathscr{H})$ in the weak topology. It follows from the arguments of Section 1.6 that the theory which considers all states over $\mathscr{B}(\mathscr{H})$ and the one which considers only the normal states are physically equivalent, and we can use either theory according to our convenience. On the other hand, the set of all the states introduced in Definition 2.2 is compact in the physical topology, it forms the biggest set of states among the physically equivalent theories and in that sense Definition 2.2 is a reasonable definition of states.

Among all states, the vector states given in (2.16) are characterized by the following theorem.

**Theorem 2.8**
The state $\psi$ defined in eq. (2.16) by a vector $\Psi$ in $\mathscr{H}$ is a pure state over $\mathscr{B}(\mathscr{H})$. Conversely, any normal pure state over $\mathscr{B}(\mathscr{H})$ is such a state.

Since a general normal state is of the form (2.18), all eigenvalues $\lambda_\nu$ of $\rho$ except for one have to be 0 in order that such a state is pure, and therefore such a state has to be of the form (2.16). Thus the second half of Theorem 2.8 is clear.

The proof of the first part is also simple, so let us show it in the following.

**Lemma 2.9**
Any state $\varphi$ satisfies the following Cauchy–Schwarz inequality. For any two elements $A, B$ in $\mathscr{B}(\mathscr{H})$

$$|\varphi(A^* B)|^2 \leq \varphi(A^* A)\varphi(B^* B). \tag{2.19}$$

**Proof** Due to the positivity and linearity of $\varphi$

$$0 \le \varphi(C^*C) = \bar{z}z\varphi(A^*A) + \bar{z}\varphi(A^*B) + z\varphi(B^*A) + \varphi(B^*B) \tag{2.20}$$

holds for $C = zA + B$ and any complex number $z$. This implies the property $\varphi(A^*B) = \overline{\varphi(B^*A)}$ and Lemma 2.9. □

Let us assume that the $\psi$ in eq. (2.16) is of the form

$$\psi(A) = \lambda\varphi_1(A) + (1-\lambda)\varphi_2(A) \tag{2.21}$$

for $0 < \lambda < 1$. If we can show that $\varphi_1 = \psi$, then it would follow that $\psi$ is a pure state. Since $\mathbf{1} - E_\Psi$ is a positive operator, we get

$$\varphi_1(\mathbf{1} - E_\Psi) \ge 0, \qquad \varphi_2(\mathbf{1} - E_\Psi) \ge 0,$$

due to the positivity of states. Therefore, setting $A = \mathbf{1} - E_\Psi$ in (2.21) and using

$$\psi(\mathbf{1} - E_\Psi) = (\Psi, (\mathbf{1} - E_\Psi)\Psi) = 0,$$

we obtain from the above positivity

$$\varphi_1(\mathbf{1} - E_\Psi) = 0 \qquad (\text{and } \varphi_2(\mathbf{1} - E_\Psi) = 0).$$

Then by (2.19) and $(\mathbf{1} - E_\Psi)^*(\mathbf{1} - E_\Psi) = \mathbf{1} - E_\Psi$ we obtain

$$\varphi_1(A(\mathbf{1} - E_\Psi)) = 0 = \varphi_1((\mathbf{1} - E_\Psi)A).$$

Substituting

$$\mathbf{1} = E_\Psi + (\mathbf{1} - E_\Psi), \qquad B = \mathbf{1}B\mathbf{1}$$

we obtain

$$\varphi_1(B) = \varphi_1(E_\Psi B E_\Psi).$$

On the other hand, by definition of $E_\Psi$,

$$E_\Psi B E_\Psi \Phi = (\Psi, \Phi) E_\Psi B \Psi = (\Psi, \Phi)(\Psi, B\Psi)\Psi = (\Psi, B\Psi) E_\Psi \Phi$$

holds for any $\Phi \in \mathscr{H}$, and hence it follows that

$$E_\Psi B E_\Psi = (\Psi, B\Psi) E_\Psi = \psi(B) E_\Psi.$$

Therefore

$$\varphi_1(B) = \psi(B)\varphi_1(E_\Psi)$$

holds for any $B$. If we set $B = \mathbf{1}$, we get $\varphi_1(E_\Psi) = 1$. Therefore $\varphi_1 = \psi$. □

To summarize this section, the necessary condition for a physical state led us to the introduction of the concept of a finite additive measure $\varphi$ on the set of questions $\mathscr{P}(\mathscr{H})$ (Theorem 2.1). For the case that the dimension of $\mathscr{H}$ is 3 or larger, it is in one-to-one correspondence with the state $\bar{\varphi}$ over $\mathscr{B}(\mathscr{H})$, defined as a positive normalized linear functional over $\mathscr{B}(\mathscr{H})$ (Definition 2.2), $\bar{\varphi}$ being the unique extension of $\varphi$ and $\varphi$ being the restriction of $\bar{\varphi}$ (Theorem 2.3). In particular, the completely additive measures (Definition 2.4) and the normal states (Definition 2.5) are in one-to-one correspondence (Theorem 2.6) and can be represented by density matrices (Theorem 2.7). The vector states which we usually use in quantum mechanics are characterized as normal pure states.

The essential part of the proof of Theorem 2.3 and Theorem 2.6 has been given by A.M. Gleason in 1957 where the case of three dimensions is crucial. Consequently, Theorem 2.3 and Theorem 2.6 are called *Gleason's theorem*. The question of whether the same theorem holds not just for $\mathscr{B}(\mathscr{H})$ but also for a general von Neumann algebra (see Appendix B, Section B.2) was a longstanding problem until the middle of the 1980s and was solved by E. Christensen and F.J. Yeadon. As in the discussion of $\mathscr{B}(\mathscr{H})$ where the case of dimension 2 is excluded, the corresponding situation is also excluded in the general case (i.e. the theorem holds if the algebra has no type $I_2$ part). For details on these points including the proof, see the references [1].

When the dimension of $\mathscr{H}$ is 2, it is easy to see that a finitely additive measure on $\mathscr{P}(\mathscr{H})$ cannot necessarily be extended to a state on $\mathscr{B}(\mathscr{H})$.

## 2.2 Algebraic viewpoint

In usual quantum mechanics discussed in the previous section the observables are represented by self-adjoint operators on a Hilbert space $\mathscr{H}$, and the algebra $\mathscr{B}(\mathscr{H})$ generated by them plays a central mathematical role. In this book we will not touch the question of why observables are represented by operators and why the algebraic laws of composition appear in physics. We develop the theory under the assumption of quantum mechanics that observables are represented by linear operators on a Hilbert space and, as a consequence, there exist algebraic laws of composition among these operators.

When performing computations with operators it is desirable that we can take limits and perform algebraic operations. Namely, the set of operators under consideration should be closed under the algebraic calculus as well as in an appropriate topology. Also non self-adjoint operators emerge from the self-adjoint operators representing the observables (even for $A^* = A$, $B^* = B$ if $AB \neq BA$ then $(AB)^* \neq AB$), and, taking all operators which emerge from the algebraic operations, we obtain a set which is closed under the $*$ operation. Thus the natural object of our discussion will be an operator algebra which we now define as follows.

**Definition 2.10**
A set $\mathfrak{A}$ of bounded linear operators on a Hilbert space is called a (concrete) $C^*$ *algebra* if the following conditions are satisfied.

1. $\mathfrak{A}$ is a $*$ algebra. That is, for any $A, B \in \mathfrak{A}$ and for any complex number $c, d$,

$$cA + dB \in \mathfrak{A}, \quad AB \in \mathfrak{A}, \quad A^* \in \mathfrak{A}$$

2. $\mathfrak{A}$ is closed in the norm topology, i.e. if a sequence $A_n \in \mathfrak{A}$ has a limit operator $A$ satisfying

$$\lim \|A_n - A\| = 0, \tag{2.22}$$

then $A \in \mathfrak{A}$.

The norm $\|B\|$ of an operator $B$ is defined by the following equation:

$$\|B\| = \sup\left\{\frac{\|B\Psi\|}{\|\Psi\|}; \Psi \in \mathscr{H}, \Psi \neq 0\right\}. \tag{2.23}$$

**Definition 2.11**
A set of bounded linear operators $\mathfrak{M}$ on a Hilbert space $\mathscr{H}$ satisfying the following conditions is called a *von Neumann algebra*:

1. $\mathfrak{M}$ is a $*$ algebra.
2. $\mathfrak{M}$ is closed in the weak operator topology. Namely, if a net of operators $A_\nu \in \mathfrak{A}$ has a *weak limit* $A$ satisfying

$$\lim(\Psi, A_\nu\Phi) = (\Psi, A\Phi) \tag{2.24}$$

for any vector $\Psi, \Phi \in \mathscr{H}$, then $A \in \mathfrak{M}$. (We write $A = w - \lim A_\nu$.)
3. $\mathbf{1} \in \mathfrak{M}$.

A concrete $C^*$ algebra $\mathfrak{A}$ as well as a von Neumann algebra $\mathfrak{M}$ on $\mathscr{H}$ are $*$ subalgebras of $\mathscr{B}(\mathscr{H})$. A $*$ subalgebra that is closed in the norm topology is called a $C^*$ algebra, and one that is closed in the weak topology is called a $W^*$ *algebra*, and if the latter contains the identity operator $\mathbf{1}$ on $\mathscr{H}$ it is called a von Neumann algebra.

The condition that the identity operator $\mathbf{1}$ is included in the algebra is simply to exclude the following situation. Let $\mathfrak{M}$ be a $W^*$ algebra on $\mathscr{H}$, and suppose that there exists a subspace $\mathscr{K}$ in $\mathscr{H}$ such that all operators in $\mathfrak{M}$ are 0 on the orthogonal complement $\mathscr{K}^\perp$ of $\mathscr{K}$ and the projection operator $\mathscr{P}(\mathscr{K})$ to the space $\mathscr{K}$ is in $\mathfrak{M}$. In this case, $\mathfrak{M}$ becomes a von Neumann algebra if it is restricted to $\mathscr{K}$. In particular if $\mathscr{K}^\perp = 0$ ($\mathscr{K} = \mathscr{H}$) then the $W^*$ algebra is a von Neumann algebra.

Originally, a $W^*$ algebra is abstractly defined as a $*$ algebra which is $*$ isomorphic to a von Neumann algebra. For the case that it is a $*$ subalgebra of $\mathscr{B}(\mathscr{H})$, it is a theorem that if we exclude the 0 part (restricting to $\mathscr{K}$) then the $W^*$ algebra becomes a von Neumann algebra.

When considering von Neumann algebras, the following terminology is important. For a self-adjoint set of operators on $\mathscr{H}$ (self-adjoint means that for $Q \in S$ also $Q^* \in S$) the *commutant algebra* $S'$ of $S$ is defined as

$$S' = \{Q \in \mathscr{B}(\mathscr{H});\ \forall Q_1 \in S \Rightarrow [Q_1, Q] = 0\}$$

$S'$ is a von Neumann algebra, and if $\mathfrak{M}$ is a von Neumann algebra then $\mathfrak{M} = (\mathfrak{M}')'$.

Since the norm topology is stronger than the weak topology (if the norm limit $A$ of a net $A_\nu$ exists in the sense of (2.22), it is equal to the weak limit in the sense of (2.24)), being closed in the weak topology implies being closed in the norm topology. Thus, a von Neumann algebra is a $C^*$ algebra. So, let us continue to explain more about $C^*$ algebras. At the end of this section, we will discuss a little for when a $C^*$ algebra is to be used and when a von Neumann algebra is to be used.

Any $*$ algebra which is isomorphic to a concrete $C^*$ algebra is called a *$C^*$ algebra.* (For $C^*$ algebras and von Neumann algebras we refer to Appendix B.) The reason why we discuss such abstract $C^*$ algebras is that we consider many different representations. A representation is defined as follows.

**Definition 2.12**
Given a $C^*$ algebra $\mathfrak{A}$ and a Hilbert space $\mathscr{L}$, a $*$ homomorphism $\pi$ which maps $\mathfrak{A}$ into $\mathscr{B}(\mathscr{L})$ is called a *representation* of $\mathfrak{A}$ on $\mathscr{L}$.

The $*$ homomorphism in this definition means that $\pi$ has the following properties:

$$\begin{aligned} \pi(cA + dB) &= c\pi(A) + d\pi(B) \qquad (A, B \in \mathfrak{A},\ c, d \in \mathbf{C}) \\ \pi(AB) &= \pi(A)\pi(B) \\ \pi(A^*) &= \pi(A)^* \end{aligned}$$

If $\pi(A) = 0$ implies $A = 0$, then $\pi$ is called a $*$ *isomorphism* and the representation $\pi$ is said to be *faithful*. By the definition of a $C^*$ algebra, there always exists a faithful representation.

In faithful representations $\pi$, the norm $\|\pi(A)\|$ does not depend on the representation $\pi$ but is determined by $A$ and is defined to be the norm $\|A\|$. For a general representation $\pi$

$$\|\pi(A)\| \leq \|A\|. \tag{2.25}$$

holds. Therefore, any representation of a $C^*$ algebra is automatically continuous.

Suppose we are given a $C^*$ algebra $\mathfrak{A}$, a representation $\pi_1$ on $\mathscr{H}_1$ and a representation $\pi_2$ on $\mathscr{H}_2$. If there exists a unitary operator $U$ from $\mathscr{H}_1$ to $\mathscr{H}_2$ satisfying

$$U\pi_1(A) = \pi_2(A)U \qquad \text{(for any element } A \text{ of } \mathfrak{A}), \tag{2.26}$$

then the representations $\pi_1$ and $\pi_2$ are said to be *unitarily equivalent*. Speaking of the same representations usually means unitarily equivalent representations.

A linear map $U$ (not necessarily a unitary map) from $\mathscr{H}_1$ to $\mathscr{H}_2$ satisfying (2.26) is called an *intertwining map* of the two representations $\pi_1$ and $\pi_2$. If such a map does not exist except for 0, then the two representations $\pi_1$ and $\pi_2$ are called *disjoint*. If $\pi_1$ and $\pi_2$ are said to be totally different representations, it means that they are disjoint.

In the usual quantum mechanics discussed in the previous section we usually fix a Hilbert space, represent the observables on this Hilbert space by self-adjoint linear operators, and describe the physical pure states by the vector states defined as expectation values with respect to the vectors in this Hilbert space. In contrast, we will use an approach where the observables are represented by the elements of an (abstract) $C^*$ algebra, and different representations of the $C^*$ algebra are used according to different physical situations (states). Such an approach is called the *algebraic viewpoint*. The advantages of this approach will become clear in the course of reading this book. The concept of a state in this context will be explained in detail in the next section. In concrete terms, we will consider vector states of various different representations in general, and the algebraic viewpoint is natural and useful in order to understand all of them in a unified way.

We now introduce various terminologies concerning representations.

**Definition 2.13**

Suppose we are given a family of Hilbert spaces $\mathscr{H}_\nu$ indexed by the elements $\nu$ of a set $I$ (finite or infinite). Then the Hilbert space $\mathscr{H}$ defined by the following conditions is called the *direct sum* of $\{\mathscr{H}_\nu\}_{\nu\in I}$.

1. A vector $\Psi$ in $\mathscr{H}$ is an indexed set $\{\Psi_\nu\}_{\nu\in I}$ of vectors $\Psi_\nu$ in $\mathscr{H}_\nu$ satisfying the following condition:

$$\sum \|\Psi_\nu\|^2 < \infty.$$

2. A linear combination of vectors in $\mathscr{H}$ is given by the following formula

$$c\Psi + d\Phi = \{c\Psi_\nu + d\Phi_\nu\}_{\nu\in I}.$$

3. The inner product of vectors in $\mathscr{H}$ is given by

$$(\Psi, \Phi) = \sum_\nu (\Psi_\nu, \Phi_\nu) \quad \text{(absolute convergence due to condition (1))}.$$

**Symbols** $\mathscr{H}$ and its vectors $\Psi$ are represented by the following symbols:

$$\mathscr{H} = \oplus_\nu \mathscr{H}_\nu, \quad \Psi = \oplus_\nu \Psi_\nu$$

If $I = \{1, 2, \ldots, n\}$ or $I = N$ {the set of all natural numbers} we can also write

$$\mathscr{H} = \mathscr{H}_1 \oplus \mathscr{H}_2 \oplus \cdots,$$

$$\Psi = \Psi_1 \oplus \Psi_2 \oplus \cdots.$$

**Definition 2.14**
Let $\mathfrak{A}$ be a $C^*$ algebra and suppose a representation $\pi_\nu$ of $\mathfrak{A}$ on each Hilbert space $\mathscr{H}_\nu$ $(\nu \in I)$ is given. If $\mathscr{H}$ is the direct sum of the $\mathscr{H}_\nu$ then the following formula defines a representation $\pi$ of $\mathfrak{A}$ on $\mathscr{H}$ and is called the *direct sum representation* of the $\pi_\nu$. For any $A \in \mathfrak{A}$ and any $\Psi \in \mathscr{H}_\nu$,

$$\pi(A)(\oplus_\nu \Psi_\nu) = \oplus_\nu \pi_\nu(A)\Psi_\nu. \tag{2.27}$$

The direct sum representation $\pi$ is represented by the following symbol:

$$\pi = \oplus_\nu \pi_\nu.$$

**Definition 2.15**
We are given a representation $\pi$ of a $C^*$ algebra $\mathfrak{A}$ on the Hilbert space $\mathscr{H}$. If the subspace $\mathscr{K}$ of $\mathscr{H}$ satisfies the following property, it is called an *invariant subspace* of the representation $\pi$.

For any element $A$ in $\mathfrak{A}$ and any vector $\Psi$ in $\mathscr{K}$, the following holds:

$$\pi(A)\Psi \in \mathscr{K}.$$

The following calculation shows that in this case the orthogonal complement $\mathscr{K}^\perp$ of $\mathscr{K}$ is also an invariant subspace of the representation $\pi$. For $\Psi \in \mathscr{K}^\perp$ and $\Phi \in \mathscr{K}$ we get

$$(\pi(A)\Psi, \Phi) = (\Psi, \pi(A)^*\Phi) = (\Psi, \pi(A^*)\Phi) = 0$$

where the last equality follows from the fact that, since $\mathscr{K}$ is invariant, $\pi(A^*)\Phi \in \mathscr{K}$, and therefore it is orthogonal to $\Psi \in \mathscr{K}^\perp$. From this equation we see that if $\Psi \in \mathscr{K}^\perp$, then $\pi(A)\Psi \in \mathscr{K}^\perp$ and it follows that $\mathscr{K}^\perp$ is an invariant subspace.

If $\mathscr{K}$ is an invariant subspace of a representation $\pi$, then by restricting the representing operator $\pi(A)$ of $A \in \mathfrak{A}$ to $\mathscr{K}$, we obtain a representation of $\mathfrak{A}$ on $\mathscr{K}$. We call this representation the *restriction* of the representation $\pi$ to $\mathscr{K}$ and denote it by $\pi_{\mathscr{K}}$:

$$\pi_{\mathscr{K}}(A)\Psi = \pi(A)\Psi \in \mathscr{K} \qquad (\Psi \in \mathscr{K},\ A \in \mathfrak{A}),$$

Defining the map $U$ from the direct sum $\mathscr{K} \oplus \mathscr{K}^{\perp}$ to $\mathscr{H}$ by

$$U(\Phi \oplus \Psi) = \Phi + \Psi \qquad (\Phi \in \mathscr{K},\ \Psi \in \mathscr{K}^{\perp}),$$

we obtain a unitary map $U$ satisfying the following property.

$$U(\pi_{\mathscr{K}} \oplus \pi_{\mathscr{K}^{\perp}})(A) = \pi(A)U.$$

In this sense the representation $\pi$ can be identified with the direct sum

$$\pi_{\mathscr{K}} \oplus \pi_{\mathscr{K}^{\perp}}.$$

Note that the projection operator $E(\mathscr{K})$ to $\mathscr{K}$ can be written in the form $(\mathbf{1} \oplus \mathbf{0})$ as

$$U(\mathbf{1} \oplus \mathbf{0}) = E(\mathscr{K})U.$$

**Definition 2.16**
We are given a representation $\pi$ of a $C^*$ algebra $\mathfrak{A}$ on a Hilbert space $\mathscr{H}$. Then $\pi$ is said to be *non-degenerate* if the only vector $\Psi$ in the Hilbert space $\mathscr{H}$ satisfying

$$\pi(A)\Psi = 0 \tag{2.28}$$

is the zero vector $\Psi = 0$.

For a general representation $\pi$, let us denote the set of all vectors $\Psi$ satisfying (2.28) as $\mathscr{K}_0$. Then $\mathscr{K}_0$ is an invariant subspace of the representation $\pi$. Due to the definition of $\mathscr{K}_0$ by (2.28), the restriction $\pi_{\mathscr{K}_0}$ to $\mathscr{K}_0$ is the zero representation

$$\pi_{\mathscr{K}_0}(A) = 0.$$

On the other hand, denoting the orthogonal complement of $\mathscr{K}_0$ by $\mathscr{K} = (\mathscr{K}_0)^{\perp}$, we see that if a vector $\Psi$ in $\mathscr{K}$ satisfies condition (2.28), then $\Psi \in \mathscr{K}_0 \cap \mathscr{K} = \{0\}$, and hence $\Psi = 0$. Thus, the restriction $\pi_{\mathscr{K}}$ to $\mathscr{K}$ is non-degenerate and the original representation $\pi$ is unitarily equivalent to the direct sum representation $\pi_{\mathscr{K}} \oplus 0$ on $\mathscr{K} \oplus \mathscr{K}_0$. We can also define $\mathscr{K}$ as the closure of

$$\pi(\mathfrak{A})\mathscr{H} \equiv \{\pi(A)\Psi;\ A \in \mathfrak{A},\ \Psi \in \mathscr{H}\}.$$

(It is easy to see that $(\pi(\mathfrak{A})\mathscr{H})^{\perp} = \mathscr{K}_0$.)

If an element $e$ of a $C^*$ algebra $\mathfrak{A}$ satisfies

$$eA = Ae = A,$$

for all $A \in \mathfrak{A}$, $e$ is called the *identity* (or unit element) of $\mathfrak{A}$ and is denoted by $\mathbf{1}_{\mathfrak{A}}$ or $\mathbf{1}$. A $C^*$ algebra does not necessarily contain an identity. Even if it does not, there exists a net of self-adjoint elements $f_\nu$ in the $C^*$ algebra $\mathfrak{A}$ which is called an *approximate identity* [3] and which has the following properties:

1. $\|f_\nu\| \leq 1$
2. $\lim \|f_\nu A - A\| = \lim \|Af_\nu - A\| = 0 \qquad (A \in \mathfrak{A})$.

For a representation $\pi$ of a $C^*$ algebra $\mathfrak{A}$, with the above decomposition into the direct sum of the null representation part $\mathscr{K}_0$ and the non-degenerate part $\mathscr{K}$

$$\pi(\mathbf{1}_{\mathfrak{A}}) = E(\mathscr{K}) \tag{2.29}$$

if $\mathfrak{A}$ has an identity, and

$$\text{w} - \lim \pi(f_\nu) = E(\mathscr{K}) \tag{2.30}$$

is satisfied for a general $\mathfrak{A}$ with an approximate identity $\{f_\nu\}$. The latter follows from an easy calculation

$$\begin{aligned}(\pi(f_\nu)\Psi, \pi(A)\Phi) &= (\Psi, \pi(f_\nu A)\Phi)\\ &\to (\Psi, \pi(A)\Phi) = (E(\mathscr{K})\Psi, \pi(A)\Phi),\end{aligned}$$

$$(\pi(f_\nu)\Psi, \Phi_0) = 0 = (E(\mathscr{K})\Psi, \Phi_0) \qquad (\Phi_0 \in \mathscr{K}_0)$$

$$\|\pi(f_\nu)\Psi\| \leq \|\pi(f_\nu)\|\|\Psi\| \leq \|\Psi\| \qquad \text{(uniform boundedness)}.$$

In particular, for a non-degenerate representation $\pi$

$$\pi(\mathbf{1}_{\mathfrak{A}}) = \mathbf{1}_{\mathscr{H}} \tag{2.31}$$

holds if $\mathfrak{A}$ possesses an identity and, for the case of a general $\mathfrak{A}$, any approximate identity $\{f_\nu\}$ satisfies

$$\text{w} - \lim \pi(f_\nu) = 1_{\mathscr{H}}. \tag{2.32}$$

Here $\mathbf{1}_{\mathscr{H}}$ is the identity operator on $\mathscr{H}$.

### 2.3 The representation associated with a state: GNS construction

We can use Definition 2.2 of a state in quantum mechanics also for the $C^*$ algebras.

**Definition 2.17**
A normalized positive linear functional $\varphi$ defined over a $C^*$ algebra $\mathfrak{A}$ is called a *state* over A.

The definiton given above means the following: The state $\varphi$ determines the expectation value $\varphi(A)$ which is a complex number for each element $A$ in $\mathfrak{A}$, satisfying the following three conditions:

1. Linearity: for $A_1, A_2 \in \mathfrak{A}$ and $c_1, c_2 \in \mathbf{C}$
$$\varphi(c_1 A_1 + c_2 A_2) = c_1 \varphi(A_1) + c_2 \varphi(A_2).$$
2. Positivity: for $A \in \mathfrak{A}$
$$\varphi(A^* A) \geq 0.$$
3. Normalization:
$$\|\varphi\| \equiv \sup\{|\varphi(A)|;\ A \in \mathfrak{A},\quad \|A\| \leq 1\} = 1. \tag{2.33}$$

Note that if $\mathfrak{A}$ contains the unit $\mathbf{1}_{\mathfrak{A}}$, the normalization condition is equivalent to
$$\varphi(\mathbf{1}_{\mathfrak{A}}) = 1. \tag{2.34}$$
(We will come back to the main point of its proof later.)

The following important and basic theorem asserts that any state is a vector state of an appropriate representation.

**Theorem 2.18**
For any state $\varphi$ over a $C^*$ algebra $\mathfrak{A}$, there exist a Hilbert space $\mathscr{H}_\varphi$, a representation $\pi_\varphi$ of $\mathfrak{A}$ on $\mathscr{H}_\varphi$ and a unit vector $\Omega_\varphi$ in $\mathscr{H}_\varphi$, satisfying the following two conditions:

1. For any $A \in \mathfrak{A}$
$$\varphi(A) = (\Omega_\varphi, \pi_\varphi(A)\Omega_\varphi). \tag{2.35}$$
2. $\Omega_\varphi$ is a *cyclic vector* of the representation $\pi_\varphi$, i.e.
$$\pi_\varphi(\mathfrak{A})\Omega_\varphi \equiv \{\pi_\varphi(A)\Omega_\varphi;\ A \in \mathfrak{A}\}$$
is dense in $\mathscr{H}_\varphi$.

The triple $(\mathscr{H}_\varphi, \pi_\varphi, \Omega_\varphi)$ satisfying these two conditions is unique as a unitary equivalence class. Thus, if another triple of a Hilbert space $\mathscr{H}'_\varphi$, a representation $\pi'_\varphi$ of $\mathfrak{A}$ on $\mathscr{H}'_\varphi$ and a vector $\Omega'_\varphi$ in $\mathscr{H}'_\varphi$ satisfies (1) and (2), then there exists a unitary map $U$ from $\mathscr{H}_\varphi$ to $\mathscr{H}'_\varphi$ satisfying
$$U\pi_\varphi(A) = \pi'_\varphi(A)U \qquad \text{(for all elements} A \in \mathfrak{A}) \tag{2.36}$$
$$U\Omega_\varphi = \Omega'_\varphi.$$

By this theorem we establish a relation between the abstract Definition 2.17 of a state and the definition of a state as an expectation value with respect to a unit vector in the Hilbert space used in quantum mechanics. (Later we explain a bit about a relation to the density matrices.)

*Main ideas of proof of the theorem*

Assume that the theorem holds. Denoting

$$\xi_A \equiv \pi_\varphi(A)\Omega_\varphi \qquad (A \in \mathfrak{A}),$$

let us analyse their properties.

(i) As for linear calculus, we obtain

$$c\xi_A + d\xi_B = c\pi_\varphi(A)\Omega_\varphi + d\pi_\varphi(B)\Omega_\varphi = \pi_\varphi(cA + dB)\Omega_\varphi = \xi_{cA+dB},$$

namely, the map from $A \in \mathfrak{A}$ to $\xi_A \in \mathscr{H}_\varphi$ is linear.

(ii) The inner product can be represented by the state $\varphi$

$$\begin{aligned}(\xi_A, \xi_B) &= (\pi_\varphi(A)\Omega_\varphi, \pi_\varphi(B)\Omega_\varphi) = (\Omega_\varphi, \pi_\varphi(A)^*\pi_\varphi(B)\Omega_\varphi) \\ &= (\Omega_\varphi, \pi_\varphi(A^*B)\Omega_\varphi) = \varphi(A^*B). \end{aligned} \tag{2.37}$$

(iii) The equalities among vectors are given by

$$\xi_{A_1} = \xi_{A_2} \Leftrightarrow \xi_{A_1} - \xi_{A_2} = \xi_{A_1 - A_2} = 0.$$

$$\xi_A = 0 \Leftrightarrow \|\xi_A\|^2 = (\xi_A, \xi_A) = \varphi(A^*A) = 0.$$

Thus, defining

$$\ker \varphi \equiv \{A \in \mathfrak{A};\ \varphi(A^*A) = 0\}, \tag{2.38}$$

we obtain the equivalence of $\xi_A = \xi_B$ with $A - B \in \ker \varphi$.

(iv) The action of the representing operator $\pi_\varphi(A)$ is computed as follows:

$$\pi_\varphi(A)\xi_B = \pi_\varphi(A)\pi_\varphi(B)\Omega_\varphi = \pi_\varphi(AB)\Omega_\varphi = \xi_{AB}. \tag{2.39}$$

(v) By using the approximate identity $f_\nu$ of $\mathfrak{A}$ we obtain the unit vector $\Omega_\varphi$ as

$$\mathrm{w} - \lim \xi_{f_\nu} = \mathrm{w} - \lim \pi_\varphi(f_\nu)\Omega_\varphi = \Omega_\varphi. \tag{2.40}$$

(If $\mathfrak{A}$ contains the unit $e = \mathbf{1}_{\mathfrak{A}}$, then $\xi_e = \Omega_\varphi$.)

Therefore, the subset

$$\xi_{\mathfrak{A}} = \{\xi_A;\ A \in \mathfrak{A}\}$$

of $\mathscr{H}_\varphi$ becomes isomorphic as a linear space to the coset space denoted by $\mathfrak{A}/\ker\varphi$, i.e. the space $\mathfrak{A}$ with an equivalence relation such that if $A - B \in \ker\varphi$ then $A, B \in \mathfrak{A}$

are equivalent. Furthermore if we define the inner product by (2.37) and add the vectors corresponding to the limit (called completion), we should be able to construct the desired space $\mathscr{H}_\varphi$. The representation $\pi_\varphi$ and the unit vector $\Omega_\varphi$ on this Hilbert space can be defined by (2.39) and (2.40).

**Proof of the theorem** The following inequality is a fundamental tool.

**Lemma 2.19** (The Cauchy–Schwarz inequality)
If $\varphi$ is a state over the $C^*$ algebra $\mathfrak{A}$, then for any $A, B \in \mathfrak{A}$ the following inequality holds,

$$|\varphi(A^*B)|^2 \leq \varphi(A^*A)\varphi(B^*B). \tag{2.41}$$

The proof is completely the same as for Lemma 2.9.

By this inequality and using eq. (2.38), we obtain for $A \in \ker\varphi$

$$\varphi(A^*B) = \varphi(B^*A) = 0, \tag{2.42}$$

which holds for any $B \in \mathfrak{A}$. Conversely, for $B = A$ the above condition yields condition (2.38). Namely, definition (2.38) is equivalent to

$$\ker\varphi = \{A \in \mathfrak{A};\ \forall B \in \mathfrak{A} \Rightarrow \varphi(B^*A) = \varphi(A^*B) = 0\}. \tag{2.43}$$

Since the condition on $A$ in this definition is linear, it follows that $\ker\varphi$ is a linear subset of $\mathfrak{A}$. Furthermore, if $A \in \ker\varphi$ then

$$\varphi(B^*CA) = \varphi((C^*B)^*A) = 0, \quad \varphi((CA)^*B) = \varphi(A^*(C^*B)) = 0.$$

Therefore it follows that for any $C \in \mathfrak{A}$ we get $CA \in \ker\varphi$. This means that $\ker\varphi$ is a left ideal of $\mathfrak{A}$.

Thus, let us consider the quotient space

$$\mathscr{D} \equiv \mathfrak{A}/\ker\varphi. \tag{2.44}$$

Since $\ker\varphi$ is a linear subset of $\mathfrak{A}$, $\mathscr{D}$ is also a complex linear space. Represent the equivalence class of an element $A \in \mathfrak{A}$ in $\mathscr{D}$ by $\xi_A$ and define

$$(\xi_A, \xi_B) \equiv \varphi(A^*B). \tag{2.45}$$

Then by using eq. (2.42), we find that the r.h.s. depends neither on the choice of the representative $A \in \mathfrak{A}$ of $\xi_A \in \mathscr{D}$, nor on that of the representative $B \in \mathfrak{A}$ of $\xi_B \in \mathscr{D}$ (the freedom of choices is as large as $\ker\varphi$). By the linearity of $\varphi$, (2.45) is linear for $\xi_B$ and antilinear for $\xi_A$. Furthermore, by the positivity of $\varphi$, we get $(\xi_A, \xi_A) \geq 0$, and we know from the definition of $\ker\varphi$ that this becomes 0 only for $A \in \ker\varphi$, i.e. for $\xi_A = 0$. Therefore, $\mathscr{D}$ is a pre-Hilbert space. The Hilbert space which is obtained by completion is called $\mathscr{H}_\varphi$ and $\mathscr{D}$ is identified with its dense linear subset (see Appendix A.2).

We define a representation $\pi_\varphi$ of $\mathfrak{A}$ by

$$\pi_\varphi(A)\xi_B = \xi_{AB}. \tag{2.46}$$

Since $\ker \varphi$ is a left ideal, $\xi_{AB} \in \mathscr{D}$ does not depend on the choice of the representative $B \in \mathfrak{A}$ of $\xi_B \in \mathscr{D}$. By eq. (2.46) $\pi_\varphi(A)$ is a linear operator defined on $\mathscr{D}$. The first two conditions of the homomorphic property of $\pi$ required in Definition 2.12 directly follow from eq. (2.46). The last condition can be verified on $\mathscr{D}$ by using (2.45)

$$(\pi_\varphi(A^*)\xi_B, \xi_C) = \varphi((A^*B)^*C) = \varphi(B^*(AC)) = (\xi_B, \pi_\varphi(A)\xi_C).$$

Therefore, $\pi_\varphi$ is a representation on $\mathscr{D}$.

Next, let us show that $\pi_\varphi(A)$ is a bounded operator. Note that we make use of the fact that for any $A, B \in \mathfrak{A}$ there exists an element $C \in \mathfrak{A}$ which satisfies

$$\|A\|^2 B^* B - B^* A^* AB = C^* C.$$

(Even if $\mathfrak{A}$ has no identity, one can prove that there exists an element $C \in \mathfrak{A}$ which satisfies the above equation as follows. Take a faithful representation $\pi_f$ of $\mathfrak{A}$. $(\|A\|^2\mathbf{1} - \pi_f(A^*A)$ is a positive operator and has a positive square root, as an operator $C'$. We can prove that $C'\pi_f(B) \in \pi_f(\mathfrak{A})$ and there exists a $C \in \mathfrak{A}$ satisfying $C'\pi_f(B) = \pi_f(C)$ and hence the above relation.) Due to the positivity of $\varphi$, $\varphi(C^*C) \geq 0$ and therefore

$$\|\pi_\varphi(A)\xi_B\|^2 = \varphi(B^*A^*AB) \leq \|A\|^2\varphi(B^*B) = \|A\|^2\|\xi_B\|^2.$$

From this we obtain eq. (2.25) which means that $\pi_\varphi(A)$ is a bounded operator. Defining the closure of $\pi_\varphi(A)$ on the entire space of $\mathscr{H}_\varphi$ we obtain the representation of the $C^*$ algebra $\mathfrak{A}$.

For an approximate identity $f_\nu$ in $\mathfrak{A}$, we obtain

$$\|\xi_{f_\nu}\|^2 = \varphi(f_\nu^* f_\nu) \leq \|f_\nu\|^2 \leq 1,$$

$$(\xi_A, \xi_{f_\nu}) = \varphi(A^* f_\nu) \to \varphi(A^*).$$

Therefore, the limit

$$\Omega_\varphi = \mathrm{w} - \lim \xi_{f_\nu}$$

exists and the equations

$$\pi_\varphi(A)\Omega_\varphi = \mathrm{w} - \lim \pi_\varphi(A)\xi_{f_\nu} = \mathrm{w} - \lim \xi_{Af_\nu} = \xi_A.$$

$$(\Omega_\varphi, \pi_\varphi(A)\Omega_\varphi) = \lim(\xi_{f_\nu}, \xi_A) = \lim \varphi(f_\nu^* A) = \lim \varphi((A^* f_\nu)^*) = \varphi(A).$$

are satisfied. Thus we have obtained the triple we wanted to construct.

Suppose we have another triple $(\mathscr{H}'_\varphi, \pi'_\varphi, \Omega'_\varphi)$. Then define a map $U$ by

$$U\pi_\varphi(A)\Omega_\varphi = \pi'_\varphi(A)\Omega'_\varphi \qquad (A \in \mathfrak{A}). \tag{2.47}$$

This is a linear map from $\pi_\varphi(\mathfrak{A})\Omega_\varphi$ to $\pi'_\varphi(\mathfrak{A})\Omega'_\varphi$ and preserves the inner product since inner products of both spaces are determined by the same $\varphi$. Consequently the closure of $U$ is a unitary map from $\mathscr{H}_\varphi$ to $\mathscr{H}'_\varphi$. Since for $\Psi = \pi_\varphi(B)\Omega_\varphi$ the following equation holds:

$$U\pi_\varphi(A)\Psi = U\pi_\varphi(AB)\Omega_\varphi = \pi'_\varphi(AB)\Omega'_\varphi = \pi'_\varphi(A)(\pi'_\varphi(B)\Omega'_\varphi) = \pi'_\varphi(A)U\Psi.$$

eq. (2.26) is satisfied for $\pi_1 = \pi_\varphi$ and $\pi_2 = \pi'_\varphi$. Substituting $A = f_\nu$ (approximate identity) in (2.47) and taking the limit we get

$$U\Omega_\varphi = \Omega'_\varphi\ .$$

Hence, the uniqueness up to unitary equivalence is also proven. □

If $\mathfrak{A}$ contains an identity it is not necessary to use the approximate identity and the proof becomes easy. For example we can show that $(\|A\|\mathbf{1}_{\mathfrak{A}} - A^*A)^{1/2}$ is in $\mathfrak{A}$, by uniformly approximating the function $(\|A\|^2 - \lambda)^{1/2}$ in the interval $[0, \|A\|^2]$ by a polynomial $P_n(\lambda)$, and taking $C = \lim P_n(A^*A)$.

The above construction was given in a joint paper of I.M. Gelfand and M.A. Naimark in 1943, and also in a paper by I.E. Segal in 1947, and is therefore referred to, by the initials of the authors, as the *GNS construction*. Furthermore $\mathscr{H}_\varphi, \pi_\varphi, \Omega_\varphi$ are called the *cyclic representation space*, the *cyclic representation* and the *cyclic vector, associated with the state* $\varphi$, respectively. They are also called the *GNS triplet*.

Next let us investigate which kind of state we obtain if we apply the GNS construction discussed here to the normal state $\varphi$ over $\mathscr{B}(\mathscr{H})$ which is given in (2.15) in terms of the density matrix $\rho$ introduced in Section 2.1. Let $\{e_\nu\}$ be an orthonormal basis of $\mathscr{H}$ consisting of eigenvectors of $\rho$, the corresponding eigenvalues being $\lambda_\nu$. The $\lambda_\nu$ satisfy (2.18a) and the $\varphi(A)$ are represented by (2.18b). $\nu$ is the label which distinguishes the $\Psi_\nu$ and the set of all labels is denoted by $I$. Let us take the Hilbert spaces $\mathscr{H}_\nu = \mathscr{H}$ labelled by $\nu$ to be all identical and take the same representation $\pi_\nu$ of $\mathfrak{A}$ on each $\mathscr{H}_\nu$ given by $\pi_\nu(A) = A$. We denote their direct sum space and representation as

$$\mathscr{L} \equiv \oplus_\nu \mathscr{H}_\nu, \quad \pi(A) = \oplus_\nu \pi_\nu(A), \ (= \oplus_\nu A).$$

Computing the expectation value with respect to the vector

$$\Omega_\rho = \oplus_\nu \sqrt{\lambda_\nu}\Psi_\nu$$

in $\mathscr{L}$ we obtain by (2.18b)

$$(\Omega_\rho, \pi(A)\Omega_\rho) = \sum_\nu \lambda_\nu(\Psi_\nu, \pi_\nu(A)\Psi_\nu) = \sum_\nu \lambda_\nu(\Psi_\nu, A\Psi_\nu) = \mathrm{Tr}(\rho A).$$

Therefore the triple consisting of the invariant subspace of $\mathscr{L}$

$$\mathscr{H}_\rho = \overline{\pi(\mathscr{B}(\mathscr{H}))\Omega_\rho} \qquad \text{(where bar means closure)},$$

the restriction of the representation $\pi$ to $\mathscr{H}_\rho$

$$\pi_\rho(A) \equiv \pi(A)|_{\mathscr{H}_\rho},$$

and the vector $\Omega_\rho$ satisfies the two conditions of Theorem 2.18 and thus, due to the uniqueness assertion of Theorem 2.18 we have obtained the GNS triple.

While we saw in this way that also the mixtures given by the density matrix can be represented as vector states of an appropriate representation, the difference from the vector states given by the vectors of the original $\mathscr{H}$ is that the latter ones are pure states. The following is a fundamental theorem which gives a distinction between the two.

**Theorem 2.20**
The following conditions about a state $\varphi$ over a $C^*$ algebra $\mathfrak{A}$ and the cyclic representation $\pi_\varphi$ associated with $\varphi$ are equivalent.

1. $\varphi$ is a pure state.
2. The representation $\pi_\varphi$ is irreducible, i.e. the only invariant subspaces of $\pi_\varphi$ are $\mathscr{H}_\varphi$ and $\{0\}$.
3. The linear operator on $\mathscr{H}_\varphi$ which commutes with all $\pi_\varphi(A)$ $(A \in \mathfrak{A})$ is a complex multiple of the identity operator, i.e.

$$\pi_\varphi(\mathfrak{A}') \equiv \{B \in \mathscr{B}(\mathscr{H}_\varphi); \forall A \in \mathfrak{A} \Rightarrow B\pi_\varphi(A) = \pi_\varphi(A)B\} = C\mathbf{1}.$$

4. If two vectors $\Psi, \Psi'$ in $\mathscr{H}_\varphi$ represent the same vector state (i.e. for all $A \in \mathfrak{A}$ we get $(\Psi, \pi_\varphi(A)\Psi) = (\Psi', \pi_\varphi(A)\Psi')$, then $\Psi'$ is proportional to $\Psi$. Namely, $\Psi' = \lambda\Psi$ for $\lambda \in \mathbf{C}$, $|\lambda| = 1$.

**Outline of the proof** (2) $\Leftrightarrow$ (3): If the subspace $\mathscr{K}$ is invariant, then since

$$\pi_\varphi(A)E(\mathscr{K})\Psi \in \mathscr{K} \qquad E(\mathscr{K})\pi_\varphi(A)E(\mathscr{K})\Psi = \pi_\varphi(A)E(\mathscr{K})\Psi$$

holds for any $\Psi$, the following equation is satisfied:

$$E(\mathscr{K})\pi_\varphi(A)E(\mathscr{K}) = \pi_\varphi(A)E(\mathscr{K}).$$

If we substitute $A^*$ into $A$ and take the $*$ on both sides we obtain

$$E(\mathscr{K})\pi_\varphi(A)E(\mathscr{K}) = E(\mathscr{K})\pi_\varphi(A).$$

Combining the two equations we get $E(\mathscr{K}) \in \pi_\varphi(\mathfrak{A})'$. So if condition (3) holds, then $E(\mathscr{K}) = 0$ or $E(\mathscr{K}) = 1$ and thus condition (2) is satisfied. Conversely, with the condition that $A = A^* \in \pi_\varphi(\mathfrak{A})'$, the spectral projection operator $E$ of $A$ is also in $\pi_\varphi(\mathfrak{A})'$ (in the weak topology we can approximate $E$ by a polynomial in $A$) and $E\mathscr{H}_\varphi$ becomes an invariant subspace. If condition (2) is satisfied then either $E = 0$ or $E = 1$. It follows that $A = \lambda\mathbf{1}$ ($\lambda \in \mathbf{C}$). For a general $A \in \pi_\varphi(\mathfrak{A})'$, the conclusion for the self-adjoint case implies $A + A^* = \lambda\mathbf{1}$ and $\mathrm{i}(A - A^*) = \lambda'\mathbf{1}$ and hence $2A = (\lambda - \mathrm{i}\lambda')\mathbf{1}$. Therefore condition (3) is satisfied.

(1) $\Leftrightarrow$ (3): For any projection operator $E \in \pi_\varphi(\mathfrak{A})'$, let

$$\Omega_1 = E\Omega_\varphi, \qquad \Omega_2 = (1 - E)\Omega_\varphi.$$

$$\lambda_1 = \|\Omega_1\|^2, \qquad \lambda_2\|\Omega_2\|^2.$$

If $\lambda_1 = 0$, then

$$E\pi_\varphi(A)\Omega_\varphi = \pi_\varphi(A)E\Omega_\varphi = 0$$

holds for any $A \in \mathfrak{A}$ and, since $\pi_\varphi(\mathfrak{A})\Omega_\varphi$ is dense in $\mathscr{H}_\varphi$ it follows that $E = 0$. Similarly, if $\lambda_2 = 1 - \lambda_1$ becomes 0 then $E = \mathbf{1}$. For the cases $\lambda_1 \neq 0, 1$, let

$$\varphi_i(A) = (\Omega_i, \pi_\varphi(A)\Omega_i)/\lambda_i \qquad (A \in \mathfrak{A}, \quad i = 1, 2).$$

Then we obtain

$$(\Omega_i, \pi_\varphi(A)\Omega_i) = (\Omega_\varphi, E_i\pi_\varphi(A)E_i\Omega_\varphi) = (\Omega_\varphi, \pi_\varphi(A)E_i\Omega_\varphi)$$

since $E_1 \equiv E$ and $E_2 \equiv 1 - E$ commute with $\pi_\varphi(A)$. Therefore the following equation is satisfied:

$$\varphi(A) = \lambda_1\varphi_1(A) + \lambda_2\varphi_2(A). \tag{2.48}$$

If (1) holds we obtain $\varphi_1 = \varphi_2 = \varphi$ because $\varphi_1$ as well as $\varphi_2$ are states over $\mathfrak{A}$. Consequently,

$$(\lambda^{-1}E\Omega_\varphi, \pi_\varphi(A)\Omega_\varphi) = \varphi_1(A) = \varphi(A) = (\Omega_\varphi, \pi_\varphi(A)\Omega_\varphi).$$

Due to the fact that $\pi_\varphi(A)\Omega_\varphi$ is dense we get $\lambda^{-1}E\Omega_\varphi = \Omega_\varphi$. Therefore, we obtain $\lambda^{-1}E = 1$, i.e. $E = 1$ by the same argument as above. Thus, (3) is satisfied. Conversely, suppose we are given the convex decomposition (2.48) with respect to the states $\varphi_1, \varphi_2$. Defining the map $T$ from $\mathscr{H}_\varphi$ to $\mathscr{H}_{\varphi_1}$ by

$$T\pi_\varphi(A)\Omega_\varphi = \pi_{\varphi_1}(A)\Omega_{\varphi_1}, \tag{2.49}$$

we get from the positivity of the state $\varphi_2$

$$\varphi(A^*A) \geq \lambda_1\varphi_1(A^*A)$$

and therefore

$$\|T\pi_\varphi(A)\Omega_\varphi\|^2 = \varphi_1(A^*A) \leq \|\pi_\varphi(A)\Omega_\varphi\|^2/\lambda_1. \tag{2.50}$$

From this it follows that (1′) if $\Psi = \pi_\varphi(A)\Omega_\varphi = 0$ then $T\Psi = 0$. Therefore if $\Psi_1 = \Psi_2$ then $T\Psi_1 - T\Psi_2 = T(\Psi_1 - \Psi_2) = 0$. (2′) Since the r.h.s. of (2.49) is linear in $A$, it follows that $T$ is a linear map. (3′) By the inequality (2.50) it follows that $T$ is bounded ($\|T\| \leq \lambda_1^{-1}$), and its closure $\bar{T}$ is a bounded linear map from $\mathscr{H}_\varphi$ to $\mathscr{H}_{\varphi_1}$. (4′) For $\Psi = \pi_\varphi(B)\Omega_\varphi$ we obtain

$$T\pi_\varphi(A)\Psi = T\pi_\varphi(AB)\Omega_\varphi = \pi_{\varphi_1}(A)\pi_{\varphi_1}(B)\Omega_{\varphi_1} = \pi_{\varphi_1}(A)T\Psi,$$

i.e. $\bar{T}$ is the non-zero intertwining map between $\pi_\varphi$ and $\pi_{\varphi_1}$

$$\bar{T}\pi_\varphi(A) = \pi_{\varphi_1}(A)\bar{T}.$$

Taking the $*$ of the above equation, we see that the intertwining map between $\pi_{\varphi_1}$ and $\pi_\varphi$ is given by $T^*$

$$\pi_\varphi(A)T^* = T^*\pi_{\varphi_1}(A).$$

Therefore $T^*\bar{T}$ is non-zero and an element of $\pi_\varphi(\mathfrak{A})'$. If (3) holds then we get the condition that $T^*\bar{T} = \lambda\mathbf{1}$, $(\lambda > 0)$. Therefore

$$\varphi_1(A) = (T\Omega_\varphi, T\pi_\varphi(A)\Omega_\varphi) = (\Omega_\varphi, T^*T\pi_\varphi(A)\Omega_\varphi) = \lambda\varphi(A).$$

Using the normalization condition $\|\varphi_1\| = \|\varphi\| = 1$, we obtain $\lambda = 1$ and $\varphi_1 = \varphi$. Hence we have shown (1).

(3) $\Leftrightarrow$ (4): The statement that $\Psi_1$ and $\Psi_2$ give the same vector state is equivalent to the following statement: A linear map $T$ which is defined by

$$T(\pi_\varphi(A)\Psi_1 + \Phi) = \pi_\varphi(A)\Psi_2, \quad \text{for any } \Phi \in (\pi_\varphi(\mathfrak{A})\Psi_1)^\perp,$$

satisfies the condition that its closure $\bar{T}$ is in $\pi_\varphi(\mathfrak{A})'$, and $T^*\bar{T}$ is a projection operator (onto $\overline{\pi_\varphi(\mathfrak{A})\Psi_1}$). Therefore the proof can be performed similarly to that of (3) $\Leftrightarrow$ (1). □

**Remark** Since the algebra $\mathscr{B}(\mathscr{H}) = \mathfrak{A}$ considered in Section 2.1 satisfies the condition (3) in the above theorem all vector states become pure states (see Theorem 2.8). Contrary to this, even for the same $\mathfrak{A} = \mathscr{B}(\mathscr{H})$ the representation $\pi_\varphi$ associated with the state $\varphi$ defined by a density matrix $\rho$ is, in general, not irreducible and the vector state $\varphi$ in this representation is not necessarily a pure state.

For a $C^*$ algebra $\mathfrak{A}$ one can consider the set of all states given by the density matrix $\rho$ on $\mathscr{H}$

$$\varphi(A) = \mathrm{Tr}(\rho\pi(A)) \tag{2.51}$$

as the states which are naturally associated to the representation $\pi$ on a Hilbert space $\mathscr{H}$ (the states usually used by physicists). We denote this set of states by $S_\pi$.

This $S_\pi$ is closed with respect to taking a mixture of states. Thus, $S_\pi$ is a convex subset of the set $S(\mathfrak{A})$ of all states on $\mathfrak{A}$. Furthermore, it is complete in the topology determined by the norm

$$\|\varphi - \psi\| \equiv \sup\{|\varphi(A) - \psi(A)|;\ A \in \mathfrak{A},\quad \|A\| \leq 1\}, \tag{2.52}$$

and hence it is a closed subset in $S(\mathfrak{A})$ with respect to the norm topology.

As for topologies on the state space $S(\mathfrak{A})$, besides the norm topology given above we can also consider the physical topology introduced in Section 1.6. It is the topology determined by the requirement that $\varphi_\alpha$ converges to $\varphi$ if and only if $\varphi_\alpha(A) \to \varphi(A)$ for any $A \in \mathfrak{A}$, (also called the weak $*$ topology). In Section 1.6 we gave a definition that $\pi_1$ and $\pi_2$ are physically equivalent if the closures $\bar{S}_{\pi_1}$ and $\bar{S}_{\pi_2}$ in this topology coincide. For this the following criterion in terms of the kernel of the representation $\pi$

$$\ker\pi \equiv \{A \in \mathfrak{A};\ \pi(A) = 0\}$$

is known.

**Theorem 2.21** [4]
Suppose we are given a unital $C^*$ algebra. The necessary and sufficient condition that the representation $\pi_1$ physically includes the representation $\pi_2$ (i.e. $\bar{S}_{\pi_1} \supset \bar{S}_{\pi_2}$) is

$$\ker\pi_1 \subset \ker\pi_2.$$

In particular, for the representations $\pi_1$ and $\pi_2$ to be physically equivalent (i.e. $\bar{S}_{\pi_1} = \bar{S}_{\pi_2}$) it is necessary and sufficient that

$$\ker\pi_1 = \ker\pi_2.$$

As an example, for a faithful representation $\pi$, $S_\pi$ is dense in the set of all states over $\mathfrak{A}$ in the physical topology.

## 2.4 The symmetries of quantum mechanics

As observables we consider the set of all self-adjoint operators in $\mathscr{B}(\mathscr{H})$ whose spectrum consists of a finite number of points (tentatively we denote this set by $\mathscr{A}$). As states we consider the set of all normal states (tentatively denoted by $\Sigma$).

According to the definition in section 1.8, a symmetry $s$ is represented as a bijection $\varphi \in \Sigma \to s\varphi \in \Sigma$ on $\Sigma$ and a bijection $Q \in \mathscr{A} \to sQ \in \mathscr{A}$ on $\mathscr{A}$ satisfying the following condition:

$$s\varphi(sQ) = \varphi(Q), \quad sf(Q) = f(sQ). \tag{2.53}$$

**Lemma 2.22**
$s$ preserves the convex combinations (mixtures) of $\Sigma$ and gives a bijection of the normal pure states.

**Proof** If $\varphi = \sum \lambda_i \varphi_i$ then by the first equation of (2.53) we get

$$s\varphi(sQ) = \varphi(Q) = \sum \lambda_i \varphi_i(Q) = \sum \lambda_i s\varphi_i(sQ).$$

Since $s$ is a surjection on $\mathscr{A}$ this equation can be understood as

$$s\varphi = \sum \lambda_i s\varphi_i,$$

i.e. $s$ preserves the convex combinations. The normal pure states are characterized as the extremal points of $\Sigma$ and since the extremal points can be defined by convex combinations alone, $s$ is a bijection of the extremal points in $\Sigma$.

Since the normal pure states $\varphi$ are vector states they can be represented by unit vectors $\Phi$ in $\mathscr{H}$. However, the representing vector $\Phi$ is not unique because all vectors in the class

$$\boldsymbol{\Phi} \equiv \{e^{i\theta}\Phi;\ 0 \leq \theta < 2\pi\}$$

represent the same $\varphi$. We call $\boldsymbol{\Phi}$ a *unit ray*. (Geometrically $\boldsymbol{\Phi}$ is a unit circle. However, originally in the real linear space, $\mathbf{R}\Phi$ (real multiples of $\Phi$) is a line, and generalizing this to the complex vector space we call $\mathbf{C}\Phi$ a ray; when restricted to the unit vector it is called a unit ray.)

If the vectors of two unit rays $\boldsymbol{\Phi}$ and $\boldsymbol{\Psi}$ are orthogonal, the corresponding states $\varphi$ and $\psi$ are also called *orthogonal*. An equivalent condition is that if the projection operator (question) $E$ satisfying $\varphi(E) = 1$ and $\psi(E) = 0$ exists, then $\varphi$ and $\psi$ are orthogonal. To see this, we note that

$$\varphi(\mathbf{1} - E) = \|(\mathbf{1} - E)\Phi\|^2 = 0, \quad \psi(E) = \|E\Psi\|^2 = 0$$

implies

$$(\mathbf{1} - E)\Phi = 0, \quad E\Psi = 0$$

and therefore we obtain

$$(\Psi, \Phi) = (\Psi, (1 - E)\Phi) + (\Psi, E\Phi) = 0 + (E\Psi, \Phi) = 0.$$

Conversely, if $(\Psi, \Phi) = 0$ then the projection operator $E$ to the one dimensional subspace $C\Phi$ (also written as $|\Phi><\Phi|$) satisfies the condition. □

The orthogonality between $\varphi$ and $\psi$ is denoted as $\varphi \perp \psi$.

**Lemma 2.23**
$s$ preserves the orthogonal relation between pure states. That is, $\varphi \perp \psi$ and $s\varphi \perp s\psi$ are equivalent.

**Proof** An operator $E \in \mathscr{A}$ (which implies $E^* = E$) being a projection is characterized by $E^2 = E$. Thus by the second equation of (2.53), $sE$ is also a projection operator. Furthermore, if $\varphi(E) = 1$ and $\psi(E) = 0$, then

$$s\varphi(sE) = 1, \qquad s\psi(sE) = 0.$$

Therefore $\varphi \perp \psi$ implies $s\varphi \perp s\psi$. Since $s^{-1}$ is also a symmetry, the converse also holds. □

Let $\varphi$ and $\psi$ be the states corresponding to the unit rays $\mathbf{\Phi}$ and $\mathbf{\Psi}$, respectively. Then

$$|(\Phi, \Psi)|^2 \equiv \langle \varphi, \psi \rangle \tag{2.54}$$

is determined only by $\varphi$ and $\psi$, and is called the *transition probability*.

**Lemma 2.24**
For any normal pure states $\varphi$ and $\psi$

$$\langle s\varphi, s\psi \rangle = \langle \varphi, \psi \rangle.$$

**Proof** The projection operator to the one dimensional subspace $\mathbf{C}\Phi$

$$E_\varphi \equiv E(\mathbf{C}\Phi)(= |\Phi><\Phi|) \in \mathscr{A}$$

can be characterized as an element in $\mathscr{A}$ which satisfies the three conditions $E_\varphi^2 = E_\varphi$, $\varphi(E_\varphi) = 1$ and $\varphi'(E_\varphi) = 0$ for any $\varphi' \perp \varphi$. Since these conditions are preserved by $s$ we get

$$sE_\varphi = E_{s\varphi}$$

On the other hand since

$$\langle \varphi, \psi \rangle = \psi(E_\varphi) \tag{2.55}$$

is satisfied we get

$$\langle s\varphi, s\psi \rangle = s\psi(E_{s\varphi}) = s\psi(sE_\varphi) = \psi(E_\varphi) = \langle \varphi, \psi \rangle.$$ □

The above discussion shows that the symmetry $s$ is represented as a bijection which conserves the transition probability of the unit rays in $\mathscr{H}$. Such bijections are characterized by *Wigner's theorem* [5]:

**Theorem 2.25**
If the dimension of $\mathscr{H}$ is 3 or larger any bijection $s$ which conserves the transition probability of the unit rays in $\mathscr{H}$ can be represented by a unitary or antiunitary map $U$ on $\mathscr{H}$ as

$$s\mathbf{\Phi} = U\mathbf{\Phi}. \tag{2.56}$$

Whether $U$ is unitary or antiunitary is determined by $s$ itself. Since the proof of this theorem is a little bit complicated, we shall give it at the end of this section.

If (2.56) holds, then by the first equation of (2.53)

$$(\Phi, Q\Phi) = (U\Phi, sQU\Phi) = \begin{cases} (\Phi, U^*(sQ)U\Phi) \\ (U^*(sQ)U\Phi, \Phi), \end{cases}$$

where on the r.h.s. the upper line corresponds to the case of unitary $U$ and the lower line to antiunitary $U$, the difference being due to the definition of the $U^*$. By the above equations

$$\begin{aligned} Q &= U^*(sQ)U, \quad \text{i.e.} \quad sQ = UQU^* \quad &\text{(unitary)},\\ Q^* &= U^*(sQ)U, \quad \text{i.e.} \quad sQ = UQ^*U^* \quad &\text{(antiunitary)}. \end{aligned}$$

Since $Q^* = Q$ for $Q \in \mathscr{A}$ in the second line $Q^*$ can be replaced by $Q$. However, if we extend our discussion to a general element $Q$ in $\mathscr{B}(\mathscr{H})$ and require $s\varphi(sQ) = \varphi(Q)$, then we should keep $Q^*$. On the other hand we may as well preserve the relation

$$sQ = UQU^*$$

and, for the antiunitary $U$ require instead

$$s\varphi(sQ) = \overline{\varphi(Q)}.$$

If $Q = Q^*$ then $\varphi(Q)$ is real and thus the above equation coincides with the first equation in (2.53). In this case the second equation of (2.53) has to be modified as

$$sf(Q) = \bar{f}(sQ).$$

However, if we restrict $f$ to real functions this modification is not necessary. We can choose between these two possibilities according to our convenience when extending a symmetry map $s$, physically given for the self-adjoint operators, to general operators.

As a summary, the symmetries $s$ occurring in the usual quantum mechanics can be represented by unitary or antiunitary operators $U_s$ on the Hilbert space $\mathscr{H}$ as

$$\varphi(A) = (\Phi, A\Phi) \to s\varphi(A) = (U_s\Phi, AU_s\Phi), \tag{2.57}$$

$$sQ = \begin{cases} U_sQU_s^* & \text{for } U_s \text{ unitary,} \quad (2.58)\\ U_sQ^*U_s^* & \text{for } U_s \text{ antiunitary.} \quad (2.59) \end{cases}$$

For a given symmetry $s$ the ambiguity in the choice of $U_s$ is limited to the choice within the *unitary ray*

$$\mathbf{U}_s = \{e^{i\theta}U_s; 0 \leq \theta < 2\pi\}. \tag{2.60}$$

To arrive at this conclusion starting from the condition (2.53), there are several other methods besides the above described Wigner's method of looking at the transition probability, as well as their generalizations.

By the proof of Lemma 2.23, $s$ defines a bijection on $\mathscr{P}(\mathscr{H})$. $\mathscr{P}(\mathscr{H})$ has the following logic structure of an *orthocomplemented lattice* (also called *quantum logic*). Let $E_i \in \mathscr{P}(\mathscr{H})$.

(i) We define $E_1 \leq E_2$ if $E_2E_1 = E_1$. This is equivalent to $E_1\mathscr{H} \subseteq E_2\mathscr{H}$, thus $\mathscr{P}(\mathscr{H})$ is an ordered set. (As a logical relation, this is the *implication* that if $E_1$ is true, then $E_2$ is true.)

(ii) There exists a least upper bound $E_1 \vee E_2$ (the least element $E \in \mathscr{P}(\mathscr{H})$ which satisfies $E \geq E_1$, $E \geq E_2$) and a greatest lower bound $E_1 \wedge E_2$ (the greatest element $E \in \mathscr{P}(\mathscr{H})$ which satisfies $E \leq E_1$, $E \leq E_2$), and $\mathscr{P}(\mathscr{H})$ is a lattice. (The *join* $E_1 \vee E_2$ is the *logical sum* and the *meet* $E_1 \wedge E_2$ is the *logical product*.)

(iii) $\mathscr{P}(\mathscr{H})$ contains the greatest element 1, and the least element 0.

(iv) If we take $E^\perp = 1 - E$ as the *orthocomplement* of $E$, then $\mathscr{P}(\mathscr{H})$ is an *orthocomplemented lattice*. It means that the following properties hold: $E \vee E^\perp = 1$, $E \wedge E^\perp = 0$, $(E^\perp)^\perp = E$, $E_1 \leq E_2 \to E_1^\perp \geq E_2^\perp$ ($E^\perp$ is the *negation* of $E$).

**Lemma 2.26**
The symmetry $s$ gives an *orthoautomorphism* of the orthocomplemented lattice $\mathscr{P}(\mathscr{H})$, i.e. $s$ is a bijection from $\mathscr{P}(\mathscr{H})$ to $\mathscr{P}(\mathscr{H})$, satisfying

$$s(E_1 \vee E_2) = sE_1 \vee sE_2, \quad s(E_1 \wedge E_2) = sE_1 \wedge sE_2, \tag{2.61}$$

$$s(E^\perp) = (sE)^\perp. \tag{2.62}$$

**Proof** $E \le F$ is equivalent to the statement that if $\varphi(E) = 1$ then $\varphi(F) = 1$ (for any $\varphi \in \Sigma$) by the following argument. As has been proven for $\varphi(A) = (\Phi, A\Phi)$ in the discussion between Lemma 2.22 and 2.23, $\varphi(E) = 1$ is equivalent to $E\Phi = \Phi$, i.e. it is equivalent to $\Phi \in E\mathscr{H}$. Thus the condition that $\varphi(E) = 1$ implies $\varphi(F) = 1$ is equivalent to the condition that $\Phi \in E\mathscr{H}$ implies $\Phi \in F\mathscr{H}$, which is equivalent to $E\mathscr{H} \subset F\mathscr{H}$ or $E \le F$.

Since $\varphi(E) = 1$ and $s\varphi(sE) = 1$ are equivalent, we obtain

$$E \le F \Leftrightarrow sE \le sF. \tag{2.63}$$

Therefore (2.61) is satisfied. Note that $F = E_1 \vee E_2$ is characterized by two properties: (1) $F \ge E_1$, $F \ge E_2$ and (2) if $F' \ge E_1$ and $F' \ge E_2$, then $F' \ge F$. These conditions then imply the same set of conditions for $sF$, $sE_1$, and $sE_2$ and hence $sF = sE_1 \vee sE_2$. Similarly for the second equation of (2.61).

By

$$\begin{aligned} s\varphi(s(cA + dB)) &= \varphi(cA + dB) = c\varphi(A) + d\varphi(B) = cs\varphi(sA) + ds\varphi(sB) \\ &= s\varphi(csA + dsB) \end{aligned}$$

the linearity of $s$

$$s(cA + dB) = csA + dsB$$

is obtained. Since the largest element is conserved due to (2.63), we obtain

$$s\mathbf{1} = \mathbf{1}.$$

Thus (2.62) holds by

$$s(E^{\perp}) = s(\mathbf{1} - E) = s\mathbf{1} - sE = \mathbf{1} - sE = (sE)^{\perp}$$

□

Concerning the orthoautomorphism of $\mathscr{P}(\mathscr{H})$ we have *Dye's theorem* [6]:

**Theorem 2.27**
If the dimension of $\mathscr{H}$ is 3 or larger, any orthoautomorphism of $\mathscr{P}(\mathscr{H})$ can be uniquely extended to a $*$ automorphism or a $*$ antiautomorphism of $\mathscr{B}(\mathscr{H})$.

A $*$ automorphism means a bijection onto itself preserving linear combinations, product and $*$ operation. In the case of a $*$ antiautomorphism the order of the product is reversed instead of being preserved:

$$s(AB) = s(B)s(A). \tag{2.64}$$

The relation with the conclusion (2.58) and (2.59) of Wigner's theorem is given by the following theorem:

**Theorem 2.28**
A $*$ automorphism $s$ of $\mathscr{B}(\mathscr{H})$ can be represented as a unitary transformation by a unitary element $U$ in $\mathscr{B}(\mathscr{H})$ as

$$sQ = UQU^*.$$

A $*$ antiautomorphism $s$ of $\mathscr{B}(\mathscr{H})$ can be represented as an antiunitary transformation by an antiunitary operator $U$ on $\mathscr{H}$ as

$$sQ = UQ^*U^*.$$

Dye's theorem (2.27) has been proven by Dye for the projection operator lattice $\mathscr{P}(\mathfrak{M})$ of a general von Neumann algebra $\mathfrak{M}$. There, a premise on $\mathfrak{M}$ is that it does not contain any type $I_2$ part. In this case, there exists a central projection operator $E$ (central means that $E$ commutes with all elements in $\mathfrak{M}$) such that

$$\mathfrak{M} = \mathfrak{M}_1 + \mathfrak{M}_2, \quad \mathfrak{M}_1 = \mathfrak{M}E, \quad \mathfrak{M}_2 = \mathfrak{M}(1 - E), \tag{2.65}$$

$sE = E$ and $s$ is decomposed into the direct sum of a $*$ automorphism $s_1$ of $\mathfrak{M}_1$ and a $*$ antiautomorphism $s_2$ of $\mathfrak{M}_2$.

Now let us consider the following three operations: The sum of observables is physically determined by the relation

$$\varphi(A + B) = \varphi(A) + \varphi(B)$$

as already discussed in Chapter 1. The multiplication of an observable by a number and the following operation, called the *Jordan product*,

$$A \circ B = (1/2)(AB + BA) = (1/2)((A + B)^2 - A^2 - B^2)$$

are physically determined by using functions of observables. A subset of $\mathscr{B}(\mathscr{H})$ which is closed under these three kinds of operations is called a Jordan algebra, and the bijection which preserves the operations of the Jordan algebra is called a Jordan isomorphism. Note that the above operations can be directly determined by observed quantities in contrast to the product $AB$. The following theorem is by R.V. Kadison [7]:

**Theorem 2.29**
A Jordan isomorphism $s$ of a von Neumann algebra $\mathfrak{M}$ can be decomposed into a direct sum of a $*$-automorphism $s_1$ of $\mathfrak{M}_1$ and a $*$-antiautomorphism $s_2$ of $\mathfrak{M}_2$ in the form of (2.65).

In the case of $\mathscr{B}(\mathscr{H})$, a central projection $E$ is either 0 or 1, and hence a Jordan isomorphism is either $s_1$ or $s_2$.

We can also characterize a symmetry as a map of states. Let $\Sigma_0$ be the set of all normal states over $\mathscr{B}(\mathscr{H})$.

**Lemma 2.30**
A symmetry $s$ preserves convex combinations (mixtures) of states.

**Proof** The same as for Lemma 2.22.

The following theorem is also due to Kadison [8]:

**Theorem 2.31**
Let $s$ be a bijection from the set $\Sigma_0$ of normal states over $\mathscr{B}(\mathscr{H})$ onto $\Sigma_0$. If $s$ preserves the convex combinations, then there exists either a $*$-automorphism or a $*$-antiautomorphism $s$ of $\mathscr{B}(\mathscr{H})$ such that

$$s\varphi(sQ) = \varphi(Q) \qquad (\text{i.e. } s\varphi(Q) = \varphi(s^{-1}Q)). \tag{2.66}$$

If the density matrix of $\varphi$ is $\rho$, then the density matrix of $s\varphi$ is given by $s\rho$.

The conclusion (2.66) also holds if we take a general von Neumann algebra instead of $\mathscr{B}(\mathscr{H})$. Furthermore, for the case of a $C^*$ algebra $\mathfrak{A}$ the same conclusion holds if $s$ is uniformly continuous in the weak $*$ topology of the states. (For example, $\Sigma_0$ can be taken as the set of all states on $\mathfrak{A}$.)

*A proof of Wigner's theorem 2.25* [5]
(1) First we prove a lemma which will be repeatedly used in the following:

**Lemma 2.32**
(i) If $\mathbf{\Phi}_1 \perp \mathbf{\Phi}_2$, then $s\mathbf{\Phi}_1 \perp s\mathbf{\Phi}_2$.
(ii) If $\Phi_i \in \mathbf{\Phi}_i$ $(i=1,2)$, $\Phi_1 \perp \Phi_2$, $c_1c_2 \neq 0$, $\Phi = c_1\Phi_1 + c_2\Phi_2$, $\Phi' \in s\mathbf{\Phi}$ ($\mathbf{\Phi}$ is the unit ray including the unit vector $\Phi$), then there exist a $\Phi_1' \in s\mathbf{\Phi}_1$ and a $\Phi_2' \in s\mathbf{\Phi}_2$ satisfying

$$\Phi' = c_1\Phi_1' + c_2\Phi_2'. \tag{2.67}$$

**Proof** (i) holds by Lemma 2.23. (ii) Fix $\mathbf{\Phi}_1$ and $\mathbf{\Phi}_2$ in (ii) and consider the set of all $\mathbf{\Phi}$ for varying $c_1$ and $c_2$. This set is characterized as the set of all unit rays which are orthogonal to any ray which is orthogonal to $\mathbf{\Phi}_1$ and to $\mathbf{\Phi}_2$. By (i) $s$ preserves this characterization, thus the set of all $s\mathbf{\Phi}$ is identical with the set of all $\mathbf{\Phi}'$ given in (2.67). The absolute value of the coefficients can be determined by the conservation of the transition probability:

$$|c_i|^2 = |(\Phi_i, \Phi)|^2 = |(\Phi_i', \Phi')|^2$$

Thus, the following equation is satisfied for any $\Phi' \in s\mathbf{\Phi}$, $\Phi_1' \in s\mathbf{\Phi}_1$ and $\Phi_2' \in s\mathbf{\Phi}_2$:

$$\Phi' = c_1'\Phi_1' + c_2'\Phi_2', \qquad |c_1'| = |c_1|, \, |c_2'| = |c_2|.$$

Hence, with an appropriate choice of $\Phi_1'$ and $\Phi_2'$ (2.67) is satisfied. □

(2) If the $U$ in the conclusion of the theorem exists then it is almost determined as we show in the following.

Fixing a $\Phi_0 \in \mathscr{H}$, then choose a $\Phi_0' \in s\mathbf{\Phi}_0$ and define

$$U\Phi_0 = \Phi_0' \tag{2.68}$$

The choice of $\Phi_0'$ leaves a freedom of only a factor $e^{i\theta}$.

If $\mathbf{\Psi} \perp \mathbf{\Phi}_0$ is not satisfied, then choose a $\Psi \in \mathbf{\Psi}$ and a $\Psi' \in s\mathbf{\Psi}$ such that

$$(\Psi, \Phi_0) > 0, \quad (\Psi', \Phi_0') > 0. \tag{2.69}$$

Such $\Psi$ and $\Psi'$ exist, and they are unique. We then define

$$U\Psi = \Psi'. \tag{2.70}$$

If $\mathbf{\Psi} \perp \mathbf{\Phi}_0$, choose a $\Psi \in \mathbf{\Psi}$. Then since $(\Phi_0 + \Psi, \Phi_0) = 1 > 0$, we define

$$U\Psi = \sqrt{2}U\{2^{-1/2}(\Phi_0 + \Psi)\} - \Phi_0'. \tag{2.71}$$

If $U$ exists, then (2.71) has to be satisfied as well. Thus $U$ is completely determined up to a factor $e^{i\theta}$. We now show that $U$ defined as above satisfies all properties required for it.

(3) First we show that $U\Psi \in s\mathbf{\Psi}$. We only need to be concerned about the case of (2.71). Since the first term on the r.h.s. of (2.71) is a linear combination of vectors $s\mathbf{\Phi}_0$ and $s\mathbf{\Psi}$ due to Lemma 2.32, $U\Psi$ also has this structure. $s\mathbf{\Phi}_0$ includes $\Phi_0'$ and is orthogonal to $s\mathbf{\Psi}$. Therefore if $U\Psi$ is shown to be orthogonal to $\Phi_0'$, then it follows that $U\Psi$ does not contain any component of $s\mathbf{\Phi}_0$, and hence it is in $s\mathbf{\Psi}$.

To this end we compute $(\Phi_0', U\Psi)$. Using the conservation of the transition probability and the positivity condition (2.67) we get

$$(U\{2^{-1/2}(\Phi_0 + \Psi)\}, \Phi_0') = 2^{-1/2}(\Phi_0 + \Psi, \Phi_0) = 2^{-1/2}.$$

Therefore we obtain

$$(U\Psi, \Phi_0') = \sqrt{2}\,2^{-1/2} - (\Phi_0', \Phi_0') = 0.$$

(4) So far we just chose one vector $\Phi$ from each ray $\mathbf{\Phi}$, determined the action of $U$ and confirmed that $U\Psi \in s\mathbf{\Psi}$. Next let us show that either

$$(U\Psi_1, U\Psi_2) = (\Psi_1, \Psi_2) \tag{2.72a}$$

or

$$(U\Psi_1, U\Psi_2) = (\Psi_2, \Psi_1) \tag{2.72b}$$

holds. Therefore consider $\Psi \perp \Phi_0$, and set

$$\Psi(\lambda, \theta) = (1-\lambda^2)^{1/2}\Phi_0 + \lambda e^{i\theta}\Psi \qquad (0 < \lambda < 1). \tag{2.73}$$

By Lemma 2.32 and eq. (2.69) we obtain

$$U\Psi(\lambda, \theta) = (1-\lambda^2)^{1/2}\Phi_0' + \lambda e^{i\omega}\Psi'$$

Here we freely fix $\Psi' \in s\mathbf{\Psi}$ and $\omega$ may depend on $\lambda$ and $\theta$. Next let us compare the following expressions:

$$(\Psi(\lambda_1,\theta_1), \Psi(\lambda_2,\theta_2)) = (1-\lambda_1^2)^{1/2}(1-\lambda_2^2)^{1/2} + \lambda_1\lambda_2 e^{i(\theta_2-\theta_1)}$$
$$(U\Psi(\lambda_1,\theta_1), U\Psi(\lambda_2,\theta_2)) = (1-\lambda_1^2)^{1/2}(1-\lambda_2^2)^{1/2} + \lambda_1\lambda_2 e^{i(\omega_2-\omega_1)}$$

Since the absolute values of both equations are equal, due to the conservation of the transition probability we get

$$\cos(\theta_2 - \theta_1) = \cos(\omega_2 - \omega_1). \tag{2.74}$$

Fixing in the above equation $\theta_1 = 0$, $\lambda_1 = \lambda_0$, and calling the corresponding $\omega_1$ as $\omega_0$, we seek a value of $\omega$ such that $\omega = \omega_2$, with the following solution:

$$\omega = \omega_0 + \sigma\theta \pmod{2\pi}.$$

where $\sigma$ is either 1 or $-1$ and may depend on $\lambda$ and $\theta$. Substituting this into $\theta_1$ and $\theta_2$ in eq. (2.74), we find that $\sigma$ does not depend on $\theta$ nor $\lambda$. If we substitute it into the above expression for the inner product, then one of the two equations in (2.72) holds for all $\lambda$ and $\theta$. Note that even if $\lambda = 0$ for $\Psi_1$ or $\Psi_2$, the two equations in (2.72) become identical due to the definition of $U$, and are both satisfied.

Now, in order to investigate the $\Psi$ dependence of $\sigma$, consider $\Phi \perp \Psi$, $\Phi \perp \Phi_0$, and

$$\Phi(\lambda, \theta, \mu) = \mu\Psi(\lambda, \theta) + (1-\mu^2)^{1/2}\Phi \qquad (0 < \mu \le 1)$$

By Lemma 2.32 and eq. (2.69) we obtain

$$U\Phi(\lambda, \theta, \mu) = \mu U\Psi(\lambda, \theta) + (1-\mu^2)^{1/2}\Phi' \qquad \Phi' \in s\mathbf{\Phi}$$

Comparing the expressions

$$(\Psi(\lambda_1,\theta_1),\Phi(\lambda,\theta,\mu)) = \mu(\Psi(\lambda_1,\theta_1),\Psi(\lambda,\theta))$$
$$(U\Psi(\lambda_1,\theta_1),U\Phi(\lambda,\theta,\mu)) = \mu(U\Psi(\lambda_1,\theta_1),U\Psi(\lambda,\theta))$$

we see that either eq. (2.72a) or eq. (2.72b) holds, the choice between them being the same as before.

If unit rays $\mathbf{\Psi}_1$ and $\mathbf{\Psi}_2$ are neither equal to $\mathbf{\Phi}_0$ nor orthogonal to $\mathbf{\Phi}_0$, and if

$$(\Psi_1,\Phi_0) > 0, \ \ (\Psi_2,\Phi_0) > 0,$$

then we can always write $\Psi_1 = \Psi(\lambda,\theta)$ and $\Psi_2 = \Phi(\lambda,\theta,\mu)$. Therefore we know that in this case one of the two equations in (2.72) is satisfied. Concerning the question which of the two equations (2.72a) or (2.72b) holds we know that if we fix $\Psi_1$ (take it as $\Psi(\lambda,\theta)$), then the choice is the same for $\Psi_2$ and, if we fix $\Psi_2$ (take it as $\Psi(\lambda,\theta)$) the same choice holds for $\Psi_1$.

Now if the $\mathbf{\Psi}_1$, $\mathbf{\Psi}_2$, $\mathbf{\Psi}'_1$, and $\mathbf{\Psi}'_2$ are all not equal to $\mathbf{\Phi}'_0$, then there exists a unit vector $\Psi_3$ neither in $\mathbf{\Phi}_0$ nor orthogonal to it, such that the inner products $(\Psi_3,\Psi_1)$ and $(\Psi_3,\Psi'_1)$ are not real. Then, for each pair $\{\Psi_1,\Psi_2\}$, $\{\Psi_1,\Psi_3\}$, $\{\Psi'_1,\Psi_3\}$, and $\{\Psi'_1,\Psi'_2\}$ the choice is the same and from this it follows that the choice is the same for any pairs $\{\Psi_1,\Psi_2\}$ and $\{\Psi'_1,\Psi'_2\}$.

Since the r.h.s. of the definition in (2.71) is a linear combination of vectors the inner product of which satisfies concurrently either (2.72a) or (2.72b), the same result is obtained for (2.71) for $\Psi$ orthogonal to $\Phi_0$.

From the discussion above we conclude that either (2.72a) or (2.72b) holds concurrently for all vectors for which $U$ is defined.

(5) Having defined $U$ for one vector from each ray, we define $U$ for a general vector by

$$U(c\Psi) = cU\Psi$$

for the case that (2.72a) holds and, by

$$U(c\Psi) = \bar{c}U\Psi$$

for the case that (2.72b) holds. Then either (2.72a) or (2.72b) is satisfied for any vector. Since $s$ is a surjection $U$ is also a surjection and it is either unitary or antiunitary. Finally, any operator $U$ satisfying (2.72a) for all vectors is a linear operator and any operator $U$ satisfying (2.72b) for all vectors is an antilinear operator. □

**Remark** From the construction of $U$, the unitary ray $\mathbf{U}$ is uniquely determined by $s$.

So far in our discussion we considered only one symmetry. Given two symmetries $r$ and $s$, we can consider the product $sr$ as a symmetry in the sense of a product of transformations of the states and observables. The corresponding unitary or antiunitary rays satisfy the following relation:

$$\mathbf{U}_{sr} = \mathbf{U}_s \mathbf{U}_r. \tag{2.75}$$

There are cases where we can decide whether $U$ is unitary or antiunitary by using this relation. We give two examples.

(1) If we have the relation $s = r^2$ between the symmetries $s$ and $r$, then

$$\mathbf{U}_s = (\mathbf{U}_r)^2$$

and hence $U_s$ is unitary, independently of whether $U_r$ is unitary or antiunitary.

For the case that the symmetries form a connected Lie group $G$, any element $s \in G$ in an appropriate neighbourhood of the identity satisfies the above properties and hence $U_s$ for $s$ in this neighbourhood is unitary. Since any element of $G$ can be represented as a product of a finite number of elements in this neighbourhood, we can represent all elements of $G$ by unitary rays.

(2) The symmetry of the time translation can be represented as

$$U(t) = \mathrm{e}^{\mathrm{i}tH}$$

by a one parameter group of unitary operators. As will be explained in section 3.6, the generator $H$ is interpreted as the energy and is required to be positive definite ($H \geq 0$). Under this condition the following relations are required for the space inversion $P(x \to -x)$ and the time inversion $T(t \to -t)$:

$$PU(t)P^* = \lambda(t)U(t), \quad TU(t)T^* = \mu(t)U(-t). \tag{2.76}$$

Here $\lambda(t)$ as well as $\mu(t)$ are indeterminate numbers with absolute value 1. In general

$$VU(t)V^* = \mathrm{e}^{\mathrm{i}VHV^*} \quad \text{(for unitary } V\text{)},$$
$$VU(t)V^* = \mathrm{e}^{-\mathrm{i}VHV^*} \quad \text{(for antiunitary } V\text{)},$$

holds. Since the spectrum of $VHV^*$ is identical with that of $H$, independently of whether $V$ is unitary or antiunitary, we can conclude by comparing both sides in (2.76) as to whether the spectra of the generators are bounded from above or from below ($\lambda(t)$ and $\mu(t)$ are merely shifting the spectrum), that $P$ must be unitary and $T$ must be antiunitary. (Due to a physical requirement $H$ is unbounded.)

We can apply the same argument to the choice between a $*$-automorphism and a $*$-antiautomorphism. For example if the symmetries form a connected Lie group then they are represented by $*$-automorphisms.

## 2.5 Symmetries from the algebraic point of view

From the algebraic point of view where the set of all observables is represented by a certain $C^*$ algebra $\mathfrak{A}$, a symmetry as discussed in the previous section is represented by a $*$-automorphism or a $*$-antiautomorphism. In particular, for the case of a connected Lie group $G$, there corresponds to each $g \in G$ an automorphism $\alpha_g$ of $\mathfrak{A}$ satisfying

$$\alpha_{gh} = \alpha_g \alpha_h. \tag{2.77}$$

In the previous section we discussed the case where the unitary operator $U_g$ corresponds to a symmetry $g$, and where the observable $Q$ transforms according to the equation

$$\alpha_g(Q) = (\mathrm{Ad}\, U_g)Q \equiv U_g Q U_g^*.$$

Note that $Q$ is concretely given as an operator on a Hilbert space and $U_g$ acts on the same Hilbert space. Ad $U$ denotes the unitary transformation of an observable by $U$.

Let us consider the case where we can construct the $U_g$ corresponding to a given $\alpha_g$.

If we convert the automorphism $\alpha_g$ into a map of the state $\varphi$ we get

$$(\alpha'_g \varphi)(Q) \equiv \varphi(\alpha_g Q).$$

From the condition $(g\varphi)(gQ) = \varphi(Q)$, the map of the state $\varphi$ by the symmetry $g$ can be represented by using $\alpha'_{g^{-1}}$. As a special case, a state which does not change under $g$ (i.e. a symmetric state) is called an *invariant state* of $g$. The condition on $\varphi$ is

$$\varphi(\alpha_g Q) = \varphi(Q), \tag{2.78}$$

which has to be satisfied for all $Q \in \mathfrak{A}$.

**Theorem 2.33**
If $\varphi$ is an invariant state with respect to the symmetry $g$, then there exists a unitary operator $U_g$ on the cyclic representation space $\mathscr{H}_\varphi$ associated with $\varphi$ satisfying the following condition:

$$U_g \Omega_\varphi = \Omega_\varphi, \quad U_g \pi_\varphi(Q) U_g^* = \pi_\varphi(\alpha_g(Q)) \tag{2.79}$$

(for $Q \in \mathfrak{A}$) and it is unique.

**Proof** On the subset $\pi_\varphi(\mathfrak{A})\Omega_\varphi$ which is dense in $\mathscr{H}_\varphi$ we define $U_g$ as

$$U_g \pi_\varphi(Q)\Omega_\varphi = \pi_\varphi(\alpha_g Q)\Omega_\varphi. \tag{2.80}$$

The range of $U_g$ is also the dense set $\pi_\varphi(\mathfrak{A})\Omega_\varphi$, and, by

$$\begin{aligned}(\pi_\varphi(\alpha_g Q_1)\Omega_\varphi, \pi_\varphi(\alpha_g Q_2)\Omega_\varphi) &= \varphi(\alpha_g(Q_1^* Q_2)) = \varphi(Q_1^* Q_2)\\ &= (\pi_\varphi(Q_1)\Omega_\varphi, \pi_\varphi(Q_2)\Omega_\varphi)\end{aligned}$$

$U_g$ is isometric. (The second equality holds due to the invariance of $\varphi$.) Therefore the closure of $U_g$ is an isometric operator with a dense range, i.e. it is a unitary operator.

If $\mathbf{1} \in \mathfrak{A}$, then we take $Q = \mathbf{1}$. Since $\alpha_g Q$ is also $\mathbf{1}$ we get the first equation of (2.79). If $\mathbf{1} \notin \mathfrak{A}$, we take an approximate identity $f_\nu$ as $Q$. Then $\alpha_g Q$ also becomes an approximate identity. Taking the limit with respect to $\nu$, we obtain the first equation of (2.79). By definition we get

$$\begin{aligned} U_g \pi_\varphi(Q) U_g^* \{\pi_\varphi(\alpha_g Q')\Omega_\varphi\} &= U_g \pi_\varphi(Q) U_g^* \{U_g \pi_\varphi(Q')\Omega_\varphi\} = U_g \pi_\varphi(Q)\pi_\varphi(Q')\Omega_\varphi \\ &= U_g \pi_\varphi(QQ')\Omega_\varphi = \pi_\varphi(\alpha_g(QQ'))\Omega_\varphi \\ &= \pi_\varphi(\alpha_g Q)\pi_\varphi(\alpha_g Q')\Omega_\varphi \end{aligned}$$

Since $\alpha_g(\mathfrak{A}) = \mathfrak{A}$ and $\pi_\varphi(\mathfrak{A})\Omega_\varphi$ is dense in $\mathscr{H}_\varphi$, the set of $\pi_\varphi(\alpha_g Q')\Omega_\varphi$ (with $Q'$ varying over all elements in $\mathfrak{A}$) is dense in $\mathscr{H}_\varphi$. Consequently we get the second equation of (2.79), i.e. we obtain a unitary operator $U_g$ which satisfies (2.79).

Conversely, if a unitary operator $U_g$ satisfies (2.79), then

$$U_g \pi_\varphi(Q)\Omega_\varphi = U_g \pi_\varphi(Q) U_g^* U_g \Omega_\varphi = \pi_\varphi(\alpha_g Q)\Omega_\varphi$$

and thus (2.80) is satisfied. Since $\pi_\varphi(\mathfrak{A})\Omega_\varphi$ is dense, such a $U_g$ is unique. □

If the symmetries form a group $G$ and if (2.77) holds, then by the definition (2.80) $U_g$ becomes a representation ($U_g U_h = U_{gh}$) of this group $G$.

Finally we add an explanation about the continuity.

A physically natural continuity is the continuity of

$$g \in G \to \varphi(\alpha_g Q),$$

for each state $\varphi$ and observable $Q$. For a $C^*$ algebra $\mathfrak{A}$ the topology of $Q \in \mathfrak{A}$ determined by $\varphi(Q)$ is the weak topology in the Banach space (i.e. for the states $\varphi_1, \ldots, \varphi_n$ and positive numbers $\varepsilon_1, \ldots, \varepsilon_n$ a neighbourhood of $Q \in \mathfrak{A}$ is given by

$$\{Q' \in \mathfrak{A};\ |\varphi_i(Q' - Q)| < \varepsilon_i\ (i = 1, \ldots, n)\}.$$

Therefore, the above continuity of $\varphi(\alpha_g Q)$ means that the map $g \in G \to \alpha_g Q \in \mathfrak{A}$ is continuous in this topology.

If $\mathfrak{A}$ is a von Neumann algebra we may restrict the states $\varphi$ to the normal states (which may be called the $\sigma$-weak operator topology), or we may either take the weak or the strong operator topology. (For $\alpha_g(Q)$ these three choices are equivalent due to the uniform boundedness of $\|\alpha_g(Q)\| = \|Q\|$ and $\alpha_g(Q)^*\alpha_g(Q) = \alpha_g(Q^*Q)$.)

On the other hand the continuity of

$$g \in G \to \alpha_g Q \in \mathfrak{A} \tag{2.81}$$

in the norm topology, a natural topology of a $C^*$ algebra, is also useful and often assumed. Actually this continuity is often realized by adopting an appropriate $\mathfrak{A}$.

For example, suppose $\mathfrak{A}$ is a von Neumann algebra and (2.81) is continuous in the $\sigma$ weak operator topology or in the weak operator topology. Then for $Q \in \mathfrak{A}$ and a $C^\infty$ function $f$ with compact support on $G$ we can define the integral

$$Q(f) = \int_G \alpha_h(Q) f(h)\, \mathrm{d}h$$

as an operator in $\mathfrak{A}$ satisfying

$$\begin{aligned} \alpha_g Q(f) &= \int_G \alpha_g \alpha_h(Q) f(h)\, \mathrm{d}h = \int_G \alpha_{gh}(Q) f(h)\, \mathrm{d}h \\ &= \int_G \alpha_{h'}(Q) f(g^{-1}h')\, \mathrm{d}h' = Q(l_g h). \end{aligned}$$

Here $\mathrm{d}h$ is the left invariant measure on $G$ and the transformation of the variable $h \to h' = gh$ is made. $l_g$ is the left regular representation

$$(l_g f)(h) = f(g^{-1}h).$$

Since $\alpha_h$ is an automorphism and $\|\alpha_g(Q)\| = \|Q\|$, we obtain

$$\begin{aligned} \|\alpha_g(Q(f)) - \alpha_{g'}(Q(f))\| &\le \int_G \|\alpha_h(Q)\|\, |l_g f(h) - l_{g'} f(h)|\, \mathrm{d}h \\ &= \|Q\|\, \|l_g f - l_{g'} f\|_1. \end{aligned}$$

Since the r.h.s. converges to 0 as $g \to g'$, $g \in G \to \alpha_g(Q(f))$ is continuous in the norm topology of $\mathfrak{A}$. Consider the $C^*$ algebra $\mathfrak{A}_1$ generated by $Q(f)$ with all $Q \in \mathfrak{A}$ and all $f$. Since $\alpha_h \mathfrak{A}_1 \subset \mathfrak{A}_1$ holds for $h = g$ and $g^{-1}$, we get $\alpha_g \mathfrak{A}_1 = \mathfrak{A}_1$. Hence $\alpha_g$ provides an automophism in $\mathfrak{A}_1$ for which (2.81) becomes continuous in the norm topology.

On the other hand, $\alpha_e Q = Q$ for the identity $e$ in $G$. Hence $\alpha_h Q \to Q$ as $h \to e$ holds in the topology in $\mathfrak{A}$ for which (2.81) is continuous. Thus for any net of functions $f(h) \ge 0$ which satisfies $\int f(h) = 1$ and the support of which tends to $e$ we get

$$\varphi(Q(f)) = \int \varphi(\alpha_h Q) f(h)\, \mathrm{d}h \to \varphi(Q).$$

for any states $\varphi$ used to define the topology of $\mathfrak{A}$. Therefore $\mathfrak{A}_1$ is dense in $\mathfrak{A}$ with respect to the topology considered in the beginning.

To describe the continuity of $U_g$ given in Theorem 2.33, we introduce the following symbol. For a given state $\varphi$ let

$$S_0(\varphi) \equiv \{\psi \in S(\mathfrak{A}); \exists Q \in \mathfrak{A}, \psi(\cdot) = \varphi(Q^* \cdot Q)\}.$$

In other words, we consider a whole set of vector states given by $\pi_\varphi(Q)\Omega_\varphi$ in the representation $\pi_\varphi$. ($S(\mathfrak{A})$ is the set of all states on $\mathfrak{A}$.)

**Corollary 2.34**
Given a state $\varphi$ invariant with respect to $\alpha_G$, the continuity of $\psi(\alpha_g Q)$ for all $Q \in \mathfrak{A}$ and $\psi \in S_0(\varphi)$ is equivalent to the condition that $U_g$ is continuous in the strong operator topology.

**Proof** To deduce the second statement from the first one, we use the following decomposition

$$\varphi(Q'\alpha_g Q) = \sum_{n=0}^{3} i^{-n}\varphi_n(\alpha_g Q)/4, \tag{2.82}$$

$$\psi_n(\cdot) \equiv \varphi((Q' + i^n\mathbf{1})^* \cdot (Q' + i^n\mathbf{1})) \qquad (n = 0, 1, 2, 3)$$

Since $\psi_n(\mathbf{1})^{-1}\psi_n \in S_0(\varphi)$, the continuity of (2.82) follows from the first condition. Hence, for $g' \to g$

$$\begin{aligned}\|\pi_\varphi(\alpha_{g'}Q)\Omega_\varphi - \pi_\varphi(\alpha_g Q)\Omega_\varphi\|^2 &= \varphi(\alpha_{g'}(Q)^*\alpha_{g'}Q) + \varphi(\alpha_g(Q)^*\alpha_g Q) \\ &\quad - \varphi(\alpha_{g'}(Q)^*\alpha_g Q) - \varphi(\alpha_g(Q)^*\alpha_{g'}Q) \\ &= 2\varphi(Q^*Q) - \varphi(Q^*\alpha_{g'^{-1}g}Q) - \varphi(Q^*\alpha_{g^{-1}g'}Q) \to 0,\end{aligned}$$

and by (2.80) $U_g$ is strongly continuous on $\pi_\varphi(\mathfrak{A})\Omega_\varphi$. By the fact that $\pi_\varphi(\mathfrak{A})\Omega_\varphi$ is dense in $\mathscr{H}_\varphi$ and by the uniform boundedness of $\|U_g\| = 1$, it follows that $U_g$ is strongly continuous.

Conversely, if $U_g$ is strongly continuous, then (2.80) is strongly continuous and the continuity of

$$\varphi(Q'^*\alpha_g(Q)Q') = \varphi(\alpha_{g^{-1}}(Q')^*Q\alpha_{g^{-1}}Q') = (\Psi_g, \pi_\varphi(Q)\Psi_g)$$

$$\Psi_g \equiv \pi_\varphi(\alpha_{g^{-1}}Q')\Omega_\varphi$$

follows. □

# 3
# THE RELATIVISTIC SYMMETRY

In this chapter we discuss the theory of special relativity as a quantum mechanical symmetry. A particle is understood as an irreducible unitary representation of the inhomogeneous Lorentz group.

## 3.1 Minkowski space

One of the key points in the theory of special relativity is that time and space are treated on an equal footing, and they are treated together as the four-dimensional space $M$ formed by the set of all space–time points

$$x = (x^{\mu})_{\mu=0,1,2,3} = (x^0, \mathbf{x}), \tag{3.1}$$

where $x^0$ represents the time and $\mathbf{x} = (x^j)_{j=1,2,3}$ is a 3 dimensional vector representing a spatial point.

In the system of units where the speed of light is 1, the bilinear form given by

$$(x, y) = x^0 y^0 - \mathbf{x} \cdot \mathbf{y} = \sum g_{\mu\nu} x^{\mu} y^{\nu} \tag{3.2}$$

plays an important role in the theory of special relativity. The metric tensor in the above equation is

$$g_{00} = -g_{11} = -g_{22} = -g_{33} = 1, \qquad g_{\mu\nu} = 0 \qquad (\mu \neq \nu). \tag{3.3}$$

In matrix form, we denote

$$g = (g_{\mu\nu}) = \begin{pmatrix} 1 & 0 & 0 & 0 \\ 0 & -1 & 0 & 0 \\ 0 & 0 & -1 & 0 \\ 0 & 0 & 0 & -1 \end{pmatrix}. \tag{3.4}$$

The 4 dimensional vectors $x$ in $M$ are classified as follows:

(i) A vector $x$ is called *timelike* if $(x, x) > 0$ ($|x^0| > |\mathbf{x}|$). In particular it is called *positive timelike* and *negative timelike*, respectively, according as $x^0 > 0$ and $x^0 < 0$.

(ii) A vector $x$ is called *lightlike* if $(x, x) = 0$ ($|x^0| = |\mathbf{x}|$). In particular it is called *positive lightlike*, *zero* and *negative lightlike*, respectively, according as $x^0 > 0$, $x^0 = 0$ and $x^0 < 0$.

(iii) A vector $x$ is called *spacelike* if $(x, x) < 0$ ($|x^0| < |\mathbf{x}|$).

The set of all lightlike vectors is called the *light cone*. The following sets of vectors

$$V_+ = \{x \in M;\ (x, x) > 0,\ x^0 > 0\} = \{\text{positive timelike vector}\}$$
$$\bar{V} = \{x \in M;\ (x, x) \geq 0,\ x^0 \geq 0\}$$

$$V_- = \{x \in M;\ (x,x) > 0,\ x^0 < 0\} = \{\text{negative timelike vector}\}$$
$$\bar{V}_- = \{x \in M;\ (x,x) \geq 0,\ x^0 \leq 0\}$$

are called the *open future cone*, the *closed future cone*, the *open past cone* and the *closed past cone*, respectively. The set of all spacelike vectors is called the *side cone*.

This classification has the following physical meaning. If $x - y$ is spacelike for two spacetime points $x, y$, then an event at a point $x$ and an event at a point $y$ cannot cause any influence on each other. In other words between $x$ and $y$ we cannot send or receive any signal. This property will be formulated in Chapter 4 as a mathematical property of the observables.

Given a subset $S$ of $M$, the set of all points which have no causal relation with $S$, i.e.

$$S' \equiv \{x \in M;\ \text{the vector } x - y \text{ is spacelike for all } y \in S\} \tag{3.5}$$

is called the *causal complement* of $S$. This notation will also be used in Chapter 4.

If $x - y$ is not spacelike, then $x$ and $y$ can have a causal relation. For example if $x - y \in \bar{V}_+$ then it is possible to send a signal from $y$ to $x$. In this case $(x - y, x - y)^{1/2}$ is called the *proper time* between $x$ and $y$. It is the time from $y$ to $x$ measured in the coordinate system in which the spatial coordinates of $x$ and $y$ are the same.

If $x - y$ is positive lightlike then a light ray starting from $y$ will pass through $x$.

## 3.2 The inhomogeneous Lorentz group

A transformation which preserves the relativistic causal relations discussed in the previous section is characterized by the following theorem.

**Theorem 3.1** [1]
Let $\kappa$ be a bijection from $M$ to $M$. The necessary and sufficient condition for $\kappa$ to satisfy the condition

$$x - y \in \bar{V}_+ \Leftrightarrow \kappa x - \kappa y \in \bar{V}_+$$

is that $\kappa$ is a linear transformation of the following form.

$$\kappa x = \lambda(\Lambda x + a), \qquad (\Lambda x)^\mu = \sum_\nu \Lambda^\mu_\nu x^\nu, \tag{3.6}$$

where $\lambda$ is a positive real number $a \in M$, and $\Lambda = (\Lambda^\mu_\nu)$ is a $4 \times 4$ matrix satisfying

$${}^t\Lambda g \Lambda = g,\ ({}^t\Lambda \text{ is the transposed matrix of } \Lambda,\ ({}^t\Lambda)^\mu_\nu = \Lambda^\nu_\mu) \tag{3.7}$$

and $\Lambda^0_0 > 0$.

In this theorem eq. (3.7) is equivalent to the invariance of the Minkowski inner product of any two points $x, y \in M$:

$$(\Lambda x, \Lambda y) = (x, y). \tag{3.8}$$

A linear transformation $\Lambda$ which satisfies (3.8) is called a *(homogeneous) Lorentz transformation*. The set of all such transformations forms a group called the *full (homogeneous) Lorentz group* and is denoted as $O(1,3)$ or $\mathscr{L}$. The image of $\bar{V}_+$ obtained by a transformation $\Lambda$ is either $\bar{V}_+$ or $\bar{V}_-$. The former case is characterized by $\Lambda^0_0 > 0$ as in the theorem above and is called an *orthochronous Lorentz transformation*. The set of all such transformations is called the *orthochronous (homogeneous) Lorentz group* and is abbreviated as $\mathscr{L}^\uparrow$. Taking the determinant on both sides of (3.7) we see that the determinant of $\Lambda$ is $\det(\Lambda) = \pm 1$. The special case $\det(\Lambda) = 1$ is called a *proper Lorentz transformation*. The set of all such transformations is called the *proper (homogeneous) Lorentz group*, denoted by $\mathscr{L}_+$. The common part of $\mathscr{L}_+$ and $\mathscr{L}^\uparrow$ is called the *restricted (homogeneous) Lorentz group* and is denoted by $\mathscr{L}^\uparrow_+$.

$\mathscr{L}^\uparrow_+$ is the connected component of the identity transformation in $O(1,3)$, and the transformations $\mathscr{L}^\uparrow_+$ can be generated by successive infinitesimal transformations. Typical transformations of $\mathscr{L}^\uparrow_+$ are the three-dimensional rotations with the time axis kept fixed:

$$\Lambda(x^0, \mathbf{x}) = (x^0, R\mathbf{x}) \quad (R \text{ represents a rotation matrix})$$

and *pure Lorentz transformations* in the direction of any unit vector $\mathbf{e}$ of the three-dimensional space given by

$$\Lambda(x^0, \mathbf{x}) = (x^0 \cosh \lambda, \mathbf{x}_\perp + (\sinh \lambda)\mathbf{x}_\parallel), \tag{3.9}$$

where $\mathbf{x}_\perp$ and $\mathbf{x}_\parallel$ are components of $\mathbf{x}$ orthogonal and parallel to $\mathbf{e}$:

$$\mathbf{x}_\parallel = (\mathbf{e} \cdot \mathbf{x})\mathbf{e}, \qquad \mathbf{x}_\perp = \mathbf{x} - \mathbf{x}_\parallel.$$

A general transformation of $\mathscr{L}^\uparrow_+$ can be represented as a product of these two types of transformations. A general transformation of $O(1,3)$ is given by a product of a transformation of $\mathscr{L}^\uparrow_+$ and one of the transformations listed below:

Identity transformation $x \to x$,
space inversion $(x^0, \mathbf{x}) \to (x^0, -\mathbf{x})$ (denoted by $P$),
time reversal $(x^0, \mathbf{x}) \to (-x^0, \mathbf{x})$ (denoted by $T$),
spacetime inversion $x \to -x$.

Among these operations the first two are orthochronous Lorentz transformations.

The part which relates to the $a \in M$ in eq. (3.6) is the transformation $x \to x + a$, called a *translation*. The set of all transformations $x \to \Lambda x + a$ given by combining translations and Lorentz transformations is called the *full inhomogeneous Lorentz group* and, as in the case of the homogeneous Lorentz group, according to the restrictions on $\Lambda$ we distinguish between the *orthochronous inhomogeneous Lorentz group*, the *proper inhomogeneous Lorentz group* and the *restricted inhomogeneous Lorentz group*. They may also be called the *Poincaré group* instead of inhomogeneous Lorentz group and are denoted by $\mathscr{P}$, $\mathscr{P}^{\uparrow}$, $\mathscr{P}_{+}$ and $\mathscr{P}_{+}^{\uparrow}$.

By Theorem 3.1, a transformation of space–time points which preserves the relativistic causal relations is given by a combination of an orthochronous inhomogeneous Lorentz transformation and a scale transformation. Further, if there is an observable which gives a scale (like a particle mass) and if we require that this observable is unchanged under the transformation, then only the transformations of $\mathscr{P}^{\uparrow}$ are allowed.

### 3.3 The relativistic symmetry in quantum mechanics

In the theory of special relativity, all transformations of $\mathscr{P}_{+}^{\uparrow}$ are assumed to be symmetries in the active sense as defined in Section 1.8. In concrete terms, a translation means a simultaneous shift of location and time of the preparation of the states of the object system as well as of the instruments which measure an observable. (In the vector $a = (a^0, \mathbf{a})$, $a^0$ corresponds to the time shift and $\mathbf{a}$ corresponds to the shift of the location.) A simultaneous (rotational) change of the direction of the object and instruments ($\mathbf{x} \to R\mathbf{x}$) corresponds to a rotation. If we put everything into a system moving with a constant velocity, this corresponds to a pure Lorentz transformation given in eq. (3.9). Note that in the unit system where the speed of light is 1, the velocity $\mathbf{v}$ of the uniformly moving system and the parameters in eq. (3.9) are related by

$$\mathbf{v} = (\tanh \lambda)\mathbf{e}.$$

Combining the above three transformations we obtain the interpretation of any transformation of $\mathscr{P}_{+}^{\uparrow}$ in the active sense. The relativistic symmetry means that when such a transformation is performed simultaneously for all the relevant systems, the measured result has to remain unchanged.

Applying Wigner's theorem given in Section 2.4 to this symmetry we obtain the unitary operator $U(g)$ implementing the transformation of the states and observables by $g$ for all $g \in \mathscr{P}_{+}^{\uparrow}$. However, since $U(g)$ and $\lambda U(g)$ generate exactly the same transformation for any complex number $\lambda$ with absolute value 1, there remains this arbitrariness in the determination of $U(g)$. (Since in a neighbourhood of the identity we can write any $g$ as $g = h^2$, we have $U(g) = \lambda(h,h)^{-1}U(h)^2$ (see eq. (3.10)) and $U(g)$ is unitary, independently of whether $U(h)$ is unitary or antiunitary.)

Since the result of two successively performed transformations $g_1$ and $g_2$ is the product $g_2 g_1$, $U(g_2)U(g_1)$ generates the transformation $g_2 g_1$ of states and observables. Due to the above mentioned arbitrariness by $\lambda$, the following equation holds:

$$U(g_2)U(g_1) = \lambda(g_2, g_1)U(g_2 g_1). \tag{3.10}$$

Note that $\lambda(g_2, g_1)$ is a complex number with absolute value 1, and is called the *phase factor*. If for each element $g$ a unitary operator $U(g)$ satisfying (3.10) is determined, $U(g)$ is called a *projective representation of the group*. The special case $\lambda(g_1, g_2) = 1$ is called a *representation of the group*. Corresponding to the group $G$ formed by the symmetries, there exists in quantum mechanics a projective representation of $G$ due to Wigner's theorem, and a transformation of states and observables by each symmetry $g$ can be represented by its representative unitary operator $U(g)$ as

$$\Psi_{g\alpha} = U(g)\Psi_\alpha, \qquad gQ = U(g)QU(g)^*, \tag{3.11}$$

where $\Psi_\alpha$ is a unitary ray representing a pure state $\alpha$, and $Q$ is (the operator representing) an observable. In particular, for relativistic symmetries, we have $G = \mathscr{P}_+^\uparrow$.

Next, consider the relation between a projective representation and a representation, i.e. let us consider $\lambda(g_2, g_1)$ given in eq. (3.10) in more detail. First we compute the product $U(g_1)U(g_2)U(g_3)$ in two different ways using associativity $(AB)C = A(BC)$. Then we find the following relation:

$$\lambda(g_1, g_2)\lambda(g_1g_2, g_3) = \lambda(g_1, g_2g_3)\lambda(g_2, g_3). \tag{3.12}$$

A $\lambda$ satisfying this relation is called a *2-cocycle* of $G$. On the other hand, using the arbitrariness in $U(g)$ we can introduce for all $g \in G$ a complex number $\mu(g)$ with absolute value 1 and define

$$U'(g) = \mu(g)U(g). \tag{3.13}$$

We may as well use $U'(g)$ instead of $U(g)$ since it also satisfies (3.11). Let us denote the factor $\lambda$ in eq. (3.10) for the case of $U'(g)$ by $\lambda'$. Then the relation between $\lambda'$ and the $\lambda$ corresponding to $U(g)$ is given by

$$\lambda'(g_1, g_2) = \partial\mu(g_1, g_2)\lambda(g_1, g_2), \tag{3.14}$$

$$\partial\mu(g_1, g_2) \equiv \frac{\mu(g_1)\mu(g_2)}{\mu(g_1g_2)}. \tag{3.15}$$

$\partial\mu$ in eq. (3.15) is called a *coboundary* and automatically satisfies (3.12) when $\partial\mu$ is substituted into $\lambda$ there. We define two cocycles $\lambda$ and $\lambda'$ to be equivalent if they are related by a coboundary as in (3.14). An equivalence class of cocycles defined in this way is called a second *cohomology class* and the set of all such equivalence classes is denoted by $H^2(G, \mathbf{T})$. ($\mathbf{T}$ represents the multiplicative group of all complex numbers with absolute value 1 viewed as the unit circle in the complex plane.) $H^2(G, \mathbf{T})$ forms a group with respect to the multiplication called the *cohomology group*.

From the above discussion we see that to each projective representation of a group there corresponds a cohomology class and, if and only if it is trivial (the equivalence class of $\lambda(g_1, g_2) = 1$) we can make the phase factor equal to 1 by redefinition of $U(g)$ as in eq. (3.13) and get a unitary representation. In particular if the cohomology group is trivial (consisting only of the equivalence class of 1), any projective representation is essentially a representation.

When the group is a Lie group, it is convenient to discuss the problem of the phase factor by dividing it into the problem about the Lie algebra and the problem about the universal covering group, as will be presented below.

Corresponding to a one parameter subgroup $g(t)$ of a Lie group $G$, $t \in \mathbf{R}(g(s)g(t) = g(s+t))$ we can define an element $\mathfrak{l}$ of the Lie algebra $\mathfrak{g}$ and write $g(t) = e^{t\mathfrak{l}}$. It is known that when considering one-parameter goups the factor can be removed (at least in the neighbourhood of $t = 0$) by a redefinition as given in (3.13); thus $U(g(t))$ can be chosen to be a one-parameter group of unitary operators (at least in the neighbourhood of $t = 0$) and can be written as

$$U(g(t)) = \exp\{\mathrm{i}t\,\mathrm{d}U(\hat{\mathfrak{l}})\}, \tag{3.16}$$

in terms of the self-adjoint operator $\mathrm{d}U(\hat{\mathfrak{l}})$. Following the conventions used in physics we have extracted here a factor i and write $\hat{\mathfrak{l}}$ instead of $\mathfrak{l}$, while in mathematics $\mathrm{d}U(\mathfrak{l}) = \mathrm{i}\,\mathrm{d}U(\hat{\mathfrak{l}})$ is used.

A Lie algebra $\mathfrak{g}$ is a real linear space with the Lie product defined. Let $\mathfrak{l}_j$ be a basis of $\mathfrak{g}$, then the Lie product is described in terms of (real) structure constants $c_{ik}^m$ by

$$[\mathfrak{l}_j, \mathfrak{l}_k] = \sum_m c_{jk}^m \mathfrak{l}_m. \tag{3.17}$$

In the case of a projective representation the following relation holds for the commutator $[x, y] = xy - yx$ of the representation operators:

$$[\mathrm{d}U(\hat{\mathfrak{l}}_j), \mathrm{d}U(\hat{\mathfrak{l}}_k)] = -\mathrm{i}\Big\{\sum c_{jk}^m\,\mathrm{d}U(\hat{\mathfrak{l}}_m) + \lambda_{jk}\Big\}. \tag{3.18}$$

Here $\lambda_{jk}$ are real numbers which appear due to the phase factor and they are antisymmetric in the indices ($\lambda_{jk} = -\lambda_{kj}$). Due to the Jacobi identity of the commutator they satisfy

$$\sum_m (c_{jk}^m\,\lambda_{ml} + c_{kl}^m\,\lambda_{mj} + c_{lj}^m\,\lambda_{mk}) = 0. \tag{3.19}$$

On the other hand, if we change the phase factor then $\lambda$ is changed as

$$\lambda'_{jk} = \lambda_{jk} - \sum c_{jk}^m\,\mu_m, \tag{3.20}$$

where $\mu_m$ can be any real number. (This corresponds to a replacement $\mathrm{d}U(\hat{\mathfrak{l}}_j) \to \mathrm{d}U(\hat{\mathfrak{l}}_j) + \mu_j = \mathrm{d}U'(\hat{\mathfrak{l}}_j)$.) In particular, if we find the relation

$$\lambda_{jk} = \sum c_{jk}^m\,\lambda_m \tag{3.21}$$

on the basis of (3.19) and antisymmetry in the indices, then by taking $\mu_m = \lambda_m$ we obtain $\lambda'_{jk} = 0$. In this case if we redefine the representation operator $U(g)$ as

$$U(\mathrm{e}^{\mathfrak{l}}) = \exp\left\{\mathrm{i}t \sum_j a^j \left(\mathrm{d}U(\hat{\mathfrak{l}}_j) + \lambda_j\right)\right\} \tag{3.22}$$

for any $\mathfrak{l} = \sum a^j \mathfrak{l}_j$, we can remove the phase factor and obtain a representation of $G$ in the neighbourhood of the identity. For a commutative Lie algebra of dimension larger than 1, the above procedure is not possible because $c_{jk}^m = 0$ and there exists a projective representation which does not become a representation (for example the canonical commutation relations). For $\mathscr{P}_+^\uparrow$ which represents the relativistic invariance, the above described procedure works. Let us explain this in more detail.

With an appropriate choice of the sign we can take the following 10 elements as a basis for the Lie algebra of $\mathscr{P}_+^\uparrow$: $\mathfrak{m}_{23}, \mathfrak{m}_{31}, \mathfrak{m}_{12}$ corresponding to rotations around the $x$-, $y$- and $z$-axes, $\mathfrak{m}_{01}, \mathfrak{m}_{02}, \mathfrak{m}_{03}$ corresponding to pure Lorentz transformations along the $x$-, $y$- and $z$-axes, and $\mathfrak{p}_0, \mathfrak{p}_1, \mathfrak{p}_2, \mathfrak{p}_3$ corresponding to translations along the time axis, $x$-, $y$- and $z$-axes, respectively. The commutation relations are

$$[\mathfrak{p}_\mu, \mathfrak{p}_\nu] = 0, \qquad [\mathfrak{m}_{\mu\nu}, \mathfrak{p}_\rho] = g_{\mu\rho}\mathfrak{p}_\nu - g_{\nu\rho}\mathfrak{p}_\mu, \tag{3.23}$$

$$[\mathfrak{m}_{\mu\nu}, \mathfrak{m}_{\rho\sigma}] = g_{\mu\rho}\mathfrak{m}_{\nu\sigma} + g_{\nu\sigma}\mathfrak{m}_{\mu\rho} - g_{\mu\sigma}\mathfrak{m}_{\nu\rho} - g_{\nu\rho}\mathfrak{m}_{\mu\sigma}, \tag{3.24}$$

where the indices run over $0, 1, 2, 3$ and $g_{\mu\nu}$ is the Minkowski metric.

We can prove that $\lambda_{\mu\nu}$ in the commutator $[\mathfrak{p}_\mu, \mathfrak{p}_\nu]$ is zero due to its antisymmetry and eq. (3.19). If we take $\mu_\sigma = \sum g^{\mu\rho}\lambda_{\mu\sigma\rho}/2$, then the $\lambda_{\mu\nu\rho}$ in the commutator of $[\mathfrak{m}_{\mu\nu}, \mathfrak{p}_\rho]$ can be written in the form of (3.21) corresponding to the redefinition of $\mathrm{d}U(\mathfrak{p}_\sigma)$. If we take $\mu_{\alpha\beta} = \sum \lambda_{\beta\mu\nu\alpha}g^{\mu\nu}$ then the $\lambda_{\mu\nu\rho\sigma}$ in the commutator $[\mathfrak{m}_{\mu\nu}, \mathfrak{m}_{\rho\sigma}]$ can also be written in the form of (3.21) corresponding to the redefinition of $\mathrm{d}U(\mathfrak{m}_{\alpha\beta})$. In this way all phase factors can be removed on the Lie algebra level.

Next, let $\tilde{G}$ be the universal covering group of the connected group $G$. Thus, there is a group homomorphism $\pi$ (covering map) from $\tilde{G}$ to $G$ such that $\pi$ is a bijection in a neighbourhood of the identity, and $\tilde{G}$ is connected and simply connected. Suppose we are given a unitary representation of $G$ in a neighbourhood of the identity, then this can be considered as a unitary representation of $\tilde{G}$ in a neighbourhood of the identity (the intersection of the above two neighbourhoods where $\pi$ is bijective and the unitary representation of $G$ is defined). Now consider an element $g = g_1 g_2 \cdots g_n$ with $g_1, g_2, \ldots, g_n$ lying in the neighbourhood of the identity, and define

$$U(g) \equiv U(g_1)U(g_2)\cdots U(g_n)$$

in terms of the known $U(g_j)$. Since $\tilde{G}$ is simply connected, it can be shown that $U(g)$ is determined, independently of how $g$ is represented as a product of the $g_1, g_2, \ldots, g_n$, and furthermore, that it is a representation of $\tilde{G}$. This means that a representation of a group $G$ in the neighbourhood of the identity can be extended to a representation of $\tilde{G}$.

For the case that $G = \mathscr{P}_+^\uparrow$, $\tilde{G}$ is obtained by replacing $\mathscr{L}_+^\uparrow$ in $\mathscr{P}_+^\uparrow$ by its universal covering group $SL(2, C)$. $SL(2, C)$ is a multiplicative group of complex $2 \times 2$ matrices with determinant 1, and the covering map $\pi$ from $SL(2, C)$ to $\mathscr{L}_+^\uparrow$ can be described as follows. Given a vector $x$ as in (3.1) we define the $2 \times 2$ matrices $\tilde{x}$ by

$$\tilde{x} = \sum x^\mu \sigma_\mu$$
$$\sigma_0 = \begin{pmatrix} 1 & 0 \\ 0 & 1 \end{pmatrix}, \quad \sigma_1 = \begin{pmatrix} 0 & 1 \\ 1 & 0 \end{pmatrix}, \quad \sigma_2 = \begin{pmatrix} 0 & -\mathrm{i} \\ \mathrm{i} & 0 \end{pmatrix}, \quad \sigma_3 = \begin{pmatrix} 1 & 0 \\ 0 & -1 \end{pmatrix}. \tag{3.25}$$

The formula which yields the components $x^\mu$ from $\tilde{x}$ is

$$x^\mu = \tfrac{1}{2}\,\mathrm{Tr}(\tilde{x}\sigma_\mu), \tag{3.26}$$

where Tr is the trace of the matrix (summation over diagonal elements). The $\pi(A) \in \mathscr{L}_+^\uparrow$ corresponding to an $A \in SU(2C)$ is determined by

$$\widetilde{\pi(A)x} = A\tilde{x}A^*. \tag{3.27}$$

The map $\pi$ is $2:1$, i.e. one $\Lambda \in \mathscr{L}_+^\uparrow$ determines two elements $\pm A(\Lambda) \in SU(2C)$. For example, for the $A$ corresponding to the identity in $\mathscr{L}_+^\uparrow$, we get $A = \pm\mathbf{1}$.

The elements of the universal covering group $\tilde{\mathscr{P}}_+^\uparrow$ of $\mathscr{P}_+^\uparrow$ are specified by a pair $(A, a)$, where $A \in SL(2, C)$ and $a \in \mathbf{R}^4$. Their product is given by

$$(a_1, A_1)(a_2, A_2) = (a_1 + \pi(A_1)a_2, A_1, A_2). \tag{3.28}$$

Due to the relativistic invariance there exists a unitary representation of $\tilde{\mathscr{P}}_+^\uparrow$

$$(a, A) \in \tilde{\mathscr{P}}_+^\uparrow \to U(a, A),$$

which gives a projective representation of $\mathscr{P}_+^\uparrow$. From the latter condition

$$u = U(0, -\mathbf{1}) \tag{3.29}$$

is a multiple of the identity operator by a complex number, and by $u^2 = U(0, \mathbf{1}) = \mathbf{1}$ we get

$$u = \pm\mathbf{1}.$$

In the case $u = \mathbf{1}$, we obtain a representation of $\mathscr{P}_+^\uparrow$. In the case $u = -\mathbf{1}$ we obtain the double-valued representation of $\mathscr{P}_+^\uparrow$. It is called double-valued because two unitary operators $\pm U(a, A)$ correspond to each element $(a, \Lambda) \in \mathscr{P}_+^\uparrow$ $(\pi(A) = \Lambda)$. $u$ itself is called the *univalence superselection rule*.

To summarize, the relativistic invariance in quantum mechanics mathematically reduces to a representation of the universal covering group $\tilde{\mathscr{P}}_+^\uparrow$ of the restricted inhomogeneous Lorentz group $\mathscr{P}_+^\uparrow$. There are two cases corresponding to $u = \pm\mathbf{1}$ which are the unitary representation and the double-valued representation of $\mathscr{P}_+^\uparrow$, respectively. (Mathematically it is also possible to consider the direct sum of these two types of representations.) Note that the assumption is made that the representations are continuous in the strong operator topology (which is equivalent to being continuous in the weak topology).

### 3.4 Irreducible representations and one-particle states

A characteristic feature of quantum mechanics is that we consider the superposition of pure states (which are different from mixtures) corresponding to linear combinations of vectors in a Hilbert space. Relativistic invariance requires a unitary representation of $\tilde{\mathscr{P}}^{\uparrow}_{+}$. The minimal unit which satisfies these two features is the concept of an *irreducible representation*, defined by the mutually equivalent conditions in the theorem below (cf. Theorem 2.20).

**Theorem 3.2**
For a unitary representation $g \in G \to U(g)$ of a group $G$ the following conditions are equivalent.

1. There is no non-trivial invariant subspace, i.e. if a subspace $\mathscr{K}$ (closed linear subset) of the representation space has the invariance property that $U(g)\psi \in \mathscr{K}$ for any $g \in G$ and $\psi \in \mathscr{K}$, then either $\mathscr{K} = \{0\}$ or $\mathscr{K}$ is the whole representation space.
2. Any bounded linear operator which commutes with all $U(g)$, $g \in G$, is a multiple of the identity operator $\mathbf{1}$ by a complex number.
3. Any unit ray is specified by the expectation value of all $U(g)$. Thus, given two vectors $\Psi, \Phi$ which satisfy $(\Psi, U(g)\Psi) = (\Phi, U(g)\Phi)$ for all $g \in G$, there exists a complex number $z$ with $|z| = 1$ such that $\Psi = z\Phi$.

The irreducible unitary representations of $\tilde{\mathscr{P}}^{\uparrow}_{+}$ are all known [2]. It is natural to classify them in the following two steps.

I. *Classification by the energy-momentum spectrum* Take as a subgroup of $\tilde{\mathscr{P}}^{\uparrow}_{+}$ the group $\mathbf{R}^4$ of the set of all translations $(a, \mathbf{1})$ and let us consider their unitary representation $T(a) = U((a, \mathbf{1}))$. By a general theory (we may call it the SNAG theorem [3], where the initials stand for Stone, Naimark, Ambrose and Godement), this can be written as

$$T(a) = \exp \mathrm{i}(P, a), \tag{3.30}$$

$$(P, a) = a^0 P^0 - a^1 P^1 - a^2 P^2 - a^3 P^3. \tag{3.31}$$

With respect to the commuting family of (unbounded) self-adjoint operators

$$P = (P^0, P^1, P^2, P^3), \tag{3.32}$$

the spectral resolution of the Hilbert space $\mathscr{H}$ is given as

$$\mathscr{H} = \int \mathscr{H}\, \mathrm{d}\mu(p) \ni \Psi = \int \Psi_p\, \mathrm{d}\mu(p), \qquad \Psi_p \in \mathscr{H}_p, \qquad (3.33)$$
$$(p = (p^0, p^1, p^2, p^3),\ p^\mu \in \mathbf{R}),$$

$$(P^\mu \Psi)(p) = p^\mu \Psi(p), \qquad (3.34)$$

$$(T(a)\Psi(p) = \mathrm{e}^{\mathrm{i}(p,a)}\Psi(p). \qquad (3.35)$$

A spectral projection operator for a measurable subset $\Delta$ of $\mathbf{R}^4$ is defined as

$$E(\Delta)\mathscr{H} = \int_\Delta \mathscr{H} \mathrm{d}\mu(p) = \{\Psi \in \mathscr{H};\quad \Psi(p) = 0 \text{ for } p \notin \Delta\}. \qquad (3.36)$$

Using this, we also have the following representation:

$$T(a) = \int \mathrm{e}^{\mathrm{i}(p,a)} E(\mathrm{d}p). \qquad (3.37)$$

By the multiplication rule of $\tilde{\mathscr{P}}_+^\uparrow$ we find

$$U(A)T(a)U(A)^* = T(\pi(A)a) \qquad (U(A) = U(0, A)). \qquad (3.38)$$

From the point of view of the spectrum, this means

$$U(A)E(\Delta)U(A)^* = E(\pi(A)\Delta), \qquad (3.39)$$

$$U(A)E(\Delta)\mathscr{H} = E(\pi(A)\Delta)\mathscr{H} \qquad (\Lambda\Delta = \{\Lambda p; p \in \Delta\}), \qquad (3.40)$$

and $U(A)$ generates the Lorentz transformation $p \to \pi(A)p$.

The set of points $p$ which are mapped to each other by transformations of the restricted Lorentz group $\mathscr{L}_+^\uparrow$ is called an *orbit*. We distinguish the following types of orbits ($m \geq 0$).

$$m_+\text{: } \{p; (p,p) = m^2,\ p^0 > 0\}$$
$$m_-\text{: } \{p; (p,p) = m^2,\ p^0 < 0\}$$
$$0\text{: one point } p = 0$$
$$im\text{: } \{p; (p,p) = -m^2\}.$$

Since a representation $U(A)$ of $\mathscr{L}_+^\uparrow$ transforms the spectrum according to (3.40), the support of the spectral measure $\mu$ has to be invariant under $\mathscr{L}_+^\uparrow$. Therefore, if one point of the orbit is included, then the whole orbit is included. If the support of the spectral measure contains two orbits or more, then there exists a measurable set $\Delta$, invariant under $\mathscr{L}_+^\uparrow$, such that the $\mu$ measure of $\Delta$ as well as that of its complement $\Delta^c$ are non-zero. In this case $E(\Delta)\mathscr{H}$ becomes a non-trivial invariant subspace of $\tilde{\mathscr{P}}_+^\uparrow$, and the representation is not irreducible. Therefore for any irreducible unitary representation of $\tilde{\mathscr{P}}^{\uparrow}_{+}$, the spectrum of $P^{\mu}$ is concentrated on one orbit and it is classified by this orbit.

According to the discussion in Section 3.6, we can interpret $P^0$ as the energy and $P^1, P^2, P^3$ as the momentum vector. Thus, the parameter $m$ of the orbit $m_+$ has the meaning of a mass.

For a fixed orbit, the equivalence class of the measure $\mu$ is uniquely determined and, except for the case of the trivial 0 orbit, can be given as follows

$$\mathrm{d}\mu(\mathbf{p}) = (2|p^0|)^{-1}\mathrm{d}^3\mathbf{p}, \qquad \mathrm{d}^3\mathbf{p} \equiv \mathrm{d}p^1\mathrm{d}p^2\mathrm{d}p^3, \tag{3.41}$$

where we have taken $\mathbf{p} = (p^1, p^2, p^3)$ as the variables. Note that in the case of the orbit $im$, it has two parts $p^0 \gtrless 0$. The above expression is obtained from $\delta((p,p) - m^2)\mathrm{d}^4p$ and thus the measure is invariant with respect to the transformation $p \to \Lambda p$, $(\Lambda \in \mathscr{L}^{\uparrow}_{+})$.

The dimension of $\mathscr{H}_p$ for all points $p$ of the orbit has to be uniform and if the dimension is 1 (spin 0), the irreducible unitary representation is determined as follows:

$$\mathscr{H} = \left\{ \Psi(\mathbf{p}); \|\Psi\|^2 \equiv \int |\Psi(\mathbf{p})|^2 \mathrm{d}\mu(\mathbf{p}) < \infty \right\}, \tag{3.42}$$

$$[U((a,\Lambda))\Psi](\mathbf{p}) = \mathrm{e}^{\mathrm{i}(a,p)}\Psi(\Lambda^{-1}\mathbf{p}), \tag{3.43}$$

where $\Lambda\mathbf{p}$ represents the spatial part of $\Lambda p$, and

$$p = (p^0, \mathbf{p}), \qquad p^0 = \sqrt{\mathbf{p}\cdot\mathbf{p} + m^2}. \tag{3.44}$$

The unitarity of this representation follows from the $\mathscr{L}^{\uparrow}_{+}$ invariance of the measure $\mathrm{d}\mu$.

II. *Classification by the spin* For a fixed point $p$ on the orbit, the subgroup under which $p$ is invariant,

$$G(p) \equiv \{A \in SL(2,C); \pi(A)p = p\}, \tag{3.45}$$

is called the *little group* (of the orbit). The action of $U(A) = U((0,A))$ can be decomposed into two parts, the map from $\mathscr{H}_p$ to $\mathscr{H}_{\pi(A)p}$ and the action of the little group on $\mathscr{H}_p$. The little groups corresponding to different points $p$ on the same orbit are mutually isomorphic by

$$G(\pi(B)p) = BG(p)B^{-1} \tag{3.46}$$

and the action of $G(p)$ on $\mathscr{H}_p$ can be reduced to the action at a specific point $p = p_0$. Thus, fixing a standard point $p_0$ on each orbit, and choosing $L(p) \in SL(2,C)$ for each point $p$ on the orbit such that

$$\pi(L(p))p_0 = p$$

holds, we identify $\mathscr{H}_p$ with the image of $\mathscr{H}_{p_0}$ through $U(L(p))$. Considering the $U(A)$ corresponding to $A \in G(p_0)$ as an irreducible unitary representation $D(A)$ of $G(p_0)$ on $\mathscr{K} = \mathscr{H}_{p_0}$, the following structure of the representation $U$ is determined:

$$\mathscr{H} = \left\{ \Psi(\mathbf{p}) \in \mathscr{K}; \|\Psi\|^2 \equiv \int \|\Psi(\mathbf{p})\|^2_{\mathscr{K}} \, \mathrm{d}\mu(\mathbf{p}) < \infty \right\}, \tag{3.47}$$

$$\begin{aligned} [U((a,A))\Psi](\mathbf{p}) &= \mathrm{e}^{\mathrm{i}(a,p)}[D(R(A,p))\Psi](\pi(A)^{-1}p) \\ R(A,p) &= L(p)^{-1}AL(\pi(A)^{-1}p) \in G(p_0). \end{aligned} \tag{3.48}$$

Here we omit the subtle discussion concerning the measure zero sets.

According to the above discussion, an irreducible unitary representation of $\tilde{\mathscr{P}}^{\uparrow}_{+}$ can be classified by an orbit and an irreducible unitary representation $D$ of the little group.

In the case of the orbit $m_+$ of positive mass $m > 0$, with $p_0 = (m, 0, 0, 0)$ the little group is the universal covering group of the three dimensional proper rotation group.

$$SU(2) = \{\mathrm{e}^{\mathrm{i}\mathbf{x}\cdot\boldsymbol{\sigma}}; \mathbf{x} \in \mathbf{R}^3\}.$$

Here $\sigma = (\sigma_1, \sigma_2, \sigma_3)$ are the Pauli spin matrices defined in (3.25) and $\mathrm{e}^{\mathrm{i}\mathbf{x}\cdot\boldsymbol{\sigma}}$ gives all $2 \times 2$ unitary matrices with determinant 1. $\pi(\mathrm{e}^{\mathrm{i}\mathbf{x}\cdot\boldsymbol{\sigma}})$ is a rotation by an angle $-2|\mathbf{x}|$ around the $\mathbf{x}$ direction. Its irreducible unitary representations can be completely classified by the parameter

$$j = 0, \tfrac{1}{2}, 1, \tfrac{3}{2}, \ldots$$

which gives the eigenvalues $j(j+1)$ of the square of the angular momentum. $j$ is called the *spin*. Using the terminology of the Lie algebra of $\tilde{\mathscr{P}}^{\uparrow}_{+}$, among polynomials in the enveloping algebra of this Lie algebra, we have the following polynomials commuting with all elements:

$$(\mathfrak{p}, \mathfrak{p}) = \sum g^{\mu\nu} \mathfrak{p}_\mu \mathfrak{p}_\nu, \tag{3.49}$$

$$(\mathfrak{w}, \mathfrak{w}) = \sum g_{\mu\nu} \mathfrak{w}^\mu \mathfrak{w}^\nu, \quad \mathfrak{w}^\kappa = \frac{1}{2} \sum_{\lambda,\mu,\nu} \varepsilon^{\kappa\lambda\mu\nu} \mathfrak{p}_\lambda \mathfrak{w}_{\mu\nu}, \tag{3.50}$$

where the $\epsilon^{\kappa\lambda\mu\nu}$ are completely antisymmetric with respect to permutation of the indices, and $\epsilon^{0123} = 1$. For the above representation, the eigenvalue of the first polynomial is $-m^2$ and the eigenvalue of the latter one is $-m^2 j(j+1)$. (Following the convention used in physics we include the i in the elements of the Lie algebra, thus the value of $(\mathfrak{p}, \mathfrak{p})$ is not $m^2$ but $-m^2$.) For the case that spin $j$ is an integer we get a representation of $\mathscr{P}^{\uparrow}_{+}$.

Let us also discuss briefly the other orbits.

For the orbit of mass zero $0_+$ (the same for $0_-$), we take

$$p_0 = (1, 0, 0, 1).$$

The little group is

$$\left\{ A = A(z,\theta) = \begin{pmatrix} \alpha & z\bar{\alpha} \\ 0 & \bar{\alpha} \end{pmatrix}, \ \alpha = e^{i\theta}, \ z \in \mathbf{C} \right\}$$

and is isomorphic to the double covering group of the two-dimensional Euclidean group. ($A(z, 0)$ represents a translation $w \longrightarrow w + z$ on the complex plane and $A(0, \theta)$ represents a rotation by an angle $2\theta$.) Among its irreducible unitary representations, the one-dimensional representation is represented as

$$U(A(z,\theta)) = e^{2ik\theta}, \qquad k = 0, \pm\tfrac{1}{2}, \pm 1, \ldots$$

$|k|$ is called the *spin*, and the sign of $k$ is called the *helicity*. All other irreducible representations of the little group are infinite dimensional and are not used in physics; therefore we do not mention them here.

The little group of the orbit 0 is isomorphic to $\mathscr{L}_+^\uparrow$ and all representations are infinite dimensional, except for the trivial representation $U(g) = \mathbf{1}$. Since in physics we usually use only the trivial representation, we do not mention the other representations here.

In the case of the orbit $im$ we take $p_0 = (0, 0, 0, 1)$. Then the little group becomes the three-dimensional restricted Lorentz group. All its irreducible representations are infinite dimensional, except for the trivial one. This orbit possesses states of negative energy, called tachyons. Since we are not going to discuss such objects in this book we also do not go into details here.

We have discussed the irreducible unitary representations of $\tilde{\mathscr{P}}_+^\uparrow$ as the minimal units possessing relativistic invariance within the framework of quantum theory. They are interpreted as the formulation of a relativistic one particle system. This means that kinematically it is not possible to decompose them further into subunits. (In this sense a bound state in a many particle system is considered as a one particle system, as long as it is stable.)

Among these one particle systems the ones which correspond to positive mass are classified completely by the spin $j$, i.e. the properties under rotation in the rest frame (corresponding to $p^0$), and the representation with the mass $m > 0$ and the spin $j$ is given by (3.47) and (3.48), where $D$ is the spin $j$ representation of $SU(2)$, the group representing rotations.

## 3.5 Free particle system and Fock space

In this section we consider a many particle system consisting of a collection of one particle systems as described in the previous section, when there is no interaction among the particles, i.e. a system of free particles.

Let us consider two particles described in the Hilbert spaces $\mathscr{H}_i$ $(i = 1, 2)$, without mutual interaction. The observables of the two particle system are generated by the operator algebras $B(\mathscr{H}_1)$ and $B(\mathscr{H}_2)$ of the observables of each individual particle. If these observables of particles 1 and 2 are simultaneously measurable (which is a formulation of the assumption that there is no interaction), i.e. if these observables mutually commute, then the corresponding von Neumann algebra is given by the tensor product of the algebras (see Appendix A)

$$B(\mathscr{H}_1) \otimes B(\mathscr{H}_2) = B(\mathscr{H}_1 \otimes \mathscr{H}_2),$$

which is the algebra of all (bounded linear) operators on the tensor product $\mathscr{H}_1 \otimes \mathscr{H}_2$ of Hilbert spaces.

From the point of view of states, if we require a product state $\varphi_1 \otimes \varphi_2$ of two normal pure states $\varphi_i (i = 1, 2)$ over $B(\mathscr{H}_i)$ to satisfy

$$(\varphi_1 \otimes \varphi_2)(A_1 \otimes A_2) = \varphi_1(A_1)\varphi_2(A_2) \tag{3.51}$$

for $A_i \in B(\mathscr{H}_i)$, then this product state is necessarily a normal pure state over $B(\mathscr{H}_1 \otimes \mathscr{H}_2)$ and becomes a vector state given by the tensor product $\psi_1 \otimes \psi_2 \in \mathscr{H}_1 \otimes \mathscr{H}_2$ of the vectors $\psi_i$ in $\mathscr{H}_i$ which represents each state $\varphi_i$ as a vector state. Equation (3.51) gives the relation which characterizes independent events in probability theory.

Similarly, we will describe a many particle system consisting of a finite number $i = 1, 2, \ldots, n$ of one particle systems without mutual interaction by using the tensor product

$$\bigotimes_{i=1}^{n} B(\mathscr{H}_i) = B\left(\bigotimes_{i=1}^{n} \mathscr{H}_i\right). \tag{3.52}$$

While the above scheme can be applied to the case of different types of particles, we have the additional feature that the particles cannot be distinguished for a many particle system of the same kind of particles. In the case that all particles $i = 1, \ldots, n$ in (3.52) are identical, define the operator representing the permutation $P$ of particles by

$$S_P(\Phi_{P(1)} \otimes \cdots \otimes \Phi_{P(n)}) = \Phi_1 \otimes \cdots \otimes \Phi_n.$$

We then obtain a representation of the permutation group $(S_{P_1} S_{P_2} = S_{P_1 P_2})$. As for the observables, we take all those operators which commute with all $S_P$ and we consider their irreducible representations. In this way we obtain inequivalent irreducible representations of observables corresponding to irreducible representations of the permutation group. However, so far only the completely symmetric and the completely antisymmetric cases are found in nature. The projection operators from the $n$th tensor product $\mathscr{H}^{\otimes n} = \mathscr{H}_1 \otimes \cdots \otimes \mathscr{H}_n$ of $\mathscr{H}_i = \mathscr{H}$ to the subspace of each irreducible representation are given by

$$S_n^+ = (n!)^{-1} \sum S_P, \qquad \text{(completely symmetric)}, \tag{3.53}$$

$$S_n^- = (n!)^{-1} \sum (\text{sign } P) S_P \qquad \text{(completely antisymmetric)}, \tag{3.54}$$

(sign $P$ is 1 for even permutations and $-1$ for odd permutations). These representations have the characteristic property that even if we restrict them to an $m$-particle subsystem ($m < n$), no other type of symmetry appears.

In order to consider an arbitrary number $n$ of particles we take the direct sum of spaces for different $n$:

$$F_\pm(\mathscr{H}) = \bigoplus_n F_\pm^n(\mathscr{H}), \quad F_\pm^n(\mathscr{H}) = S_n^\pm(\mathscr{H}^{\otimes n}). \tag{3.55}$$

$F_+$ and $F_-$ are called the *Boson* and *Fermion Fock* space [4], respectively. The $n$ summation runs over all natural numbers and 0. $F_\pm^0$ is a one-dimensional space representing the *vacuum state* in which no particle is present. We choose one vector $\Omega$ in $F_\pm^0$ and call it the *Fock vacuum*.

For $f \in \mathscr{H}$ we define a *creation operator* $(a^*, f)$ by

$$\begin{aligned} &(a^*, f) S_n^\pm (\Phi_1 \otimes \cdots \otimes \Phi_n) = \sqrt{n+1} S_{n+1}^\pm (f \otimes \Phi_1 \otimes \cdots \otimes \Phi_n) \\ &(a^*, f)\Omega = f. \end{aligned} \tag{3.56}$$

It is a closable linear operator defined on a dense domain

$$D_0^\pm \equiv \left\{ \Psi \in \bigoplus_{j=0}^{n} F_\pm^n(\mathscr{H}); n = 0, 1, 2, \ldots \right\}, \tag{3.57}$$

in $F_\pm(\mathscr{H})$. For $f_j \in \mathscr{H}$ ($j = 1, \ldots, n$), we introduce the symbol

$$\Phi_\pm(f_1 \cdots f_n) = (a^*, f_n) \cdots (a^*, f_n)\Omega = \sqrt{n!} S_n^\pm (f_1 \otimes \cdots \otimes f_n). \tag{3.58}$$

Then their inner product is given by

$$(\Phi_\pm(f_1 \cdots f_n), \Phi_\pm(g_1 \cdots g_m)) = \delta_{nm} \sum_P \varepsilon_P^\pm (f_1, g_{P(1)}) \cdots (f_n, g_{P(n)}), \tag{3.59}$$

where $\varepsilon_P^+ = 1$, $\varepsilon_P^- = \text{sign } P$.

The adjoint operator of $(a^*, f)$ restricted to $D_0^\pm$ is denoted as $(f, a)$ and called an *annihilation operator*. Explicitly it is given by

$$(f, a)\Phi_\pm(f_1 \cdots f_n) = \sum_{j=1}^{n} (\pm 1)^{j-1} (f, f_j) \Phi_\pm(f_1 \cdots \hat{f_j} \cdots f_n), \tag{3.60a}$$

$$(f, a)\Omega = 0, \tag{3.60b}$$

where $f_j$ is to be omitted from the variables of $\Phi_\pm$ on the r.h.s. (this omission being indicated by $\hat{f_j}$).

On $F_+(\mathscr{H})$ both creation and annihilation operators are unbounded and leave the common domain $D_0^\pm$ invariant. $(a^*,f)$ is linear in $f$, $(f,a)$ is conjugate linear in $f$, and the following relations on $D_0^+$, called the *canonical commutation relations*, hold:

$$\begin{aligned}&[(a^*,f_1),(a^*,f_2)] = [(f_1,a),(f_2,a)] = 0,\\ &[(f_1,a),(a^*,f_2)] = (f_1,f_2)\mathbf{1},\end{aligned} \tag{3.61}$$

where $[X, Y] = XY - YX$.

On $F_-(\mathscr{H})$ the creation and annihilation operators are bounded and satisfy

$$\|(a^*,f)\| = \|(f,a)\| = \|f\|(= (f,f)^{1/2}).$$

On $D_0^-$ the following relations, called the *canonical anticommutation relations*, are satisfied.

$$\begin{aligned}&[(a^*,f_1),(a^*,f_2)]_+ = [(f_1,a),(f_2,a)]_+ = 0,\\ &[(f_1,a),(a^*,f_2)]_+ = (f_1,f_2)\mathbf{1},\end{aligned} \tag{3.62}$$

where $[X, Y]_+ = XY + YX$.

Corresponding to an operator $A$ on $\mathscr{H}$ we define an operator $\Gamma(A)$ on $D_0^\pm$ by

$$\Gamma(A)S_n^\pm(f_1\otimes\cdots\otimes f_n) = S_n^\pm(Af_1\otimes\cdots\otimes Af_n).$$

On $F_\pm^n$, $\Gamma(A)$ is given by $A^{\otimes n}$. In particular, we define

$$\Gamma(A)\Omega = \Omega. \tag{3.63}$$

$\Gamma(A)$ leaves each $F_\pm^n$ invariant and it satisfies the following relations on $D_0^\pm$:

$$\Gamma(A)\Gamma(B) = \Gamma(AB), \quad \Gamma(\mathbf{1}) = \mathbf{1}. \tag{3.64}$$

Thus for a unitary operator $U$, $\Gamma(U)$ is unitary. If a unitary representation $U(g)$, $g \in G$ of the group $G$ on $\mathscr{H}$ is given, then we obtain a unitary representation $\Gamma(U(g)), g \in G$ on the corresponding Fock space $F_\pm(\mathscr{H})$. Creation and annihilation operators are transformed by $\Gamma(U)$ as

$$\Gamma(U)(a^*,f)\Gamma(U)^* = (a^*, Uf), \tag{3.65a}$$

$$\Gamma(U)(f,a)\Gamma(U)^* = (Uf,a). \tag{3.65b}$$

We define the Segal field $\phi(f)$ for $f \in \mathscr{H}$ as

$$\phi(f) = (a^*,f) + (f,a). \tag{3.66}$$

$\phi(f)$ is essentially self-adjoint. The so-called Weyl operator defined by

$$W(f) = \mathrm{e}^{\mathrm{i}\overline{\phi(f)}} \qquad (\overline{\phi(f)} \text{ is the closure of } \phi(f)) \tag{3.67}$$

is a unitary operator and satisfies the following relation, called the Weyl form of the canonical commutation relations:

$$W(f_1)W(f_2) = W(f_1 + f_2)\exp[-\mathrm{i}\,\mathrm{Im}(f_1, f_2)], \tag{3.68}$$

where $\mathrm{Im}\, z$ represents the imaginary part of the complex number $z$.

Next, we consider the Fock space related to the irreducible representation of positive mass $m > 0$ and spin 0 given in (3.42) and (3.43). For a well-behaved real function $h(x)$ on the four-dimensional space $M$ (say, a $C^\infty$-function with compact support), define its Fourier transform by

$$\tilde{h}(p) = (2\pi)^{-3/2} \int \mathrm{e}^{\mathrm{i}(p,x)} h(x)\, \mathrm{d}^4 x. \tag{3.69}$$

Consider the restriction of $\tilde{h}(p)$ to the orbit $m_+$ as a vector in a Hilbert space $\mathscr{H}$ in (3.42), and denote it by $\hat{h}$. Using (3.66) we define

$$A(h) = \int A(x)h(x)\, \mathrm{d}^4 x \equiv \phi(\hat{h})$$

$A(h)$ is real linear in $h$ (i.e. $A(r_1 h_1 + r_2 h_2) = r_1 A(h_1) + r_2 A(h_2)$ for real coefficients $r_1$ and $r_2$) and the map

$$h \to (\Psi, A(h)\Phi)$$

is a Schwartz distribution for any vectors $\Psi$, $\Phi$ in $D_0^\pm$. In this sense $A(x)$ is called an operator-valued distribution, $h(x)$ is called a test function and the integral sign was used symbolically in the defining equation for $A(h)$ above. The $A(x)$ defined here is called a *(neutral) free scalar field*. On $D_0^+$, $A(h)^* = A(h)$ holds.

The field $A(x)$ satisfies the following Klein–Gordon equation:

$$(\Box + m^2)A(x) = 0, \quad \Box = (\partial/\partial x^0)^2 - \sum_{k=1}^{3} (\partial/\partial x^k)^2 \tag{3.70}$$

and also the following commutation relations in four dimensions:

$$[A(x), A(y)] = \mathrm{i}\Delta(x - y), \tag{3.71}$$

$$\Delta(x) = -(2\pi)^{-3} \int_{m_+} \sin(p, x)\, \mathrm{d}\mathbf{p}/p^0. \tag{3.72}$$

These equations are to be interpreted in the distribution sense, namely they are satisfied after multiplication of a well-behaved function $h(x)$, integration with respect to each variable, and a formal partial integration such that the derivatives act on $h$. Therefore, (3.70) means that $A((\Box + m^2)h) = 0$ holds for any $h$, and if we compute $\tilde{h}_1$ from $h_1 = (\Box + m^2)h$, we immediately see the validity of this equation because $\tilde{h}_1$ becomes automatically 0 on the orbit $m_\pm$. (3.71) means

$$[A(h_1), A(h_2)] = 2\,\mathrm{Im} \int_{m_+} \tilde{h}_1(p)^* \tilde{h}_2(p)\, \mathrm{d}\mu(\mathbf{p})\mathbf{1}$$

where $\mathrm{d}\mu$ is the invariant measure given by (3.41). These equations hold when acting on a vector in the domain $D_0^+$ which is dense in $F_+(\mathscr{H})$.

The free neutral scalar field $A(x)$ is one example of a Wightman field to be discussed in the next chapter. We see that if $x - y$ is spacelike, we get from (3.71)

$$[A(x), A(y)] = 0. \tag{3.73}$$

(This means that if the support of $h_1$ and that of $h_2$ are mutually spacelike then

$$[A(h_1), A(h_2)]\Psi = 0, \quad \Psi \in D_0^+.)$$

We can define free fields also for the case where the spin is not 0 (see Appendix C) and, for odd half-integer spin, eq. (3.73) holds for the anticommutator $[X, Y]_+ = XY + YX$. The profound reason for the latter situation is discussed in Chapter 6.

### 3.6 Energy momentum

The energy $\Pi^0$ and the momentum vector $\mathbf{\Pi} = (\Pi^1, \Pi^2, \Pi^3)$ are required to satisfy the following conditions as self-adjoint operators representing unbounded observables.

1. They are simultaneously measurable, i.e. they have a common spectral resolution and relations completely analogous to (3.33) and (3.40) hold for $\Pi^\mu$. (We do not require that its resolution has to be the same as the resolution for $P$.)
2. They are conserved under space–time translations and transform like a four-vector under homogeneous Lorentz transformations:

$$U(a, A)\Pi^\mu U(a, A)^* = \sum_\nu \pi(A^{-1})^\mu_\nu \Pi^\nu. \tag{3.74}$$

3. They are additive under composition of independent systems, i.e. from $\Pi_i^\mu$ of each system described in $\mathscr{H}_i$, we get

$$\Pi^\mu = \sum_i \mathbf{1} \otimes \cdots \otimes \mathbf{1} \otimes \Pi_i^\mu \otimes \mathbf{1} \otimes \cdots \otimes \mathbf{1} \tag{3.75}$$

on $\otimes \mathscr{H}_i$ for the composed system.

First in the irreducible unitary representation of mass $m > 0$ and spin $j$, $\Pi^\mu$ has to be proportional to the generator $P^\mu$ of translations, as defined in Section 3.4.

$$\Pi^\mu = \alpha P^\mu. \tag{3.76}$$

The reason for this can be explained briefly as follows. Since $\Pi^\mu$ and $T(a)$ commute we get

$$\Pi^\mu = \int \Pi^\mu(p)\, \mathrm{d}\mu(\mathbf{p}).$$

Furthermore, since all $U((a, A))$ commute with the inner product $(P, \Pi)$, we can conclude by the irreducibility of the representation that

$$(P, \Pi) = \int (p, \Pi(p))\, \mathrm{d}\mu(\mathbf{p}) = \alpha m^2 \mathbf{1}$$

where $\alpha$ is a real number. Next, at $p = p_0$ we can see that $\Pi^k(p)$ $(k = 1, 2, 3)$ transforms as a three-dimensional vector under rotation. Due to the Wigner–Eckart theorem [5] of irreducible projective unitary representations of the rotational group, $\Pi^k(p)$ is proportional to the vector formed by the generators of the Lie algebra of the rotational group. Since the $\mathfrak{w}^\mu$ introduced in (3.50) are invariant under translation, we can represent them by $\int \mathfrak{w}^\mu(p)\, \mathrm{d}\mu(p)$, and thus $\Pi^k(p)$ is proportional to $\mathfrak{w}^k(p)$ at $p = p^0$. Since $(\mathfrak{p}, \mathfrak{w}) = 0$, $\mathfrak{w}^0(p) = 0$ at $p = p_0$. Therefore if we take the transformation behaviour under $U(A)$ into account, we can conclude that

$$\Pi^\mu(p) = \alpha p^\mu + \beta(p)\mathfrak{w}^\mu(p).$$

From condition (1) we obtain

$$0 = [\Pi^\mu(p), \Pi^\nu(p)] = \beta(p)^2 [\mathfrak{w}^\mu(p), \mathfrak{w}^\nu(p)]$$

and thus $\beta(p) = 0$. Hence we get (3.76).

A similar discussion applies to representations of mass 0 and spin $j$.

For a free particle system, $P^\mu$ satisfies the same equation as (3.75) on the Fock space due to condition (3). Hence (3.76) holds in the Fock space of each particle. At this stage $\alpha$ may differ depending on the particle.

The rest of the discussion in the following proceeds under some suitable assumptions. First we assume that if we wait sufficiently long, then all stable particles move straight and the mutual distance becomes infinitely large (with probability 1), thus the system approaches a free particle system. (This feature is explained in Chapter 5 when we discuss scattering theory.) In this asymptotic situation we can apply the above discussion and (3.76) holds for each stable particle. We also assume a similar feature for the time $t \to -\infty$. Finally in real phenomena, any stable particle can be produced from other particles by pair creation. In such a process, $P^\mu$ is invariant under time translation due to the relativistic invariance and

$\Pi^\mu$ satisfies the same property by condition (2). Thus, if we compute $P$ and $\Pi$ at $t = +\infty$ and at $t = -\infty$ by (3.76) for each particle and condition (3), we find that if the pair creation of another kind of particle occurs in a system of one kind of particles, then the $\alpha$ for both kinds of particles must be identical.

If we find by the above discussion that $\alpha$ is a common constant for all particles, then (3.76) generally holds due to the argument of the scattering states at $t = +\infty$ and condition (3). $\alpha$ is normally taken to be $\hbar^{-1}$, i.e. $2\pi$ divided by the Planck constant, and we take the unit system where this constant is 1. Then we can call $P^\mu$ the *energy momentum*.

# 4
# LOCAL OBSERVABLES

In this chapter we consider for each space–time domain the set of all those observables which can be measured there and we discuss its basic properties, especially in connection with the vacuum state.

## 4.1 General properties of local observables

Each measurement of an observable is performed in some specific limited space–time domain. For example, an observable measured in a spatially limited region $A$ of the laboratory during a limited time period $T$ can be considered as an observable measured in the space–time domain $T \times A$. Thus, it is possible to define the set of all observables $\mathcal{O}(D)$ which can be measured in each space–time domain $D$. The esssence of field theory is to investigate the relation between space–time points and observables. The main point of this book is to investigate in particular the theory of the set $\mathcal{O}(D)$ of observables determined by each space–time domain $D$. As a starting point we present in this section the basic properties satisfied by $\mathcal{O}(D)$ as axioms and explain their physical meaning.

The discussion is based on the relativistic quantum theory described in Chapters 2 and 3. Therefore $\mathcal{O}(D)$ is a subset of the $C^*$ algebra $\mathfrak{A}$ formed by the observables, and the relativistic symmetry is given by the inhomogeneous Lorentz group $\mathscr{P}_+^\uparrow$, each transformation $g \in \mathscr{P}_+^\uparrow$ being represented by an automorphism $\alpha_g$ of $\mathfrak{A}$. (The Hilbert space and the unitary representation on it appear later when we discuss states.) $\alpha_g(Q)$ is assumed to be continuous in $g$.

Concerning $\mathcal{O}(D)$ we make the following four basic assumptions.

**Axioms for local observables** ($D$ is a finite space–time domain).

1. *Monotone property*: If $D_1 \supset D_2$ then $\mathcal{O}(D_1) \supset \mathcal{O}(D_2)$.
2. *Covariance*: For $g = (a, \Lambda) \in \mathscr{P}_+^\uparrow$, $\alpha_g \mathcal{O}(D) = \mathcal{O}(gD)$, where $gD = \{\Lambda x + a; x \in D\}$.
3. *Locality*: If $D_1$ and $D_2$ are spacelike separated, then $\mathcal{O}(D_1)$ and $\mathcal{O}(D_2)$ commute.
4. *Generating property*: $\cup_D \mathcal{O}(D)$ generates the algebra $\mathfrak{A}$ of observables as a $C^*$ algebra.

In the following we explain the meaning of these axioms.

1. An observable which can be measured in a certain domain $D_2$ will be considered to be also measurable in a bigger domain $D_1$. Therefore this is a natural assumption.

2. It follows from the discussion about symmetries given in Section 1.8 and from the meaning of relativistic symmetry given in Section 3.3 that if an observable measurable in a space–time domain $D$ is transformed by $\alpha_g$, we obtain an observable which is measurable in the transformed space–time domain $gD$. Therefore $\alpha_g \mathcal{O}(D) \subset \mathcal{O}(gD)$. It then follows that

$$\alpha_{g^{-1}} \mathcal{O}(gD) \subset \mathcal{O}(g^{-1}gD) = \mathcal{O}(D)$$

holds and we obtain $\alpha_g \mathcal{O}(D) = \mathcal{O}(gD)$.

3. This assumption concerns the relativistic causality mentioned in Section 3.1. The statement that $D_1$ and $D_2$ are spacelike separated means that $x - y$ is spacelike, for any point $x$ in $D_1$ and any point $y$ in $D_2$. By the definition given in Section 3.1, this is equivalent to the statement that $D_2$ is contained in the causal complement $D_1'$ of $D_1$. According to the theory of special relativity, an event in $D_1$ and an event in $D_2$ do not interfere each other. Thus, if we perform a measurement in both regions $D_1$ and $D_2$ we get the same result as if the measurement were performed in each region separately. This means that a measurement of an observable $Q_1 \in \mathcal{O}(D_1)$ in $D_1$ and a measurement of an observable $Q_2 \in \mathcal{O}(D_2)$ in $D_2$ can be performed simultaneously, and therefore according to the discussion in Section 1.3 and eq. (2.1), the observables $Q_1$ and $Q_2$ commute. Hence we arrive at the assumption of locality which states that the operators in $\mathcal{O}(D_1)$ commute with those in $\mathcal{O}(D_2)$.

4. This is the assumption that we only consider observables measurable in a bounded space–time domain. The statement that a $C^*$ algebra is generated means that we consider a number of operators corresponding to the observables measurable in bounded space–time domains, we construct sequences of polynomials with complex coefficients from these operators and take limits in the norm topology; taking together all operators obtained in this manner, we obtain the $C^*$ algebra mentioned above.

An operator corresponding to an observable is usually self-adjoint. Hence, there are no other assumptions on $\mathcal{O}(D)$ than that it is a set of self-adjoint operators in a $C^*$ algebra $\mathfrak{A}$. Note that we can consider the $C^*$ subalgebra $\mathfrak{A}(D)$ generated by $\mathcal{O}(D)$, and for $\mathfrak{A}(D)$ the above axioms hold in exactly the same form:

1. Monotone property: If $D_1 \supset D_2$ then $\mathfrak{A}(D_1) \supset \mathfrak{A}(D_2)$.
2. Covariance: $\alpha_g \mathfrak{A}(D) = \mathfrak{A}(gD)$.
3. Locality: If $D_1' \supset D_2$, then $\mathfrak{A}(D_1)' \supset \mathfrak{A}(D_2)$.
4. Generating property: $\cup_D \mathfrak{A}(D)$ is dense in $\mathfrak{A}$ with respect to the norm topology.

$\mathfrak{A}(D)'$ denotes the set of all operators commuting with all operators in $\mathfrak{A}(D)$, where we only consider operators in the algebra $\mathfrak{A}$ in the present situation. Since all $\mathfrak{A}(D)$ are $C^*$ algebras, the generating property now becomes the condition that we obtain $\mathfrak{A}$ if all norm limits are added.

Usually an operator in $\mathfrak{A}(D)$ is also called a *local observable* in the domain $D$ (including also operators which are not self-adjoint or not normal). The set $\{\mathfrak{A}(D)\}$ satisfying the above four properties is called a *local* $C^*$ *system*.

For the mathematical deduction it is often convenient if each local algebra $\mathfrak{A}(D)$ is a von Neumann algebra. A von Neumann algebra can be obtained by considering an appropriate representation $\pi$ of a local $C^*$ system (for example, the representation associated with a vacuum state discussed in the next section), and defining the von Neumann algebra $W(D)$ generated by $\pi(\mathfrak{A}(D))$ for each $D$. Then $W(D)$ satisfies assumptions (1)–(3). In this case $\mathfrak{A}$ is the $C^*$ algebra generated by all $W(D)$. If each local algebra $\mathfrak{A}(D)$ is a von Neumann algebra, then we speak of a *local $W^*$ system*. In this case it is appropriate to assume for the continuity of $\alpha_g$ that for each $D$ the map $Q \in \mathfrak{A}(D) \to \alpha_g(Q)$ is continuous in the weak operator topology of the von Neumann algebra instead of the norm topology. (This is equivalent to continuity in the strong operator topology.)

As for the shape of the bounded domain $D$, the double cone

$$V_q^p \equiv \{x; p - x \in V_+, x - q \in V_+\} \tag{4.1}$$

is considered to be the simplest choice. It is the intersection of an open past cone $p + V_-$ with vertex $p$ and an open future cone $q + V_+$ with vertex $q$, where $p - q$ is positive timelike. Thus the double cone is the union of two circular cones with vertices $p$ and $q$, and the common base which is a sphere of radius $|p - q|/2$. The centre of this sphere is the intersection of the segment $\overline{pq}$ with a hyperplane bisecting this segment orthogonally.

## 4.2 The vacuum state

In order to define the vacuum state, let us first explain the notion of energy-increasing and energy-decreasing operators.

Let the Fourier transform of a $C^\infty$-function $\tilde{g}$ be

$$g(x) = (2\pi)^{-4} \int \tilde{g}(p) \mathrm{e}^{-\mathrm{i}(p,x)} \,\mathrm{d}^4 p. \tag{4.2}$$

For each four-dimensional bounded closed set $\Delta$, the operator

$$Q(g) = \int \alpha_{(x,1)}(Q) g(x) \,\mathrm{d}^4 x \tag{4.3}$$

constructed from any element $Q \in \mathfrak{A}$ and any $C^\infty$-function $\tilde{g}$ with support in $\Delta$ will be called an *operator which increases the energy momentum by* $\Delta$. Here we use the notation $(p, x) = p^0 x^0 - \mathbf{p} \cdot \mathbf{x}$ introduced in Section 3.1. The origin of the name is as follows.

Consider a representation $\pi$ of $\mathfrak{A}$ to which the discussion in Chapter 3 can be applied, and assume that the relativistic symmetry $\tilde{\mathscr{P}}_+^\uparrow$ is represented by a unitary representation $U(a, A)$ as

$$\pi(\alpha_{(a,\Lambda)}(Q)) = U(a, A)\pi(Q)U(a, A)^*, \tag{4.4}$$

where $\Lambda = \Lambda(A)$. Furthermore, assume that the energy-momentum interpretation of the generators of translations $U(a, \mathbf{1})$ given in Section 3.6 applies, and take the unit system where the proportionality constant becomes 1. Then the following lemma holds:

**Lemma 4.1**
If the energy momentum spectrum of a vector $\xi$ in the Hilbert space is a closed set $F$, then the energy-momentum spectrum of $Q(g)\xi$ is contained in $F+\Delta$.

If the spectral decomposition of the translation $U(a, \mathbf{1})$ is given by

$$U(a,\mathbf{1}) = \int e^{i(a,p)} E(d^4p), \tag{4.5}$$

then the support of the measure $\|E(d^4p)\xi\|^2$ is the energy-momentum spectrum of $\xi$. A formal computation using (4.3), (4.4) and the above spectral decomposition yields

$$Q(g)\xi = \int E(d^4p)Q\tilde{g}(p-P)\xi = \int E(d^4p)Q\tilde{g}(p-q)E(d^4q)\xi.$$

Then the region of the $p$ integration is given by the sum of the support $F$ of $E(d^4q)\xi$ and the support of $\tilde{g}$ which is in $\Delta$, and it contains the energy-momentum spectrum of $Q(g)\xi$ by definition. It is easy to make this a mathematical proof, but we do not give it explicitly here.

From this lemma the above interpretation of $Q(g)$ follows. For the definition of the vacuum we still need another lemma.

**Lemma 4.2**
Given a $C^\infty$-function $\tilde{g}$, the support of which is bounded and has no intersection with $\bar{V}_+$, there exist finitely many positive timelike unit vectors $e_i$, $(i=1,\dots,n)$ and $C^\infty$-functions $\tilde{g}_i$, $(i=1,\dots,n)$ with respective support in

$$\{p;\ (p,e_i)<0\} \equiv M_-(e) \tag{4.6}$$

such that

$$\tilde{g} = \tilde{g}_1 + \cdots + \tilde{g}_n. \tag{4.7}$$

**Outline of the proof** The complement of $\bar{V}_+$ is the union of $M_-(e)$ over all positive timelike unit vectors $e$. $M_-(e)$ is an open set and since by assumption the support of $\tilde{g}$ is compact, it is contained in the union of $M_-(e_i)$ for a finite number of $e_i$ $(i=1,\dots,n)$. Hence, there exist positive-valued $C^\infty$-functions $h_i$ with respective support in $M_-(e_i)$ satisfying

$$\sum h_i(p) = 1$$

on the support of $\tilde{g}$ (resolution of 1). Now, taking $\tilde{g}_i = \tilde{g}h_i$, the lemma is satisfied. □

In a coordinate system where the time axis lies in direction of $e_i$, $(p, e_i)$ is the $0th$ component of $p$ and can be interpreted as the energy. Denoting the Fourier transform (4.2) of $\tilde{g}_i$ by $g_i$, we can interpret $Q(g_i)$ as the operator which decreases the energy in such a coordinate system.

For each state $\varphi$, the set of all operators annihilating $\varphi$

$$\ker\varphi \equiv \{Q \in \mathfrak{A}, \varphi(Q^*Q) = 0\} \tag{4.8}$$

is a left ideal of $\mathfrak{A}$ by the discussion in Section 2.3.

**Definition 4.3**
A state $\varphi$ is called a *vacuum* if $Q \in \ker\varphi$ for any operator $Q \in \mathfrak{A}$ which decreases the energy in some coordinate system (depending on $Q$).

The meaning of the above definition is that the vacuum is the state of lowest energy (and therefore stable) under any local perturbation.

**Corollary 4.4**
$\varphi$ is the vacuum state if and only if $Q(g)$ defined in (4.2) and (4.3) is in $\ker\varphi$ for any $Q \in \mathfrak{A}$ and for any $C^\infty$-function $\tilde{g}$, the support of which is bounded and has no intersection with the closed future cone $\bar{V}_+$.

This follows immediately from Lemma 4.2.

**Theorem 4.5**
(1) Any vacuum state is translation invariant, i.e. for any $Q \in \mathfrak{A}$

$$\varphi(\alpha_{(a,1)}Q) = \varphi(Q) .$$

(2) On the GNS representation space of the vacuum state $\varphi$, a unitary representation $T_\varphi$ of the translation group satisfying

$$T_\varphi(a)\pi_\varphi(Q)\Omega_\varphi \equiv \pi_\varphi(\alpha_{(a,1)}Q)\Omega_\varphi \tag{4.9a}$$

can be defined and the spectral (projection operator valued) measure $E_\varphi$ defined by

$$T_\varphi(a) = \int \mathrm{e}^{\mathrm{i}(p,a)} E_\varphi(\mathrm{d}^4 p) \tag{4.9b}$$

has its support contained in $\bar{V}_+$.

**Outline of the proof** (1) The function

$$h(x) = \varphi(\alpha_{(x,1)}Q)$$

is a bounded ($|h(x)| \leq \|Q\|$) continuous (under assumed continuity of $\alpha_g(Q)$) function. It is enough to show that this function is constant.

Let us consider a $C^\infty$-function $\tilde{g}(p)$ with a bounded support not containing the origin. Similarly to the proof of Lemma 4.2 there exist finitely many positive timelike unit vectors $e_i$, signs $\sigma_i$ ($+$ or $-$) and $C^\infty$-functions $g_i$, ($i = 1, \ldots, n$) with respective support in

$$M_{\sigma_i}(e_i) = \{p; \sigma_i(p, e_i) > 0\}$$

such that we can decompose $\tilde{g}$ as

$$\tilde{g} = \tilde{g}_1 + \cdots + \tilde{g}_n$$

If $\sigma_i = -$ then $Q(g_i) \in \ker\varphi$ and, if $\sigma_i = +$ then $Q(g_i)^* = Q^*(\bar{g}_i) \in \ker\varphi$. In either case we get

$$\int h(x)g_i(x)\,\mathrm{d}^4x = \varphi(Q(g_i)) \; (= \overline{\varphi(Q(g_i)^*)}) = 0$$

Here $g_i$ is the Fourier transform of $\tilde{g}_i$ given by (4.2). By summing over $i$, we see that the Fourier transform $g$ given by (4.2) satisfies

$$\int h(x)g(x)\,\mathrm{d}^4x = 0$$

if the support of $\tilde{g}$ does not include the origin. This shows that the support of the Fourier transform $\tilde{h}$ of $h$ as a distribution is the origin, hence $\tilde{h}$ is given by a finite sum of $\delta$ functions and their finite order derivatives. Consequently, $h(x)$ has to be a polynomial in the variable $x$. Due to the boundedness of $h$, $h(x)$ is a constant.

(2) By Theorem 2.33 and Corollary 2.34 of Section 2.5 the existence of a continuous unitary representation $T_\varphi$ of the translation group satisfying (4.9) follows. If the support of $\tilde{g}$ has no common point with $\bar{V}_+$, then by Corollary 4.4 we get

$$\int T_\varphi(a)g(a)\,\mathrm{d}a\pi_\varphi(Q)\Omega_\varphi = \pi_\varphi(Q(g))\Omega_\varphi = 0$$

Thus for such a $\tilde{g}$

$$\int T_\varphi(a)g(a)\,\mathrm{d}a = \int \tilde{g}(p)E_\varphi(\mathrm{d}^4p) = 0$$

holds and hence the support of $E_\varphi$ is contained in $\bar{V}_+$. □

### 4.3 Irreducibility

In this section we assume that a vacuum state $\varphi$ is not merely translation invariant but also invariant under $\mathscr{P}_+^\uparrow$. By the following theorem the analysis of the vacuum state essentially reduces to the case where $\pi_\varphi$ is an irreducible representation and the translation invariant vectors form a one-dimensional subspace $\mathbf{C}\Omega_\varphi$ (these two conditions are equivalent).

> **Theorem 4.6** [1]
> 1. For a $\mathscr{P}_+^\uparrow$ invariant vacuum state $\varphi$, the commutant $\pi_\varphi(\mathfrak{A})'$ of the cyclic representation associated with $\varphi$ is a commutative algebra and it commutes with $T_\varphi(a)$.
> 2. For a vacuum state the following conditions are equivalent. ($\mathscr{P}_+^\uparrow$ invariance is not assumed.)
>    (a) $\pi_\varphi(\mathfrak{A})''$ is a factor ($\pi_\varphi(\mathfrak{A})'' \cap \pi_\varphi(\mathfrak{A})' = \mathbf{C}1$).
>    (b) $\pi_\varphi(\mathfrak{A})$ is irreducible ($\pi_\varphi(\mathfrak{A})' = \mathbf{C}1$).
>    (c) Any translation invariant vector is proportional to $\Omega_\varphi$.

**Remark** If $\pi_\varphi(\mathfrak{A})'$ is a commutative algebra as in (1), it coincides with the centre of $\pi_\varphi(\mathfrak{A})''$ because $\pi_\varphi(\mathfrak{A})'' \supset \pi_\varphi(\mathfrak{A})'$. And since $T_\varphi(a) \in \pi_\varphi(\mathfrak{A})''$ by (1), both $\pi_\varphi(\mathfrak{A})$ and $T_\varphi(a)$ are decomposed by the central decomposition. In particular, if $\mathfrak{A}$ is separable, $\varphi$ can be written as a direct integral of those vacuum states $\psi$, for which $\pi_\psi(\mathfrak{A})''$ is a factor (a von Neumann algebra with the trivial centre $\mathbf{C}1$) and $\pi_\varphi$, $\Omega_\varphi$, and $T_\varphi$ can also be written as direct integrals of $\pi_\psi$, $\Omega_\psi$, and $T_\psi$. In this decomposition we can apply to each $\psi$ condition (2) of the theorem and thus, the representation $\pi_\psi$ is irreducible and the vectors invariant under $T_\psi(a)$, $a \in \mathbf{R}^4$, are limited to vectors proportional to $\Omega_\psi$ (uniqueness of the vacuum in the representation space). These properties are assumed in the following analysis. The results so obtained can be applied to a general vacuum state through the direct integral of the central decomposition.

**Outline of the proof** The proof of (1) consists of five steps.

(i) For $S \in \pi_\varphi(\mathfrak{A})'$ we show that $[S, T_\varphi(a)] = 0$. Let $Q \in \mathfrak{A}$ and

$$Q(x) = \alpha_{(x,1)}Q, \quad \pi_\varphi(Q(x)) = T_\varphi(x)\pi_\varphi(Q)T_\varphi(x)^*$$

Since $S \in \pi_\varphi(\mathfrak{A})'$, $S$ and $\pi_\varphi(Q(x))$ commute and we get

$$(\Omega_\varphi, \pi_\varphi(Q(x))S\Omega_\varphi) = (\Omega_\varphi, S\pi_\varphi(Q(x))\Omega_\varphi)$$

Substituting the expression for $\pi_\varphi(Q(x))$ into the above formula, and using $T_\varphi(x)^*\Omega_\varphi = \Omega_\varphi$ we obtain

$$(\Omega_\varphi, \pi_\varphi(Q)T_\varphi(-x)S\Omega_\varphi) = (\Omega_\varphi, ST_\varphi(x)\pi_\varphi(Q)\Omega_\varphi)$$

Taking the Fourier transform we get a complex measure

$$(\Omega_\varphi, \pi_\varphi(Q)E_\varphi(-\mathrm{d}^4p)S\Omega_\varphi) = (\Omega_\varphi, SE_\varphi(\mathrm{d}^4p)\pi_\varphi(Q)\Omega_\varphi)$$

Since the support of the measure is contained in $\bar{V}_+$ on the r.h.s. and in $-\bar{V}_+$ on the l.h.s., it consists of a single point $\{0\}$. Hence, the original function is constant and we obtain

$$(\pi_\varphi(Q)^*\Omega_\varphi, T_\varphi(-x)S\Omega_\varphi) = (\pi_\varphi(Q)^*\Omega_\varphi, S\Omega_\rho)$$

(the r.h.s. is the value at $x = 0$). Since $\pi_\varphi(\mathfrak{A})\Omega_\varphi$ is dense, we get

$$T_\varphi(x)S\Omega_\varphi = S\Omega_\varphi, \text{ i.e. } \quad S(x)\Omega_\varphi = S\Omega_\varphi,$$

where $S(x) = T_\varphi(x)ST_\varphi(x)^*$. Due to $S \in \pi_\varphi(\mathfrak{A})'$, the following equation holds for any $Q \in \mathfrak{A}$:

$$[S(x), \pi_\varphi(Q)] = T_\varphi(x)[S, \pi_\varphi(\alpha_{(-x,1)}Q)]T_\varphi(x)^* = 0$$

Thus,

$$S(x)\pi_\varphi(Q)\Omega_\varphi = \pi_\varphi(Q)S(x)\Omega_\varphi = \pi_\varphi(Q)S\Omega_\varphi = S\pi_\varphi(Q)\Omega_\varphi$$

Again, by the fact that $\pi_\varphi(\mathfrak{A})\Omega_\varphi$ is dense,

$$S(x) = S, \quad \text{i.e. } [T_\varphi(x), S] = 0$$

is satisfied. (Note that in this part of the proof we have only used the assumption about the support of $E_\varphi$.)

(ii) Denoting the projection operator to the subspace of all translation invariant vectors by $E_0$, we prove the following equation for $x \neq 0$:

$$\text{w} - \lim_{\lambda\to\infty} T_\varphi(\lambda x) = E_0. \tag{4.10}$$

Due to the assumption of $\mathscr{P}_+^\uparrow$ invariance, there exists a unitary representation of $\mathscr{P}_+^\uparrow$ on $\mathscr{H}_\varphi$, and $T_\varphi(a)$ is a part of the representation. From the analysis of unitary representations of $\mathscr{P}_+^\uparrow$ given in Chapter 3, it follows that the Fourier transform of $(\Phi, T_\varphi(\lambda x)\Psi)$ with $\lambda \in \mathbf{R}$ becomes an absolutely continuous measure for $E_0\psi = 0$. Thus, there exists an appropriate $L_1$ function $f(l)$ such that

$$(\Phi, T_\varphi(\lambda x)\Psi) = \int \mathrm{e}^{\mathrm{i}\lambda l}f(l)\,\mathrm{d}l$$

Hence, due to the Riemann–Lebesque Lemma [2] this converges to 0 for $\lambda \to \infty$. On the other hand, since $T_\varphi(x)E_0 = E_0$, we obtain (4.10).

(iii) If $x$ is spacelike, we show for any $Q_1, Q_2 \in \mathfrak{A}$ that

$$\lim_{\lambda\to\infty} \|[Q_1, \alpha_{(\lambda x, 1)} Q_2]\| = 0 \tag{4.11}$$

is satisfied (called an asymptotic Abelian property).

By the generating property (4), for any $\varepsilon > 0$ there exist bounded space–time domains $D_1$, $D_2$ and $Q_i' \in \mathfrak{A}(D_i)$, $(i = 1, 2)$ such that $Q_i$ can be approximated as

$$\|Q_1' - Q_1\| < \varepsilon\,, \quad \|Q_2' - Q_2\| < \varepsilon.$$

If $\lambda$ is sufficiently large, $D_1$ and $D_2 + \lambda x$ become spacelike, and since $\alpha_{(\lambda x, 1)} Q_2' \in \mathfrak{A}(D_2 + \lambda x)$ we get

$$[Q_1', \alpha_{(\lambda x, 1)} Q_2'] = 0 \quad \text{(locality(3))}.$$

On the other hand, the following approximation holds

$$\begin{aligned} &\|[Q_1, \alpha_{(\lambda x, 1)} Q_2] - [Q_1', \alpha_{(\lambda x, 1)} Q_2']\| \\ &\le \|[Q_1 - Q_1', \alpha_{(\lambda x, 1)} Q_2]\| + \|[Q_1', \alpha_{(\lambda x, 1)}(Q_2 - Q_2')]\| \\ &\le 2\varepsilon\|Q_2\| + 2\varepsilon\|Q_1'\| \end{aligned}$$

Consequently, (4.11) holds.

(iv) We show that $E_0 \pi_\varphi(\mathfrak{A})'' E_0$ is a commutative algebra.

First, $T_\varphi(a) \in \pi_\varphi(\mathfrak{A})''$ by (i), and hence we obtain $E_0 \in \pi_\varphi(\mathfrak{A})''$. From this we see that $E_0 \pi_\varphi(\mathfrak{A})'' E_0$ forms an algebra ($W^*$ algebra). To show the commutativity, let $Q, Q' \in \mathfrak{A}$, then from (iii) we get

$$\lim_{\lambda\to\infty} E_0[\pi_\varphi(Q), T_\varphi(\lambda x)\pi_\varphi(Q')T_\varphi(\lambda x)^*]E_0 = 0.$$

On the other hand, using $T_\varphi(x)^* E_0 = E_0 = E_0 T_\varphi(x)$, we obtain

$$\begin{aligned} &E_0[\pi_\varphi(Q), T_\varphi(x)\pi_\varphi(Q')T_\varphi(x)^*]E_0 \\ &\quad = E_0\pi_\varphi(Q)T_\varphi(x)\pi_\varphi(Q')E_0 - E_0\pi_\varphi(Q')T_\varphi(-x)\pi_\varphi(Q')E_0. \end{aligned}$$

Hence by (ii)

$$\text{w} - \lim_{\lambda\to\infty} E_0[\pi_\varphi(Q), T_\varphi(\lambda x)\pi_\varphi(Q')T_\varphi(\lambda x)^*]E_0 = [E_0\pi_\varphi(Q)E_0, E_0\pi_\varphi(Q')E_0].$$

From these two limit expressions the commutativity follows:

$$[E_0\pi_\varphi(Q)E_0, E_0\pi_\varphi(Q')E_0] = 0.$$

By taking the limit with respect to $Q, Q'$ we get the commutativity of $E_0\pi_\varphi(\mathfrak{A})''E_0$.

(v) We show the commutativity of $\pi_\varphi(\mathfrak{A})'$.

Since $E_0 \in \pi_\varphi(\mathfrak{A})''$, we get by the general theory

$$E_0\pi_\varphi(\mathfrak{A})' = (E_0\pi_\varphi(\mathfrak{A})''E_0)'E_0 \quad \text{(the commutant on } E_0\mathscr{H}_\varphi).$$

On the other hand, since $\pi_\varphi(\mathfrak{A})\Omega_\varphi$ is dense,

$$(E_0\pi_\varphi(\mathfrak{A})''E_0)\Omega_\varphi = E_0\pi_\varphi(\mathfrak{A})''\Omega_\varphi$$

is dense in $E_0\mathscr{H}_\varphi$. Again, by the general theory $E_0\pi_\varphi(\mathfrak{A})''E_0$ is maximal abelian as an Abelian algebra with a cyclic vector (see Appendix B for the general theory). Namely

$$(E_0\pi_\varphi(\mathfrak{A})''E_0)'E_0 = E_0\pi_\varphi(\mathfrak{A})''E_0.$$

Hence

$$E_0\pi_\varphi(\mathfrak{A})' = E_0\pi_\varphi(\mathfrak{A})''E_0.$$

By (iv) the r.h.s. is commutative. If $S \in \pi_\varphi(\mathfrak{A})'$ satisfies $E_0S = 0$, then $S\Omega_\varphi = SE_0\Omega_\varphi = E_0S\Omega_\varphi = 0$ and we get

$$S\pi_\varphi(Q)\Omega_\varphi = \pi_\varphi(Q)S\Omega_\varphi = 0.$$

By the fact that $\pi_\varphi(\mathfrak{A})\Omega_\varphi$ is dense we obtain $S = 0$. Consequently, $\pi_\varphi(\mathfrak{A})'$ is isomorphic to $E_0\pi_\varphi(\mathfrak{A})'$ and therefore commutative.

The proof of (2) is given in the following order: $(c) \to (b) \to (a) \to (c)$. By the proof of (i) $E_0 \in \pi_\varphi(\mathfrak{A})''$.

If (c) holds, then $E_0$ is one-dimensional, and thus $\pi_\varphi(\mathfrak{A})'$, which is isomorphic to $E_0\pi_\varphi(\mathfrak{A})'$, is also one-dimensional and becomes $\mathbf{C}1$. Therefore, (b) holds.

If (b) holds, then $\pi_\varphi(\mathfrak{A})'' = \mathscr{B}(\mathscr{H}_\varphi)$ and thus (a) holds.

From (a) the following *clustering property* can be derived:

$$\lim_{\lambda\to\infty} \varphi(Q_1\alpha_{(\lambda x, 1)}Q_2) = \varphi(Q_1)\varphi(Q_2) \tag{4.12}$$

where $x$ is spacelike and $Q_1$ and $Q_2$ can be any elements in $\mathfrak{A}$.

We come back to its proof later. First we want to derive (c) from (4.12). Denoting the one-dimensional projection operator to $\mathbf{C}\Omega_\varphi$ by $E_\Omega$ we get

$$\varphi(Q_1)\varphi(Q_2) = (\pi_\varphi(Q_1^*)\Omega_\varphi, E_\Omega\pi_\varphi(Q_2)\Omega_\varphi).$$

On the other hand the expression on the l.h.s. of (4.12) is

$$(\Omega_\varphi, \pi_\varphi(Q_1)T_\varphi(\lambda x)\pi_\varphi(Q_2)\Omega_\varphi).$$

From the fact that $\pi_\varphi(\mathfrak{A})\Omega_\varphi$ is dense and the uniform boundedness of $\|T_\varphi(\lambda x)\| = 1$

$$\text{w} - \lim T_\varphi(\lambda x) = E_\Omega \tag{4.13}$$

is equivalent to (4.12). If we multiply $E_0$ onto (4.13), then since $\Omega \in E_0\mathscr{H}_\varphi$ the r.h.s. remains to be $E_\Omega$ while the l.h.s. becomes $E_0$ by $T_\varphi(\lambda x)E_0 = E_0$. Hence, we obtain $E_0 = E_\Omega$ and (c) holds.

The proof of the clustering property (4.12):

Let

$$Q_\varphi(x) = \pi_\varphi(\alpha_{(\lambda x,1)}Q). \tag{4.14}$$

The cluster point of $Q_\varphi(\lambda x)$

$$Q_\infty \in \bigcap_L \overline{\{Q_\varphi(\lambda x); |\lambda| \geq L\}}^w \tag{4.15}$$

exists, due to $\|Q_\varphi(x)\| \leq \|Q\|$ and the weak compactness of the unit sphere of $\mathscr{B}(\mathscr{H})$. Here $\overline{\{\ldots\}}^w$ represents the weak closure of $\{\ldots\}$.

Since $Q(\lambda x) \in \pi_\varphi(\mathfrak{A})$, we get $Q_\infty \in \pi_\varphi(\mathfrak{A})''$. On the other hand, due to (4.11), $Q_\infty$ commutes with $\pi_\varphi(Q)$ for any $Q \in \mathfrak{A}$. Therefore,

$$Q_\infty \in \pi_\varphi(\mathfrak{A})'' \bigcap \pi_\varphi(\mathfrak{A})' = \text{centre of } \pi_\varphi(\mathfrak{A})''.$$

By assumption (a) the r.h.s. is $\mathbf{C1}$, thus

$$Q_\infty = c\mathbf{1}.$$

The complex number $c$ can be determined from the fact that $Q_\infty$ is a cluster point of $Q_\varphi(\lambda x)$ as follows:

$$c = (\Omega_\varphi, Q_\infty\Omega_\varphi) = \lim(\Omega_\varphi, Q_\varphi(\lambda x)\Omega_\varphi) = \varphi(Q)$$

Namely, the set of cluster points (4.15) consists of one point $c\mathbf{1}$. Hence, the following equation is proven:

$$\mathrm{w} - \lim Q_\varphi(x) = \varphi(Q)\mathbf{1}. \tag{4.16}$$

Substituting the above equation into the l.h.s. of (4.12) written as

$$\varphi(Q_1\alpha_{(\lambda x,1)}Q_2) = (\Omega_\varphi, \pi_\varphi(Q_1)Q_{2\varphi}(\lambda x)\Omega_\varphi),$$

we obtain the clustering property (4.12). □

**Remark 1** By Theorem 4.6 (1), the central decomposition of $\varphi$ gives the vacuum state $\psi$. However we do not know whether it is $\mathscr{P}_+^\uparrow$ invariant. On the other hand, since the proof of Theorem 4.6 (2) does not make use of the $\mathscr{P}_+^\uparrow$ invariance, we can apply it to $\psi$, and $\psi$ satisfies all conditions (a), (b), (c).

**Remark 2** In order to emphasize the point of view that the vacuum is a stable state, i.e. it is the state of minimal energy (ground state) we did not require $\mathscr{P}_+^\uparrow$ invariance in the definition of the vacuum. Below we consider the vacuum state which satisfies condition (2) of Theorem 4.6 (i.e. which is pure), and the $\mathscr{P}_+^\uparrow$ invariance of $\varphi$ will be introduced in the cases where it becomes necessary.

**Remark 3** Comparing the matrix elements of both sides of eq. (4.10) between vectors in the dense set $\pi_\varphi(\mathfrak{A})\Omega_\varphi$ we obtain the condition

$$\lim \varphi(Q_1\alpha_{(\lambda x,1)}Q_2) = (\Omega_\varphi, \pi_\varphi(Q_1)E_0\pi_\varphi(Q_2)\Omega_\varphi),$$

which is equivalent to (4.10). Moreover if condition (2) in Theorem 4.6 is satisfied, $E_0$ is identical with the one-dimensional projection operator $E_\Omega$, and the above equation becomes the clustering property (4.14). Thus, the $\mathscr{P}_+^\uparrow$ invariant pure vacuum state $\varphi$ possesses the clustering property (4.14) not only with respect to spacelike $x$, but also with respect to arbitrary $x$.

### 4.4 Mass gap and exponential clustering property

In this section we show that if zero is an isolated point in the energy-momentum spectrum, i.e. if there is a gap between the energy of the vacuum vector $\Omega_\varphi$ and its orthogonal complement, then a much stronger form of the clustering property than (4.12) holds. This means that in a very good approximation the vacuum is a product state for the observables of two spacelike separated domains $D_1$ and $D_2$.

First we derive an exponential clustering property of a pure vacuum state from the spectral condition for energy and the locality by using a simple method. In Section 4.5, we will derive the best exponent for the exponential clustering property by quoting a result obtained by a more complicated method.

**Theorem 4.7**
We make the following two assumptions:
(a) For a vacuum state $\varphi$, the spectrum of $E_\varphi$ in $\Omega_\varphi^\perp$ is contained in $\{p;\ p^0 \geq m\}$ (the energy is larger than $m$).
(b) Two domains $D_1$ and $D_2$ are such that $D_1$ and $D_2 + te$ are spacelike separated ($D_1' \supset D_2 + te$) for $|t| < T$, where $e = (1, 0, 0, 0)$.
Then the following inequality holds for any $Q_1 \in \mathfrak{A}(D_1)$ and $Q_2 \in \mathfrak{A}(D_2)$.

$$|\varphi(Q_1Q_2) - \varphi(Q_1)\varphi(Q_2)| \leq Gk(mT/2) \tag{4.17}$$

$$k(a) = \tfrac{1}{2}((\pi a)^{-1/2} + (\pi a)^{-1})e^{-a}$$

$$G = [\varphi(Q_1Q_1^*)\varphi(Q_2^*Q_2)]^{1/2} + [\varphi(Q_1^*Q_1)\varphi(Q_2Q_2^*)]^{1/2}$$

**Proof** Set

$$\left.\begin{aligned}\varphi(Q_1\alpha_{(te,1)}Q_2)-\varphi(Q_1)\varphi(Q_2)&=\int e^{its}\,d\mu_1(s)\\ \varphi(\alpha_{(te,1)}(Q_2)Q_1)-\varphi(Q_1)\varphi(Q_2)&=\int e^{-its}\,d\mu_2(s)\end{aligned}\right\}\qquad(*)$$

Let us denote $E'_\varphi(\cdot)\equiv E_\varphi(\cdot)(1-E_\Omega)$. (See (4.9b) for $E_\varphi(\cdot)$.) Then

$$\mu_1(\Delta)=(\pi_\varphi(Q_1^*)\Omega_\varphi,E'_\varphi(\{p;\,p^0\in\Delta\})\pi_\varphi(Q_2)Q_\varphi),$$

$$\mu_2(\Delta)=(\pi_\varphi(Q_2^*)\Omega_\varphi,E'_\varphi(\{p;\,p^0\in\Delta\})\pi_\varphi(Q_1)\Omega_\varphi).$$

The support of $\mu_1$ and $\mu_2$ are in $[m,\infty)$, and the following estimates hold:

$$\|\mu_1\|=\int|\mu_1(ds)|\le\|\pi_\varphi(Q_1^*)\Omega_\varphi\|\cdot\|\pi_\varphi(Q_2)\Omega_\varphi\|$$

$$\|\mu_2\|=\int|\mu_2(ds)|\le\|\pi_\varphi(Q_2^*)\Omega_\varphi\|\cdot\|\pi_\varphi(Q_1)\Omega_\varphi\|$$

(The norm $\|\mu\|$ of a measure $\mu$ is the smallest positive number $\|\mu\|$ satisfying $|\int g\,d\mu|\le\|\mu\|\sup|g|$.)

By locality and condition (b)

$$\varphi(t)\equiv\varphi([Q_1,\alpha_{(te,1)}Q_2])=0$$

holds for $|t|<T$. Since $\varphi(t)$ is the difference of the two expressions in $(*)$ the equation

$$\int\varphi(t)f(t)\,dt=\int\tilde f(s)\,d\mu_1(s)-\int\tilde f(-s)\,d\mu_2(s)\qquad(**)$$

holds for any $L_1$ function $f(t)$ and its Fourier transform

$$\tilde f(s)=\int e^{its}f(t)\,dt.$$

For $f(t)$ we use the following function $f_\varepsilon(t)$:

$$f_\varepsilon(t)=f_{1\varepsilon}(t)f_2(t)\quad(\varepsilon>0,l>0),$$

$$f_{1\varepsilon}(t)=(2\pi)^{-1}(it+\varepsilon)^{-1},\quad f_2(t)=e^{-t^2/4l}.$$

By computing $\tilde f_{1\varepsilon}$ and $\tilde f_2$ we obtain the Fourier transform of $f_\varepsilon$ as follows:

$$\tilde f_\varepsilon(s)=(l/\pi)^{1/2}\int_{-\infty}^{s}e^{-lu^2+\varepsilon u-\varepsilon s}\,du$$

In the limit $\varepsilon \to +0$, $f_\varepsilon$ converges to $f_0$ in $L_1$ for $|t| \geq T$, and $\tilde{f}_\varepsilon$ converges uniformly to $\tilde{f}_0$. Now, substituting $f_\varepsilon$ into $f$ of (**) and taking the limit $\varepsilon \to +0$, we obtain eq. (**), where $f_0$ is substituted into $f$. We estimate both sides of this equation as follows:

$$\begin{aligned}\left|\int \varphi(t) f_0(t)\,\mathrm{d}t\right| &\leq \sup|\varphi(t)|\left(\int_{-\infty}^{-T} + \int_T^{\infty}\right)|f_0(t)|\,\mathrm{d}t \\ &= \sup|\varphi(t)| \int_T^{\infty} \mathrm{e}^{-t^2/4l}\,\mathrm{d}t/(\pi t) \\ &= \sup|\varphi(t)| \int_{(T^2/4l)}^{\infty} \mathrm{e}^{-t'}\,\mathrm{d}t'/(2\pi t') \\ &\leq \sup|\varphi(t)|(2l/\pi T^2)\exp(-T^2/4l)\end{aligned}$$

where $\sup|\varphi(t)| \leq G$. Furthermore,

$$\begin{aligned}\left|\int \tilde{f}_0(-s)\,\mathrm{d}\mu_2(s)\right| &\leq \|\mu_2\| \sup\{|\tilde{f}_0(-s)|; s \geq m\} \\ &= \|\mu_2\|(l/\pi)^{1/2} \int_{-\infty}^{-m} \exp(-lu^2)\,\mathrm{d}u \\ &= \|\mu_2\| \int_{lm^2}^{\infty} \mathrm{e}^{-v}\,\mathrm{d}v/(4\pi v)^{1/2} \leq \|\mu_2\|\mathrm{e}^{-lm^2}/(2m\sqrt{\pi l})\end{aligned}$$

$$\begin{aligned}\left|\int (1-\tilde{f}_0(s))\,\mathrm{d}\mu_1(s)\right| &\leq \|\mu_1\| \sup\{|1-\tilde{f}_0(s)|; s \geq m\} \\ &\leq \|\mu_1\|\mathrm{e}^{-lm^2}/(2m\sqrt{\pi l})\end{aligned}$$

(where we have used the equality $1 - \tilde{f}_0(s) = \tilde{f}_0(-s)$). Therefore,

$$\begin{aligned}&|\varphi(Q_1Q_2) - \varphi(Q_1)\varphi(Q_2) = \left|\int \mathrm{d}\mu_1(t)\right| \\ &\leq \left|\int \varphi(t) f_0(t)\,\mathrm{d}t\right| + \left|\int \tilde{f}_0(-s)\,\mathrm{d}\mu_2(s)\right| + \left|\int (1-\tilde{f}_0(s))\,\mathrm{d}\mu_1(s)\right| \\ &\leq G\{(2l/\pi T^2)\exp(-T^2/4l) + (2m\sqrt{\pi l})^{-1}\exp(-lm^2)\}.\end{aligned}$$

Thus by setting $l = T/(2m)$, we obtain

$$|\varphi(Q_1Q_2) - \varphi(Q_1)\varphi(Q_2)| \leq G\left\{\frac{1}{\pi m T} + \frac{1}{\sqrt{2\pi m T}}\right\}\mathrm{e}^{-(mT/2)}.$$

□

**Remark** In the estimate (4.17), $k(Tm/2)$ is a quantity determined by the condition on the energy spectrum (namely $m$) and the mutual relations between the two regions (i.e. $T$), and $G$ is a quantity determined by the norm of the observables $Q_1$ and $Q_2$ as

$$|G| \leq 2\|Q_1\| \cdot \|Q_2\|.$$

For the case where $D_1$ and $D_2$ are double cones with bases $S_1$ and $S_2$ lying in the constant time hypersurface $\{x;\, x^0 = t\}$, we can take as $T$ the spacelike distance $R$ between $S_1$ and $S_2$. Namely, eq. (4.17) represents the exponential clustering property with respect to the distance between the two domains $D_1$ and $D_2$.

Actually, by a more complicated discussion (using more information than the assumptions in Theorem 4.7) we can improve the exponent up to $e^{-mR}$. It is possible to give a physical interpretation for such an exponential clustering property by Yukawa's theory. Namely, the correlation at a distance $R$ is interpreted as being induced by an energy exchange (in Yukawa's theory, a particle exchange) and its effective radius is determined by (and hence the correlation tends to zero beyond the distance equal to) the reciprocal of the exchanged energy. This is because the amount of a virtual energy transfer from one region to another, which must be at least the minimal energy $m$ (the mass of the particle in Yukawa's theory), has to be within the limit allowed by the uncertainty principle of time and energy, where the time is the flight time of the energy (the virtual particle in Yukawa's theory) to the original region and hence is equal to $R$ or larger in the unit $c = 1$. Note that the Yukawa potential decreases as $e^{-mR}$ (neglecting the inverse powers of $R$).

The clustering property of the form $e^{-mR}$ is obtained in the next section by the use of the JLD representation. (JLD stands for Jost, Lehmann and Dyson.)

## 4.5 The JLD representation

**Theorem 4.8**

For any local observables $Q_1 \in \mathfrak{A}(D_1)$, $Q_2 \in \mathfrak{A}(D_2)$ and for a pure vacuum state $\varphi$, the following *JLD representation* holds as a consequence of covariance and locality of local observables and of the definition of vacuum states (the condition on the energy momentum spectrum) [3]:

$$\begin{aligned}&\varphi(Q_1 Q_2(x)) - \varphi(Q_1)\varphi(Q_2)\\ &= \int d\xi^0 \int_{S_\xi} d\boldsymbol{\xi} \int_0^\infty \rho(d\kappa, \xi)\{2\varphi(\xi^0)(\partial/\partial x^0)\Delta_\kappa^+(\xi - x) - \varphi'(\xi^0)\Delta_\kappa^+(\xi - x)\}\end{aligned} \tag{4.18}$$

Here, $\varphi(\xi^0)$ is a $C^\infty$-function introduced for the sake of mere convenience and is a function satisfying

$$\int \varphi(\xi^0)\, d\xi^0 = 1,$$

with its support in

$$\{\xi^0;\ |\xi^0 - t| < \varepsilon\}$$

for some $\varepsilon > 0$ and some $t$. $\rho(\mathrm{d}\kappa, \xi)$ is a measure in the variable $\kappa$ for each $\xi$, and is defined by the equation

$$\int_0^\infty g(\kappa)\rho(\mathrm{d}\kappa, \xi) = (2\pi)^{-4} \int g((p,p)^{1/2})\mathrm{e}^{\mathrm{i}(p,\xi)}\mu(\mathrm{d}^4 p).$$

$(p,p)$ and $(p,\xi)$ are the Minkowski inner products introduced in Section 3.1, and the measure $\mathrm{d}\mu$ is defined as a complex measure, like $\mathrm{d}\mu_1 - \mathrm{d}\mu_2$ in the previous section by

$$\varphi([Q_1, Q_2(x)]) = \int \mathrm{e}^{i(p,x)}\mu(\mathrm{d}^4 p). \tag{4.19}$$

Note that for an observable which is translated by $x$ we use the notation

$$\alpha_{(x,1)}Q = Q(x). \tag{4.20}$$

Since $(p,p) \geq 0$ on the support of the measure $\mu$ by the definition of a vacuum state $\varphi$, we can use $(p,p)^{1/2}$ in the defining equation for $\rho$. Furthermore, by the uniqueness of the state vector of the vacuum (since $\varphi$ is the pure vacuum state), the $\rho$ measure of the point 0 is zero. $\Delta_\kappa^+$ in eq. (4.18) is a distribution given by

$$\Delta_\kappa^+(x) = -\mathrm{i}(2\pi)^{-3} \int (2p^0)^{-1}\mathrm{e}^{-\mathrm{i}(p,x)}\mathrm{d}^3\mathbf{p}, \tag{4.21}$$

where $p^0 = (\mathbf{p}^2 + \kappa^2)^{1/2}$. Note that for spacelike points $x$ we can replace $\Delta_\kappa^+$ by the following function:

$$\Delta_\kappa^+(x) = -\mathrm{i}(2\pi)^{-2}\kappa K_1(\kappa s)/s, \qquad s = (-(x,x))^{1/2}. \tag{4.22}$$

$K_1$ is the modified Bessel function which decreases exponentially at infinity.

The region $S_\varepsilon$ of the $\boldsymbol{\xi}$ integration in eq. (4.18) can be determined as follows. Due to covariance and locality the l.h.s. of (4.18) vanishes on the causal complement $(D_1 - D_2)'$ of the region

$$D_1 - D_2 = \{x - y;\ x \in D_1, y \in D_2\}$$

On the constant time hypersurface $\{x; x^0 = s\}$, let the complement of the section (by the hypersurface) of $(D_1 - D_2)'$ be $\{(s, \mathbf{x}); \mathbf{x} \in S^s\}$ and the convex hull of $S^s$ be $S_0^s$. $S^s$ is the set of space coordinates of the image obtained by causally projecting $D_1 - D_2$ onto the plane $x_0 = s$ (the causal shadow). We denote the union of all $S_0^s$ for $s$ satisfying $|s - t| \leq \varepsilon$ as $S_\varepsilon$ (where $S_0 = S_0^t$), and we take

$$D = [t - \varepsilon, t + \varepsilon] \times S_\varepsilon. \tag{4.23}$$

Especially if $D_1$ and $D_2$ are double cones with bases in $x^0 = t$, $S_0$ is the base of the double cone $D_1 - D_2$ (its section by $x^0 = t$).

For a general $x$, eq. (4.18) is meaningful in the distribution sense when it is multiplied by a $C^\infty$-function $f(x)$ with compact support and then integrated over $x$. Namely, first the $x$ integration on the r.h.s. is performed in the distribution sense, and the $\kappa$ and $\xi$ integrations are performed afterwards. However, if $x$ is spacelike separated from $D$, we can use eq. (4.22) and the computation of the integration can be performed in the usual way.

The l.h.s. of eq. (4.18) can be written as follows:

$$\begin{aligned}(\Omega_\varphi, \pi_\varphi(Q_1)T_\varphi(x)\pi_\varphi(Q_2)\Omega_\varphi) - \varphi(Q_1)\varphi(Q_2)\\ = \int_{V_+} \mathrm{e}^{\mathrm{i}(p,x)}(\Omega_\varphi, \pi_\varphi(Q_1)E_\varphi(\mathrm{d}^4p)\pi_\varphi(Q_2)\Omega_\varphi).\end{aligned}$$

Note that $E_\varphi$ represents the spectral projection operator of the unitary translation group $T_\varphi$. Taking the Fourier transform on both sides of (4.18) and using the fact that the region $D$ of the $\xi$ integration is compact, we see that for each $\kappa > 0$ the support of $E_\varphi$ either includes the whole hyperbolic surface

$$\{p \in V_+; (p,p) = \kappa^2, p^0 > 0\}$$

or does not include it at all. Thus we get the following result of Borchers.

**Corollary 4.9**
For a vacuum state $\varphi$, the support of the energy momentum

$$P^\mu = \int p^\mu E_\varphi(\mathrm{d}^4p)$$

is Lorentz invariant.

In particular, if there is an energy gap as in the case of Theorem 4.7, then we also have a mass gap and the $\kappa$ integration of (4.18) starts from $m$. Under this circumstance the following estimate holds [4].

**Corollary 4.10**
If $D_1 - D_2 - x$ is spacelike under condition (a) of Theorem 4.7, the following relation holds:

$$\begin{aligned}&|\varphi(Q_1Q_2(x)) - \varphi(Q_1)\varphi(Q_2)|\\ &\quad\leq (2\pi)^{-6}G|S_\varepsilon|\{2m^2(|t-x^0|+\varepsilon)d^{-2}(-K_1'(md))\\ &\quad + 2\varepsilon^{-1}md^{-1}K_1(md)\}.\end{aligned} \tag{4.24}$$

Here $d$ is the minimal invariant distance between $D$ and $x$:

$$d^2 = \inf\{-(x-y, x-y);\ y \in D \equiv [t-\varepsilon, t+\varepsilon] \times S_\varepsilon\}.$$

$G$ is the quantity given in Theorem 4.7.

**Remark** Since both $K_1(z)$ and $K_1(z)'$ behave as $(\pi/2z)^{1/2}\mathrm{e}^{-z}$ for large $z$, eq. (4.24) gives an exponential decrease $d^{-3/2}\mathrm{e}^{-md}$ for large $d$. Here $d$ can be understood as the spatial distance between the support of $Q_1$ and that of $Q_2(x)$.

**Proof** We estimate each term in the bracket of eq. (4.18) separately. Since $\xi - x$ is spacelike according to our assumption, we use eq. (4.22), where

$$s = (-(x-\xi, x-\xi))^{1/2},$$

$$(\partial/\partial x^0)\Delta_\kappa^+(\xi - x) = (\xi^0 - x^0)s^{-1}(\partial/\partial s)\Delta_\kappa^+(\xi - x).$$

The absolute value of (4.22) is monotonically decreasing with respect to $\kappa$, due to $(zK_1(z))' = -zK_0(z) \leq 0$, and thus in the region $\kappa \geq m$ it takes its maximal value at $\kappa = m$. By the relations

$$|(\partial/\partial s)\Delta_\kappa^+| = (2\pi)^{-2}(2s^{-2}\kappa K_1(\kappa s) + s^{-1}\kappa^2 K_0(\kappa s)),$$

$$(\partial/\partial\kappa)(2s^{-2}\kappa K_1(\kappa s) + s^{-1}\kappa^2 K_0(\kappa s)) = -\kappa^2 K_1(\kappa s) < 0,$$

$|(\partial/\partial s)\Delta_\kappa^+|$ also takes its maximal value at $\kappa = m$. Hence we obtain

$$\begin{aligned} &|\varphi(Q_1 Q_2(x)) - \varphi(Q_1)\varphi(Q_2)| \\ &\leq \int \mathrm{d}\xi^0 \int_{S_\varepsilon} \mathrm{d}\boldsymbol{\xi}\, \|\rho\| \{2|\varphi(\xi^0)(\partial/\partial x^0)\Delta_m^+(\xi - x)| + |\varphi'(\xi^0)\Delta_m^+(\xi - x)|\}, \end{aligned} \tag{4.25}$$

$\|\rho\| = (2\pi)^{-4}\|\mu\|$ and we can estimate $\|\mu\|$ by $G$.

Using the monotone property with respect to $\kappa$ we get

$$(\partial/\partial s)(s^2|\Delta_\kappa^+(\xi - x)|) = \kappa s(\partial/\partial\kappa)|\Delta_\kappa^+(\xi - x)| < 0,$$

$$(\partial/\partial s)(s^3|(\partial/\partial s)\Delta_\kappa^+|) = \kappa s^2(\partial/\partial\kappa)|(\partial/\partial s)\Delta_\kappa^+| < 0,$$

and, since $s^{-2}$ and $s^{-4}$ are monotone decreasing, the following two expressions:

$$|\Delta_m^+(\xi - x)| = s^{-2}|s^2\Delta_m^+(\xi - x)|,$$

$$s^{-1}|(\partial/\partial s)\Delta_m^+(\xi - x)| = s^{-4}(s^3|(\partial/\partial s)\Delta_m^+(\xi - x)|,$$

are both monotone decreasing in $s$. Hence, we can replace the variable $s$ in eq. (4.25) by its minimal value $d$. We also use

$$|\xi^0 - x^0| \leq |\xi^0 - t| + |t - x^0| \leq \varepsilon + |t - x^0|,$$

$$\int |\varphi(\xi^0)|\, \mathrm{d}\xi^0 = 1, \qquad \int |\varphi'(\xi^0)| \mathrm{d}\xi^0 = 2 \sup \varphi(\xi^0) < 2/\varepsilon.$$

Here, we choose a function $\varphi$ which is positive and takes only one maximum value smaller than $\varepsilon^{-1}$. Substituting the above estimates into (4.25) we get (4.24). □

**Remark 1** In order to perform the above described estimate we have introduced the JLD representation containing the function $\varphi$; usually $\varphi$ is not contained in the JLD representation. For example, if $Q_2(x)$ is differentiable the JLD representation without $\varphi$ is given by

$$\begin{aligned} &\varphi(Q_1 Q_2(x)) - \varphi(Q_1)\varphi(Q_2) \\ &\quad = \int_{S_0} \mathrm{d}\boldsymbol{\xi} \int_m^\infty \{\rho'(\mathrm{d}\kappa, \xi)\Delta_\kappa^+(\xi - x) + \rho(\mathrm{d}\kappa, \xi)(\partial/\partial x^0)\Delta_\kappa^+(\xi - x)\}_{\xi^0 = t}, \end{aligned} \tag{4.26}$$

where the complex measure $\rho'$ is obtained by replacing the $\mu$ appearing in the definition of $\rho$ in Theorem 4.8 by the following $\mu'$ ($\mu'(\mathrm{d}p) = ip^0\mu(\mathrm{d}p)$)

$$\varphi([Q_1, \dot{Q}_2(x)]) = \int \mathrm{e}^{\mathrm{i}(x,p)} \mu'(\mathrm{d}^4 p).$$

Here $\dot{Q}_2(x) = (\partial/\partial x^0)Q_2(x)$ and we can estimate $\mu'$ as $\|\mu'\| < G'$, where $G'$ denotes the quantity $G$ in Theorem 4.7 in which $Q_2$ is replaced by $\dot{Q}_2 = \dot{Q}_2(0)$. A differentiable local observable can be obtained as

$$Q_2 = \int Q(\xi)\varphi(\xi)\, \mathrm{d}^4\xi. \tag{4.27}$$

If $Q \in \mathfrak{A}(D)$ and the support of $\varphi$ is in $D_1$ then $Q_2 \in \mathfrak{A}(D + D_1)$ and, if $\varphi$ is differentiable then using $\dot{\varphi}(\xi) = (\partial/\partial \xi^0)\varphi(\xi)$

$$Q_2(x) = \int Q(\xi + x)\varphi(\xi)\, \mathrm{d}^4\xi, \qquad \dot{Q}_2 = -\int Q(\xi)\dot{\varphi}(\xi) \mathrm{d}^4\xi.$$

Thus $Q_2$ is differentiable.

**Remark 2** To prove the JLD representation, the function

$$H(x, x^4) = \int \mathrm{e}^{\mathrm{i}(p,x)} \cos(x^4 (p,p)^{1/2}) \mu(\mathrm{d}^4 p)$$

is introduced. Since $H$ satisfies the five-dimensional wave equation for the variables $(x^0, \mathbf{x}, x^4)$, it can be represented by using the initial values of $H$ and $\dot{H}$ at $x^0 = t$. At $x^4 = 0$, $H$ is zero in the region $(D_1 - D_2)'$. Therefore by applying Asgeirsson's theorem about the solution of the wave equation, we obtain the result that $H$ vanishes on the plane $x^0 = s$ at $x \notin S_0^s$ and for any $x^4$. Specializing to $x^4 = 0$ we obtain the JLD representation (4.26). Multiplying $\varphi(s)$ by the representation with initial value $x^0 = s$ and integrating over $s$, we obtain the JLD representation (4.18).

## 4.6 Truncated expectation values and multiple clustering property

The clustering property discussed in the previous two sections gives the asymptotic behaviour of two observables when their supports are taken far apart in spacelike directions. In this section we discuss the clustering property of $n$ observables, when their supports are taken mutually far apart in spacelike directions.

As we have seen in the preceding two sections, the vacuum expectation values of two observables, after the contribution of the vacuum intermediate state (the quantity obtained by inserting the projection operator to the vacuum in between the two observables, i.e. the term $\varphi(Q_1)\varphi(Q_2)$) has been subtracted, vanishes exponentially in the limit of an infinite spacelike separation. In the case of $n$ observables, if we construct an expression such that the contribution from the vacuum intermediate state is automatically subtracted for any partition of the $n$ observables into two groups, we expect an exponential decrease to zero whenever we separate supports of observables far apart. Such an expression is the following *truncated* function.

For a pure vacuum state $\varphi$, the truncated expectation value $\varphi^T$ of $n$ observables $Q_1, \dots, Q_n$ is defined as follows:

For any subset

$$I = \{i_1, i_2, \dots, i_k\}, \qquad i_1 < i_2 < \cdots < i_k$$

of $(1, 2, \dots, n)$ we define

$$Q(I) = Q_{i_1} Q_{i_2} \cdots Q_{i_k}$$

and we define $\varphi^T$ as

$$\varphi^T(Q(I)) = \sum_{m=1}^{|I|} \sum_{\{I_\nu\}} (-1)^{m-1}(m-1)! \prod_{\nu=1}^{m} \varphi(Q(I_\nu)), \tag{4.28}$$

where the $I_\nu (\nu = 1, \dots, m)$ have no intersection with each other, their union is $I$, and the second summation is over all such partitions of $I$. $|I|$ denotes the number of elements in $I$. By solving this equation for $\varphi$ in terms of $\varphi^T$ we get

$$\varphi(Q(I)) = \sum_{m=1}^{|I|} \sum_{\{I_\nu\}} \prod_{\nu=1}^{m} \varphi^T(Q(I_\nu)). \tag{4.29}$$

The relation between the two can be given by using a formal power series in the commutative, nilpotent indeterminate elements $t_n$ ($n \in N$) as

$$1+\sum_I t(I)\varphi(Q(I)) = \exp\sum_I t(I)\varphi^T(Q(I))$$

$$\sum_I t(I)\varphi^T(Q(I)) = \log\left\{1+\sum_I t(I)\varphi(Q(I))\right\}$$

where $t(I) = \prod\{t_n; n \in I\}$, $t_n^2 = 0$.

**Theorem 4.11**
Suppose that for bounded domains $D_1, \ldots, D_n$ there exists a division of the indices into two sets $\{J, J^c\}$, $J \subset \{1, \ldots, n\}$ ($J^c$ is the complement of $J$), such that for any $i \in J, j \in J^c$

$$D_i - D_j + te,$$

is spacelike for $|t| < T$ ($D_i$ and $D_j$ are spacelike separated by an order of $T$ or larger), where $e = (1, 0, 0, 0)$. Then there exists a combinatorial constant $A_n$ determined only by $n$ such that for any $Q_i \in \mathfrak{A}(D_i)$

$$|\varphi^T(Q_1 \cdots Q_n)| \leq A_n\left(\prod_j \|Q_j\|\right)k(mT/2). \qquad (4.30)$$

Note that $k$ is the function given in Theorem 4.7 and it decreases exponentially. For an irreducible vacuum state $\varphi$ we make the same assumption as in Theorem 4.7(a).

**Proof** Since

$$|\varphi(Q(I))| \leq \|Q(I)\| \leq \prod_{j\in I}\|Q_j\|,$$

we obtain by (4.28)

$$|\varphi^T(Q(I))| \leq c_n \prod_{j\in I}\|Q_j\|, \qquad c_n = \sum_m (m-1)!S(|I|, m), \qquad (4.31)$$

where $S(k, m)$ denotes the total number of partitions of $k$ indices into $m$ sets (Stirling number of the 2nd kind). ($c_n = \sum_1^\infty 2^{-k}k^{n-1}$).

Now we prove (4.30) by mathematical induction on $n$. For $I = \{1, \ldots, n\}$ we split the summation of the r.h.s. of (4.29) into the following three parts.

(a) the term $m = 1$: $\varphi^T(Q_1, \ldots, Q_n)$,

(b) the sum over subpartitions of $(J, J^c)$: $\varphi^T(Q(J))\varphi^T(Q(J^c))$,

(c) other terms: $m \geq 2$ and there exists at least one $I_\nu$ which has an intersection with both $J$ and $J^c$.

Moving the (b) and (c) parts to the left-hand side of (4.29) we obtain

$$\varphi^T(Q_1, \ldots, Q_n) = \varphi(Q_1, \ldots, Q_n) - \varphi(Q(J))\varphi(Q(J^c)) - \sum{}^{(c)}$$

where $\sum^{(c)}$ is the sum of the contributions of $(c)$.

Since by assumption $D_i$ and $D_j$ are spacelike separated (at $t = 0$) for $i \in J, j \in J^c$, $Q_i$ and $Q_j$ commute. Therefore

$$\varphi(Q_1, \ldots, Q_n) = \varphi(Q(J)Q(J^c)).$$

By Theorem 4.7 we can estimate

$$|\varphi(Q_1, \ldots, Q_n) - \varphi(Q(J))\varphi(Q(J^c))| < Gk(mT/2),$$
$$G \leq 2\|Q(J)\| \cdot \|Q(J^c)\| \leq 2\prod_j \|Q_j\|.$$

Concerning each term in $\sum^{(c)}$, at least one $I_\nu$ has non-empty intersections with $J$ and $J^c$ and $|I_\nu| < n$, due to $m > 1$. By the inductive assumption we obtain

$$|\varphi^T(Q(I_\nu))| \leq A_{|I_\nu|}\left(\prod_{j \in I_\nu} \|Q_j\|\right)k(mR/2).$$

Applying (4.31) to the remaining factors $\varphi^T(Q(I_\mu))$ $(\mu \neq \nu)$ in the same term we obtain

$$\left|\prod_\mu \varphi^T(Q(I_\mu))\right| \leq \left(\prod_j \|Q_j\|\right)k(mR/2)A_{|I_\nu|}\prod_{\mu \neq \nu} c_{|I_\mu|}.$$

Therefore, if we take

$$A_n = 2 + \sum A_{|I_\nu|}\prod_{\mu \neq \nu} c_{|I_\mu|}$$

we obtain (4.30).

**Corollary 4.12**
Suppose that $|x^0| < \delta$ and $Q_j \in \mathfrak{A}(D)$, $(j = 1, \ldots, n)$. Then, for a constant $A$ determined by $D, \delta$ and $n$,

$$|\varphi^T(Q_1(x_1) \cdots Q_n(x_n))| \leq A\left(\prod_j \|Q_j\|\right)\mathrm{e}^{-mR/2} \tag{4.32}$$

holds, where $R = (n-1)^{-1}\max|\mathbf{x}_i - \mathbf{x}_j|$.

**Proof** First we show for any given $\mathbf{x}_1, \ldots, \mathbf{x}_n$ that there exists a $J$ such that $|\mathbf{x}_i - \mathbf{x}_j| \geq R$ for any $i \in J, j \in J^c$ in the splitting of $\{1, \ldots, n\}$ into $\{J, J^c\}$.

Take a pair $(i,j)$ such that $|\mathbf{x}_i - \mathbf{x}_j|$ becomes maximal, and denote the segment between $\mathbf{x}_i$ and $\mathbf{x}_j$ by $L$. A plane passing through a point $\mathbf{x}_k$ and being orthogonal to $L$ always intersects with $L$, since $|\mathbf{x}_i - \mathbf{x}_j|$ is the maximum. Denote an intersection point by $P_k$ and label them according to their order starting from one end as $P_{k_1}, \ldots, P_{k_n}$ with $k_1 = i$ and $k_n = j$, for example. Since the length of $L$ is $(n-1)R$, necessarily there exists at least one pair of neighbouring points with their distance equal to or larger than $R$. Denote this pair by $k_l$ and $k_{l+1}$ and take $J = \{k_1, \ldots, k_l\}$. Then we get

$$|\mathbf{x}_p - \mathbf{x}_q| \geq |P_{k_p} - P_{k_q}| \geq |P_{k_l} - P_{k_{l+1}}| \geq R$$

for any $p \in J$, $q \in J^c$.

Let us denote the radius of the causal image of $D - D$ onto the plane $x^0 = 0$ by $a$,

$$a = \sup\{|x^0| + |\mathbf{x}|;\ x = x_1 - x_2,\ x_1 \in D,\ x_2 \in D\}.$$

If $R > a + 2\delta$, then the support $D + x_p$ of $Q_p(x_p)$ and the support $D + x_q$ of $Q_q(x_q)$ for $p \in J$ and $q \in J^c$ have the property that $(D + x_p) - (D + x_q) + te$ becomes spacelike for $|t| < R - a - 2\delta$, and we can apply Theorem 4.11 to obtain (4.32). □

### 4.7 Additivity assumption and Reeh–Schlieder theorem

**Definition 4.13**
Let $\pi_\varphi$ be the GNS representation of a state $\varphi$. If for any double cone $K$ and any of its open coverings $K = \cup_i D_i$ the equation

$$\pi_\varphi(\mathfrak{A}(K))'' = \left(\bigcup_i \pi_\varphi(\mathfrak{A}(D_i))\right)'' \tag{4.33}$$

is satisfied, we say that *additivity* holds. If

$$\pi_\varphi(\mathfrak{A})'' = \left(\bigcup_x \pi_\varphi(\mathfrak{A}(D + x))\right)'' \tag{4.34}$$

is satisfied for any domain $D$, then we say that *weak additivity* holds.

Here, $M''$ represents the von Neumann algebra generated by $M$.

**Remark** In the case of weak additivity, (4.34) holds for any open set $D$, due to the monotone property of $\mathfrak{A}(D)$, even if we restrict $D$ to the double cone in the definition.

**Theorem 4.14 (Reeh–Schlieder theorem)** [5]
Assume weak additivity for a vacuum state $\varphi$. Then the vacuum vector $\Omega_\varphi$ is a cyclic vector of $\pi_\varphi(\mathfrak{A}(D))$ for any domain $D$, and it is a separating vector of $\pi_\varphi(\mathfrak{A}(D))''$ for any bounded domain $D$.

**Proof** For the proof of the cyclicity it is enough to show that if $\Psi \perp \pi_\varphi(\mathfrak{A}(D))\Omega_\varphi$, then $\Psi = 0$. Take a domain $D_1$ which satisfies $\bar{D}_1 \subset D$. Then for any $x$ in a sufficiently small neighbourhood $N$ of 0 we get $D_1 + x \subset D$. Since $Q_j(x_j) \in \mathfrak{A}(D_1 + x_j) \subset \mathfrak{A}(D)$ for any elements $Q_1, \ldots, Q_n$ in $\mathfrak{A}(D_1)$ and any points $x_1, \ldots, x_n$ in $N$, we obtain

$$(\Psi, Q_{1\varphi}(x_1) \cdots Q_{n\varphi}(x_n)\Omega_\varphi) = 0, \tag{4.35}$$

where we have used the notation of eq. (4.14).

As we discussed in Section 2.5, the $T_\varphi$ defined in eq. (4.9) satisfies

$$Q_{j\varphi}(x_j) = T_\varphi(x_j) Q_{j\varphi} T_\varphi(x_j)^*, \quad Q_{j\varphi} \equiv \pi_\varphi(Q_j), \tag{4.36}$$

$$T_\varphi(x) = \int e^{i(x,p)} E_\varphi(d^4p).$$

Since the support of $E_\varphi$ is contained in $\bar{V}_+$

$$T_\varphi(z) = \int e^{i(\zeta,p)} E_\varphi(d^4p)$$

can be defined for a complex vector $\zeta$ satisfying $\operatorname{Im}\zeta \in \bar{V}_+$, $T_\varphi(z)$ is continuous in the strong operator topology and it is a holomorphic function in $\zeta$ for $\operatorname{Im}\zeta \in V_+$. Therefore, for $\operatorname{Im}\zeta_j \in \bar{V}_+$ we can define the continuous function

$$(\Psi, T_\varphi(\zeta_1) Q_{1\varphi} T_\varphi(\zeta_2) Q_{2\varphi} \cdots T_\varphi(\zeta_n) Q_{n\varphi} \Omega_\varphi) \tag{4.37}$$

which is holomorphic for $\zeta_j \in V_+$. Taking in this equation $\zeta_1 = x_1$, $\zeta_k = x_k - x_{k-1}$, $(k = 2, \ldots, n)$ we obtain the l.h.s. of (4.35) by (4.36) and by $T_\varphi(x_{k-1})^* T_\varphi(x_k) = T_\varphi(x_k - x_{k-1})$. The holomorphic function (4.37) must be identically 0 since (4.35) and hence (4.37) vanish for $x_j$ in $N$. Therefore, eq. (4.35) holds for all $x_j$.

The elements of the $*$ algebra $\mathcal{B}$ generated by $\cup_x \pi_\varphi(\mathfrak{A}(D_1 + x))$ are linear combinations of monomials of the form

$$Q_{1\varphi}(x_1) \cdots Q_{n\varphi}(x_n).$$

Since eq. (4.35) holds for any $x$, we know that $\Psi$ is orthogonal to $\mathcal{B}\Omega_\varphi$. On the other hand, by the weak additivity (4.34), the closure of $\mathcal{B}$ is $\pi_\varphi(\mathfrak{A})''$. Furthermore, since $\pi_\varphi(\mathfrak{A})\Omega_\varphi$ is dense due to the GNS construction, $\mathcal{B}\Omega_\varphi$ is also dense. Therefore $\Psi = 0$. Namely, $\pi_\varphi(\mathfrak{A}(D))\Omega_\varphi$ is dense.

Next we sketch the proof of the separating property. We show for $A, B \in \pi_\varphi(\mathfrak{A}(D))''$ that if $A\Omega_\varphi = B\Omega_\varphi$, then $A = B$. For the domain $D_1$ which is spacelike to $D$, $\mathfrak{A}(D_1)$ commutes with $\mathfrak{A}(D)$ due to locality. Thus, $AC\Omega_\varphi = BC\Omega_\varphi$ holds for any $C \in \pi_\varphi(\mathfrak{A}(D_1))$. Since $\pi_\varphi(\mathfrak{A}(D_1))\Omega_\varphi$ is dense as we proved, we get $A = B$. □

**Remark 1** The clustering property suggests that for two regions which are spatially separated far enough, any irreducible vacuum state is independent (a product state) in a very good approximation; on the other hand, the Reeh–Schlieder theorem shows the aspects of non-independence, such as the feature that any vector can be approximated by multiplying an observable of a bounded domain $D$ onto the vacuum state vector.

**Remark 2** For a domain $E$ which includes a translated $(D + x)$ of any double cone $D$ (at least for one $x$ depending on $D$), we denote the corresponding $*$ algebra

$$\{\mathfrak{A}(D);\ D \subset E\}$$

by $\mathfrak{A}_0(E)$. Then even if we do not assume weak additivity, we can prove in the way as shown above that such an $E$ possesses the Reeh–Schlieder property ($\pi_\varphi(\mathfrak{A}_0(E))\Omega_\varphi$ is dense in $\mathscr{H}_\varphi$). To give an example of such an $E$, take a cone $E$ (extended to infinity, with angle at its vertex as small as one pleases).

The following theorem is due to Borchers, and having in mind the association with archaeologists who guess the events in the surrounding area from samples dug out at a single point, we may call this result the *archaeologists' causality* [6].

**Theorem 4.15**
Let $D$ be a double cone with vertices $x$ and $y$, their separation $x - y$ being timelike, and let $I$ be any open set including the open segment connecting $x$ and $y$. (Both $x$ and $y$ are excluded from the segment.) If additivity holds, then

$$\pi_\varphi(\mathfrak{A}(D))'' \subset \pi_\varphi(\mathfrak{A}(I))''.$$

**Remark** If $I \subset D$, then the equality holds. We can take $I$ as "thin" as we like.

The proof can be given with a similar method as used for the JLD representation. We skip it here [7].

### 4.8 Quantum field theory

The framework of quantum field theory which started from the canonical quantization of classical fields has been axiomatized by Wightman as follows. They are referred to as the *Wightman axioms* [8].

**Axiom 1 (quantum field)**
The operators $\phi_1(f), \dots, \phi_n(f)$ are given for each $C^\infty$-function $f$ with compact support on the Minkowski space $\mathbf{R}^4$. Each $\phi_j(f)$ and its Hermitian conjugate operator $\phi_j(f)^*$ are defined at least on a common dense linear subset $\mathfrak{D}$ of the Hilbert space $\mathscr{H}$ and $\mathfrak{D}$ satisfies

$$\phi_j(f)\mathfrak{D} \subset \mathfrak{D}, \qquad \phi_j(f)^*\mathfrak{D} \subset \mathfrak{D} \tag{4.38}$$

for any $f$ and $j = 1, \dots, n$. For any $\Phi, \Psi \in \mathfrak{D}$,

$$f \to (\Phi, \phi_j(f)\Psi)$$

is a complex valued distribution (linear and continuous with respect to $f$).

**Remark** In simple words, the quantum fields $\phi_j$ are operator valued distributions. The operators $\phi_j(f)$ are generally unbounded. Therefore it becomes necessary to state the properties of their domain $\mathfrak{D}$. By assumption (4.38), we may operate arbitrarily many $\phi_j(f)$ successively on a vector $\Psi$ in $\mathfrak{D}$.

**Axiom 2 (relativistic symmetry)**
On $\mathscr{H}$ there exists a unitary representation $U(a, A)$ of $\tilde{\mathscr{P}}_+^\uparrow$ ($a \in \mathbf{R}^4$, $A \in SL(2, C)$), satisfying

$$U(a, A)\mathfrak{D} = \mathfrak{D} \quad \text{(invariance of the common domain of the fields)}$$

$$U(a, A)\phi_j(f)U(a, A)^* = \sum S(A^{-1})_{jk}\phi_k(f_{(a,A)}), \tag{4.39}$$

$$f_{(a,A)}(x) = f(\Lambda(A)^{-1}(x - a)), \tag{4.40}$$

where the matrix $(S(A)_{j,k})$ is an $n$-dimensional representation of $A \in SL(2C)$.

**Remark** For the case $n = 1$, $S(A)$ is the identity representation and $\phi$ is called a scalar field. Usually the notation

$$\phi(f) = \int \phi(x)f(x)\,\mathrm{d}^4x$$

is used. Then (4.39) can be represented in the following form where the transformation (4.40) acts on $\phi$, after a change of integration variables

$$U(a, A)\phi(x)U(a, A)^* = \phi(\Lambda(A)x + a)$$

Generally, $S(A)$ can be written as a sum of irreducible representations. For the case of the Dirac field, $n = 4$ and $S(A)$ given by the Dirac $\gamma$ matrices is the direct sum of two two-dimensional representations which are conjugate to each other ($A$ and $\bar{A}$). If one treats one scalar field and one Dirac field at the same time, one may take $n = 5$.

**Axiom 3 (local commutativity)**
If the support of $f$ and $g$ is spacelike separated, then for any vector $\Phi$ in $\mathfrak{D}$

$$[\phi_j(f)^{(*)}, \phi_k(g)^{(*)}]_\pm \Phi = 0. \tag{4.41}$$

Here the symbol $(*)$ indicates a choice between $*$ and no $*$ and the above equation holds for either choice for each $(*)$, i.e. no $*$, one $*$ and both with $*$. The commutator and anticommutator are denoted by

$$[X, Y]_\pm \equiv XY \pm YX,$$

where the choice of the sign $\pm$ depends on the indices $j, k$.

**Remark** If we write (4.41) in distribution notation we get

$$[\phi_j(x)^{(*)}, \phi_k(y)^{(*)}]_\pm = 0 \qquad ((x - y, x - y) < 0).$$

In the framework of Axioms 1–4 it can be shown that the choice of $\pm$ in (4.41) has to be the same for $\phi_j$ and $\phi_j^*$. If $+$ holds for the case that both $j$ and $k$ are double valued representations of $\mathscr{P}_+^\uparrow$ (half odd spin), and $-$ holds for other pairs $j, k$, then we speak of *normal commutation relations*. It is known that the general case can be reduced to the normal commutation relations by the so-called *Klein transformations* [9].

**Axiom 4 (vacuum state)**
There exists a vector $\Omega$ in $\mathfrak{D}$ satisfying the following conditions:

(i) $U(a, A)\Omega = \Omega$ (invariance).

(ii) The set of all vectors obtained by acting an arbitrary polynomial $P$ of the fields on $\Omega$ is dense in $\mathscr{H}$ (cyclicity).

(iii) The spectrum of the translation group $U(a, \mathbf{1})$ on $\Omega^\perp$ is contained in

$$\bar{V}_m = \{p;\, (p, p) \geq m^2,\, p^0 > 0\} \quad (m > 0)$$

(spectrum condition).

**Remark** Here we have assumed a mass gap $m > 0$. More generally, we may consider the case $m = 0$. In that case, it is sometimes convenient to assume the uniqueness of the vacuum vector, i.e. if $U(a, A)\Phi = \Phi$ then $\Phi = c\Omega$ (where $c$ is a complex number). This assumption corresponds to considering a pure vacuum state. Usually the set of all vectors given in (ii) is taken as the domain $\mathfrak{D}$ of the fields.

*Comparison with the theory of local observables*

If we denote the set of all $C^\infty$-functions with support contained in a space–time domain $D$ as $\mathscr{D}(D)$, then the $*$ algebra $\mathscr{P}(D)$ of all polynomials in the fields $\phi_j(f)$, $f \in \mathscr{D}(D), j = 1, \ldots, n$, plays the role of the local observables on $D$. In this case the monotone property of the local observables is satisfied by definition. Covariance and locality correspond to the Axioms 2 and 3 given above. The properties of the GNS representation constructed from the vacuum state correspond to Axiom 4 (i) and (ii) and the spectrum condition with $\bar{V}_0 \equiv \bar{V}_+$. In quantum field theory the additivity and also the special case of the weak additivity used in the previous section are satisfied in the following way. For an open covering $\{D_i\}$ of the domain $D$ and for any $f \in \mathscr{D}(D)$ there exists a finite number of $\chi_i \in \mathscr{D}(D_i)$ satisfying $\sum \chi_i = 1$ on the (compact) support of $f$. If we take $f_i \equiv f\chi_i \in \mathscr{D}(D_i)$, then $\phi(f) = \sum \phi(f_i)$ and we obtain the additivity for $\mathscr{P}(D)$.

The differences to the theory of local observables are as follows.

1. The quantum fields $\phi_j(f)$ are in general unbounded operators, while for local observables only bounded operators are considered. Consequently, the domain of field operators is not the whole $\mathscr{H}$ but its dense linear subset $\mathfrak{D}$.
2. For a local observable $Q$, the $Q(x)$ which is taken to be $\alpha_{(x,1)}Q$ is a function of $x$, whereas the $\phi_j(x)$s are distributions, although their transformation under translations is the same.
3. With respect to $\tilde{L}_+^\uparrow$, the $\phi_j(x)$s give a finite-dimensional representation, while the above $Q(x)$ does not possess such a simple transformation property.
4. Concerning local commutativity, $Q(x)$ and $Q(y)$ commute if $x - y$ is spacelike to $D - D$ and the support of Q is $D$, while a quantum field $\phi_j(x)$ behaves as if its support $D$ is a single point $\{x\}$.
5. In the local commutativiy of quantum fields, both commutativity ($[X, Y]_-$) and anticommutativity ($[X, Y]_+$) of $\phi_j(x)$ and $\phi_k(y)$ for spacelike $x - y$ are considered, while in the theory of local observables only the commutativity is considered.

(3) and (4) are special properties of the quantum fields, and it may be that only in very special cases are we able to extract such quantum fields from the system of local observables. On the other hand, we can also understand that due to the requirement (4), the mathematical treatment of quantum fields is more difficult than that of $Q(x)$ with respect to points (1) and (2).

In the next section we will discuss in which case the algebra of local observables $\mathfrak{A}(D)$ can be constructed from the given quantum fields. In this consideration the difference in (5) can be understood by the sector theory treated in Chapter 6. Namely, for the case that $\phi_i(x)$ and $\phi_j(y)$ are anticommuting, the even polynomials of the quantum fields commute for spacelike separation (due to $[X_1X_2, Y]_- = X_1[X_2, Y]_+ - [X_1, Y]_+X_2$). Starting from a subalgebra satisfying such a local commutativity, we construct $\mathfrak{A}(D)$. By developing the sector theory we can derive out of it operators which are anticommuting. Since the local commutativity is based on the relativistic causality principle the anticommutativity does not fit into the fundamental axioms. In the theory of local observables we assume commutativity in the fundamental axioms and, after the theory has been developed, anticommuting fields are derived which is one of the merits of the theory of local observables.

## 4.9 From quantum fields to local observables

For simplicity, let us take one type of real scalar field $\phi(x)$ (by real we mean that $\phi(f)^*$ coincides with $\phi(\bar{f})$ on $\mathscr{D}$) satisfying the Wightman axioms given in the previous section. However, the condition $m > 0$ is not necessary and it is sufficient to assume the uniqueness of the vacuum vector.

Let us denote the $*$ algebra consisting of all complex polynomials of $\phi(f)$, $f \in \mathscr{D}(D)$ (which are unbounded operators on $\mathfrak{D}$) by $\mathscr{P}(D)$. The set of all bounded linear operators $L$ which satisfy

$$(L^*\Phi, X\Psi) = (X^*\Phi, L\Psi)$$

for any $X \in \mathscr{P}(D)$ and any $\Phi, \Psi \in \mathfrak{D}$ is called the *weak commutant* of $\mathscr{P}(D)$ and denoted by $\mathscr{P}(D)^w$. Now consider the following two conditions:

(a) $\mathscr{P}(D)^w$ is an algebra for any double cone $D$.

(b) The vacuum vector $\Omega$ is cyclic for the union of $\mathscr{P}(D')^w$ over all double cones $D$, where $D'$ is the causal complement of $D$. Thus, $\cup_D \mathscr{P}(D')^w\Omega$ is dense in $\mathscr{H}$.

Under these conditions the following result is known [10]:

**Theorem 4.16**
Assume conditions (a) and (b). For each double cone $D$ we define

$$\mathfrak{A}(D) = (\mathscr{P}(D)^w)'.$$

$\mathfrak{A}(D)$ is a von Neumann algebra, and it satisfies axioms (1)–(3) for local observables. Define $\mathfrak{A}$ by the generating condition (4). The state over $\mathfrak{A}$ determined by the vacuum vector $\Omega$ is then a pure vacuum state, satisfying the following properties:

(i) $\mathfrak{A}(D)\Omega$ is dense in $\mathscr{H}$ for any double cone $D$.

(ii) Each $X \in \mathscr{P}(D)$ has a closed extension $X_e \subset X^{\dagger *}$ affiliated with $\mathfrak{A}(D)$.

Note that $A$ is affiliated with $\mathfrak{A}(D)$ if, with the polar decomposition $A = U|A|, (|A| = (A^*A)^{1/2})$, $U$ and all spectral projection operators of $A$ are in $\mathfrak{A}(D)$, and $A^\dagger$ is the restriction of $A^*$ to $\mathfrak{D}$ (i.e. the Hermitian conjugate of $A$). $X_e \subset A$ means that the domain of $X_e$ is contained in the domain of $A$ and that $X_e = A$ holds on the domain of $X_e$.

For the premises (a) and (b) of this theorem, the following sufficient conditions are known. These conditions are verified for concrete examples.

**Theorem 4.17**
If the following generalized H-boundedness holds, then $\mathscr{P}(D)^w$ is an algebra for any open set $D$. For any $f \in \mathfrak{D}$ there exists a natural number $n$ for which $\phi(f)(1+H)^{-n}$ is a bounded operator, where $H$ is the generator of the time translation (the energy operator).

**Theorem 4.18**
If under the assumption (a) there exists a $\Psi \in \mathscr{H}$ such that the vector state $\omega_\Psi(A) = (\Psi, A\Psi)$ is a weakly correlated state over $\mathscr{P}(D_1 \cup D_2')$ (in the following sense), for any pair of open double cones $D_1, D_2$ satisfying $\bar{D}_1 \subset D_2$, then condition (b) is satisfied and the conclusion of Theorem 4.16 holds.

A state $\omega_\Psi$ is weakly correlated over $\mathscr{P}(D_1 \cup D_2')$ if there exists a constant $c$ satisfying

$$\omega_\Psi(\pi(A)) \leq c\omega_\Omega \otimes \omega_\Omega(A),$$

for all positive $A \in \mathscr{B}$, where $\omega_\Omega$ denotes the vacuum state for $\mathscr{P}(D_1)$ and for $\mathscr{P}(D_2')$, $\mathscr{B}$ denotes the tensor product of $\mathscr{P}(D_1)$ with $\mathscr{P}(D_2')$ and $\pi$ is the natural map from $\mathscr{B}$ to $\mathscr{P}(D_1 \cup D_2')$.

# 5
# SCATTERING THEORY

In this chapter we explain how particle scattering can be described by local observables. First we introduce the concept of the $S$-matrix, which is an idea common to classical and quantum mechanics. We show the construction of incoming and outgoing states which have a certain asymptotic behaviour at time $-\infty$ and $+\infty$, and we also show that the constructed states have an acceptable physical asymptotic behaviour. Then we derive the representation of the $S$-matrix given by these scattering states and discuss its properties.

## 5.1 The concept of scattering states and the $S$-matrix

First let us consider particle scattering in classical mechanics using the same symbols as in quantum mechanics. Let $\mathbf{f}(\mathbf{x})$ be the given field of force at each point $\mathbf{x}$, then the motion of one particle with mass $m$ is governed by the following equation of motion:

$$\ddot{\mathbf{x}}(t) = \mathbf{f}(\mathbf{x}(t))/m. \tag{5.1}$$

Let us view the solution $\mathbf{x}(t)$ as a particle state (in the Heisenberg representation) and denote the set of all solutions by $\mathscr{H}$.

If the field of force $\mathbf{f}(\mathbf{x})$ vanishes at large distances, the motion of a particle there will be a straight-line motion:

$$\mathbf{x}(t) = \mathbf{a} + \mathbf{v}t, \tag{5.2}$$

as a solution of $\ddot{\mathbf{x}}(t) = 0$. Although the state $\mathbf{x}(t)$ in $\mathscr{H}$ takes this form only when $\mathbf{x}(t)$ reaches the remote region, we consider the states (5.2) in which the particle moves along a straight line all the time $t$ as a means to describe asymptotic states at large distances, and we denote its totality by $\mathscr{H}_0$.

Since the initial value problem of (5.1) has a unique solution if $\mathbf{f}$ is smooth, the solution of (5.1) which coincides with (5.2) at $t < -T$ for a sufficiently large $T$ uniquely exists. For states $\varphi \in \mathscr{H}_0$ and $\Psi \in \mathscr{H}$, let the corresponding motion be represented by $\mathbf{x}_\varphi(t)$ and $\mathbf{x}_\Psi(t)$. Then to each $\varphi \in \mathscr{H}_0$, there corresponds a unique $\Psi \in \mathscr{H}$ satisfying

$$|\mathbf{x}_\varphi(t) - \mathbf{x}_\Psi(t)| \to 0 \text{ (actually} = 0) \qquad \text{for } t \to -\infty. \tag{5.3}$$

We denote this $\Psi$ as

$$\Psi = \Psi_-(\varphi).$$

$\Psi_-$ is a map from $\mathscr{H}_0$ into $\mathscr{H}$. The image of $\Psi_-$, $\mathscr{H}_- = \Psi_-\mathscr{H}_0$, does not necessarily coincide with $\mathscr{H}$. This is due to possible bound states of particles which could not escape to a remote distance, being trapped by the field of force. However, for any $\Psi \in \mathscr{H}_-$, the state $\varphi \in \mathscr{H}_0$ satisfying $\Psi = \Psi_-(\varphi)$ is uniquely determined by (5.3). We write it as

$$\varphi = \varphi_-(\Psi).$$

$\varphi_-$ is a map from $\mathscr{H}_-$ to $\mathscr{H}_0$ and is the inverse of the map $\Psi_-$.

Similarly, if we define, for $\varphi \in \mathscr{H}_0$ and $\Psi \in \mathscr{H}$,

$$\Psi = \Psi_+(\varphi), \quad \varphi = \varphi_+(\Psi)$$

by the condition

$$|\mathbf{x}_\varphi(t) - \mathbf{x}_\Psi(t)| \to 0 \qquad \text{for } t \to +\infty \tag{5.4}$$

then $\Psi_+$ is a map from $\mathscr{H}_0$ into $\mathscr{H}$ and $\varphi_+$ is a map from $\mathscr{H}_+ = \Psi_+(\mathscr{H}_0)$ to $\mathscr{H}_0$, being the inverse map of $\Psi_+$.

$\Psi_-(\varphi)$ represents the motion under the influence of the force field for the particle which came in from infinity in the state of free motion $\varphi$. $\Psi_+(\varphi)$ represents the motion under the influence of the force field for the particle which finally comes out to infinite distance in the state of free motion $\varphi$. $\varphi_-(\Psi)$ describes the state of free motion of the incoming particle, from which the state of motion $\Psi$ under the influence of the force field is realized, and $\varphi_+(\Psi)$ describes the state of free motion of the outgoing particle, which comes out of the state of motion $\Psi$. Now let us consider

$$S_0 = \varphi_+\Psi_-. \tag{5.5}$$

Then $S_0(\varphi) = \varphi_+(\Psi_-(\varphi))$ represents the final outgoing state reached under the influence of the force field by a particle with incoming state $\varphi$. Namely, it is a map describing the scattering of a particle by a field of force. In this description, incoming and outgoing states are described by a free motion and $S_0$ is a map from $\mathscr{H}_0$ to $\mathscr{H}_0$.

In the case where the field of force is a conserved force $\mathbf{f}(\mathbf{x}) = -\nabla V(\mathbf{x})$, the total energy

$$E = (m\dot{\mathbf{x}}(t)^2/2) + V(\mathbf{x}(t))$$

is a constant independent of $t$. The states of both $\mathscr{H}_+$ and $\mathscr{H}_-$ have the special property that $E > 0$ and $\mathscr{H}_+ = \mathscr{H}_-$. In this case $S_0$ becomes a bijection.

The following map is often considered instead of $S_0$:

$$S = \Psi_-\varphi_+. \tag{5.6}$$

$S$ is a map from $\mathscr{H}_+$ to $\mathscr{H}_-$ and it maps a state $\Psi_+(\varphi)$ with the outgoing state $\varphi$ to a state $\Psi_-(\varphi)$ with incoming state $\varphi$. If $\mathscr{H}_+ = \mathscr{H}_-$ we can write

$$S = (\varphi_+)^{-1}S_0(\varphi_+) = (\varphi_-)^{-1}S_0(\varphi_-).$$

Conceptually, $S_0$ may be called the *S-matrix*, but often this expression is used for $S$.

The above description holds under various different circumstances. Instead of scattering under a field of force, the classical motion of interacting particles under a conserved force determined by the mutual relative position of $n$ particles is one example. Due to the fact that some particles can form bound states, there is a possibility for the case of $n$ incoming particles that the outgoing particles may include such bound states. Thus, treating a bound state as a different kind of particle, $\mathscr{H}_0$ does not merely contain the free motion of $n$ particles but we also have to consider states of a many particle system, including the free motion of bound states. With an appropriate choice of $\mathscr{H}_0$ we can get a description of scattering states as formulated above also in this case.

In quantum mechanics $\mathscr{H}$ and $\mathscr{H}_0$ are Hilbert spaces with time evolution in each space described in terms of self-adjoint operators $H$, $H_0$ as

$$\Psi(t) = \mathrm{e}^{-\mathrm{i}tH}\Psi, \quad \varphi(t) = \mathrm{e}^{-\mathrm{i}tH_0}\varphi. \tag{5.7}$$

Conditions (5.3) and (5.4) for the asymptotic free motion become

$$\lim_{t\to\pm\infty} \|\Psi(t) - \varphi(t)\| = 0. \tag{5.8}$$

Since $\mathrm{e}^{-\mathrm{i}tH}$ and $\mathrm{e}^{-\mathrm{i}tH_0}$ are unitary, this condition can be rewritten as

$$\lim_{t\to\pm\infty} \|\Psi_\pm(\varphi) - \mathrm{e}^{\mathrm{i}tH}\mathrm{e}^{-\mathrm{i}tH_0}\varphi\| = 0$$
$$\lim_{t\to\pm\infty} \|\varphi_\pm(\Psi) - \mathrm{e}^{\mathrm{i}tH_0}\mathrm{e}^{-\mathrm{i}tH}\Psi\| = 0.$$

Usually, $H_0$ is the kinetic energy and $H$ is of the form $H = H_0 + V$. If the potential $V$ satisfies an appropriate condition in the case of the potential scattering of a single particle or the scattering of two particles by mutual interaction potential, then

$$\Psi_\pm = \lim_{t\to\pm\infty} \mathrm{e}^{\mathrm{i}tH}\mathrm{e}^{-\mathrm{i}tH_0} \tag{5.9}$$

exists, it is an isometry, the two spaces $\mathscr{H}_\pm = \Psi_\pm\mathscr{H}_0$ coincide and become the space of the absolute continuous spectrum of $H$. The limit

$$\varphi_\pm = \lim_{t\to\pm\infty} \mathrm{e}^{\mathrm{i}tH_0}\mathrm{e}^{-\mathrm{i}tH} \tag{5.10}$$

exists on $\mathscr{H}_+ = \mathscr{H}_-$ and is identical with $(\Psi_\pm)^*$. $\Psi_\pm$ are called *wave operators*. For $\varphi, \psi \in \mathscr{H}_0$ the quantity

$$(\psi, S_0\varphi) = (\Psi_+(\psi), \Psi_-(\psi)) = (\Psi_-(\psi), S\Psi_-(\varphi)) = (\Psi_+(\psi), S\Psi_+(\varphi)) \tag{5.11}$$

is called an *S-matrix element*, representing the probability amplitude that an incoming particle in the state $\varphi$ becomes an outgoing particle in the state $\psi$ (the probability is given by the square of its absolute value). Here $S_0$ and $S$ are defined by (5.5) and (5.6), and under the above described circumstances $S_0$ becomes unitary.

Under the same circumstances $\mathcal{H}_0$ and $\mathcal{H}$ are taken to be the same Hilbert space. However, in the scattering of a large number of particles it is necessary to prepare $\mathcal{H}_0$ separately from $\mathcal{H}$ because bound states have to be taken into consideration among the incoming and outgoing particles in general. Thus, it becomes necessary to take different spaces for $\mathcal{H}_0$ depending on the system. $\mathcal{H}_0$ is the space which describes the asymptotic behaviour of the scattering system at time infinity. In the following let us continue the discussion of how $\mathcal{H}_0$, $\Psi_\pm$ etc. are represented in the theory of local observables. Completely the same discussion occurs in quantum field theory. In this case $\Psi_\pm$ are usually denoted by $\Psi^{\text{out}}$ and $\Psi^{\text{in}}$.

## 5.2 Description of asymptotic states

As a physical picture we want to describe the asymptotic behaviour at time infinity by the states of free motion (which is classically a straight-line motion of constant velocity) of many non-interacting particles.

First we have to clarify what a particle is. Here let us consider a particle as the smallest unit in the framework of relativistic quantum theory taken in this book. Since the states, which are transformed to each other under the relativistic symmetry $\mathscr{P}_+^\uparrow$, are considered as the same physical object with different location, direction and velocity, the whole set of states of one particle has to be invariant under $\mathscr{P}_+^\uparrow$. Further, the quantum mechanical states are represented by unit vectors in a Hilbert space $\mathcal{H}$ and a state which can be represented by a linear combination of two vectors $\Psi_1$, $\Psi_2$ (norm to be normalized to 1 by multiplication of a positive number) is called *superposition*. If $\Psi_1$ and $\Psi_2$ represent two states of the same particle, then by the requirement of quantum mechanics their superposition also describes a state of the same particle. Furthermore, the limit of a one particle state is also a one particle state. Summarizing these three conditions, the set of all states of one particle has to be represented by the set of all unit vectors of a relativistic invariant subspace (closed linear subset). Since the relativistic symmetry can be represented by a unitary representation $U(g)$ of $\tilde{\mathscr{P}}_+^\uparrow$ ($g \in \tilde{\mathscr{P}}_+^\uparrow$), according to the discussion in Chapter 3, the condition of the smallest unit becomes the condition that there is no subspace invariant under $\tilde{\mathscr{P}}_+^\uparrow$ except $\{0\}$ and the whole space itself. This is just the condition for an irreducible unitary representation of $\tilde{\mathscr{P}}_+^\uparrow$.

Based on the above discussion, we define a one particle state as a state which can be represented by the unit vectors of an irreducible representation of $\tilde{\mathscr{P}}_+^\uparrow$. This can also be considered as a concept which includes the relativistic objects corresponding to bound states of non-relativistic quantum mechanics.

The irreducible representations of $\tilde{\mathscr{P}}_+^\uparrow$ are completely classified. One of the classification parameters is the mass $m$, which is defined as

$$m^2 = (P, P) = (P^0)^2 - (P^1)^2 - (P^2)^2 - (P^3)^2$$

using the energy momentum $P^\mu$. Since $m^2$ is positive or 0, due to the assumption of positivity of the energy (see Definition 4.3, Theorem 4.5 in Section 4.2), $m$ is also a real number which is either positive or 0. For the case $m > 0$ we obtain a complete classification by the pair $(m,j)$ of mass $m$ and spin $j$. The spin takes values $0, 1/2, 1, 3/2, \ldots$; however, for simplicity we explain in this section only the case of one type of particle with mass $m > 0$ and spin 0. Note that there is a discussion in Chapter 6 relating to the case of half-odd integer spin like $1/2, 3/2, \ldots$.

Note that in the cyclic representation associated with a pure ground state (which we will be considering in this chapter), we consider the unitary representation of $\mathscr{P}_+^\uparrow$ because of the irreducibility of the algebra generated by observables and the existence of the vacuum vector $\Omega$ invariant under $U(g)$. Therefore, only one particle states of integer spin can appear there. One particle states of half-odd integer spin may appear in other representation spaces of observables, as is discussed in Chapter 6.

Considering only one type of particle with mass $m$ and spin 0, the state of free motion of the corresponding non-interacting many particle system can be described by a Fock space as already explained in Section 3.5. In the following we show how the abstract description given in Section 3.5 can be written by using a concrete wave function.

The vector $f$ in the Fock space can be represented as a series of wave functions $f_n(p_1 \ldots p_n)$, $(n = 0, 1, 2, \ldots)$, which give the probability amplitude that the energy momenta of $n$ given particles are $p_1, \ldots, p_n$. Here $p_j$ is a four-dimensional vector variable with mass $m$ and satisfying the condition of positivity of the energy

$$(p_i, p_j) = m^2, \qquad (p_j)^0 > 0,$$

the inner product of the vectors is given by

$$(g,f) = \sum_{n=0}^{\infty} \int g_n(p_1 \ldots p_n)^* f_n(p_1 \ldots p_n)\, \mathrm{d}\mu(\mathbf{p}_1) \cdots \mathrm{d}\mu(\mathbf{p}_n), \tag{5.12}$$

and $\mathrm{d}\mu(\mathbf{p})$ is the invariant measure given in eq. (3.41).

The wave function of $n = 0$ is a complex number $f_0$ and $|f_0|^2$ represents the probability that the state $f$ is the vacuum. The $n = 1$ wave function $f_1(p_1)$ corresponds to $\Psi(\mathbf{p})$ in eq. (3.42), $\mathscr{H}_p$ of eq. (3.33) is one-dimensional since the spin is 0, and $f_1(p_1)$ is a complex valued function. Since the subspace for general $n$ is spanned by the product of $n$ one particle wave functions (tensor product), the corresponding vector is given by a function in $n$ vector variables $p_1, \ldots, p_n$, and also the measure of the inner product can be given by a product measure as in (5.12).

By the situation explained in Section 3.5, the wave function $f_n$ is restricted to those functions which are invariant under permutation of the $n$ variables $p_1, \ldots, p_n$ (completely symmetric). We will find out later that this is related to the locality of the observables. On the other hand, in the case of a particle with half-odd integer spin we consider only completely antisymmetric functions (exchange of any two variables changes the sign), and the deep reason for this will be explained in Chapter 6.

Representing the states of $\mathscr{H}_0$ by such wave functions, the $S$-matrix elements given in (5.11) can usually be represented as

$$(g, S_0 f) = \sum_{n,m} \int g_n(p'_1 \cdots p'_n)^* f_m(p_1 \cdots p_m) S_0^{nm}(p'_1 \cdots p'_n;\ p_1 \cdots p_m)$$
$$\times\, \mathrm{d}\mu(p'_1) \cdots \mathrm{d}\mu(p'_n)\, \mathrm{d}\mu(p_1) \cdots \mathrm{d}\mu(p_m). \qquad (5.13)$$

If we restrict $g_n$ and $f_n$ to rapidly decreasing $C^\infty$ class functions $\mathscr{S}$, the slowly increasing hyperfunction $S_0(\ldots;\ldots)$ can be defined by using (5.13). For example in the case of the trivial $S$-matrix $S_0 = 1$

$$S_0^{nm} = \delta_{nm} \sum_{\sigma} \prod_{j=1}^{n} \{\delta^3(\mathbf{p}'_j - p_{\sigma(j)}) 2p_j^0\},$$

where summation is over all permutations $\sigma$ of $1, \ldots, n$.

### 5.3 Construction of asymptotic states

In this section we consider the situation where one particle states with mass $m$ and spin 0 (the $[m, 0]$ representation of $\mathscr{P}_+^\uparrow$) appear in the GNS representation space $\mathscr{H}_\omega$ (to distinguish from the map $\varphi$ of Section 5.1, we use $\omega$ instead of $\varphi$ used in Chapter 4) constructed from the vacuum state $\omega$ and construct the candidate $\Psi_\pm$ introduced in Section 5.1 as a unitary map from the space $\mathscr{H}_0$, representing the asymptotic behaviour explained in Section 5.2, to $\mathscr{H}_\omega$. (It is called the *Haag–Ruelle scattering theory* [1].) In the next section we will verify that the states $\Psi_+(\varphi)$, $\Psi_-(\varphi)$ constructed mathematically in this section show certainly the asymptotic behaviour represented by $\varphi$ at future and past infinity, respectively. As a result we show that if $\mathscr{H}_\omega$ contains one particle states then it also contains $n$ particle scattering states.

Below we fix one $\mathscr{P}_+^\uparrow$ invariant pure vacuum state $\omega$ and we simply write $\mathscr{H}$ for the representation space $\mathscr{H}_\omega$, $\Omega$ for the vector representing $\omega$, and $Q$ for the representing operator $\pi_\omega(Q)$ of the observable $Q$. By the $\mathscr{P}_+^\uparrow$ invariance of $\omega$, the unitary representation $U$ of $\mathscr{P}_+^\uparrow$ satisfying

$$U(g)\Omega = \Omega, \quad U(g)QU(g)^* = gQ \qquad (g \in \mathscr{P}_+^\uparrow) \qquad (5.14)$$

is determined (see Theorem 2.33). We denote the translation group part as $T(a)$ $(= U(a, \mathbf{1}))$ (in Chapter 4 it was denoted by $T_\varphi$). The irreducible representation $[m, 0]$ of $\mathscr{P}_+^\uparrow$ (the one particle state of mass $m > 0$ and spin 0) is assumed to be contained in $\mathscr{H}$ with multiplicity 1, and the projection operator to this irreducible representation is denoted by $E_1$: $E_1\mathscr{H} = \mathscr{H}_1$. The support of the spectrum of the energy momentum $P^\mu$ generating $T(a)$ in $(\mathbf{1} - E_1)\mathscr{H}$ is assumed to be separated from $(p, p) = m^2$.

#### (a) Construction of the quasi-local one particle creation operator

Since $\mathfrak{A}\Omega$ is dense in $\mathscr{H}$ and $\mathfrak{A}$ is generated by $\{\mathfrak{A}(D);\ D\}$, an appropriate double cone $D$ and a $Q_0 \in \mathfrak{A}(D)$ satisfy

$$f_0 \equiv E_1 Q_0 \Omega \neq 0. \qquad (5.15)$$

Since $f_0$ is a one particle state, as a vector in the space of an irreducible representation $[m, 0]$ of $\mathscr{P}_+^\uparrow$ it can be represented by a wave function $f_0(\mathbf{p})$.

Next, choose a $C^\infty$ function $\tilde{f}_j(p)$ which has a compact support disjoint from the support of $P^\mu$ in $(\mathbf{1} - E_1)\mathscr{H}$ and put

$$\begin{aligned} Q_j &= \int Q_0(x) F_j(x)\, \mathrm{d}^4 x \\ Q_0(x) &= T(x) Q_0 T(x)^* \\ F_j(x) &= (2\pi)^{-4} \int \mathrm{e}^{-\mathrm{i}(p,x)} \tilde{f}_j(p)\, \mathrm{d}^4 p. \end{aligned} \tag{5.16}$$

Then

$$(\mathbf{1} - E_1) Q_j \Omega = (\mathbf{1} - E_1) \tilde{f}_j(P) Q_0 \Omega = 0.$$

Here $\tilde{f}_j(P)$ denotes the function

$$\tilde{f}_j(P) = \int \tilde{f}_j(p) E(\mathrm{d}^4 p)$$

of $P^\mu = \int p^\mu E(\mathrm{d}^4 p)$ and $(\mathbf{1} - E_1)\tilde{f}_j(P) = 0$ by assumption. Thus

$$f_j \equiv Q_j \Omega = E_1 Q_j \Omega = \tilde{f}_j(P) E_1 Q_0 \Omega = \tilde{f}_j(P) f_0 \tag{5.17}$$

is a one particle state and its wave function is given by

$$f_j(\mathbf{p}) = \tilde{f}_j(p) f_0(\mathbf{p}), \qquad p^0 = (\mathbf{p} \cdot \mathbf{p} + m^2)^{1/2}.$$

We have attached a suffix $j = 1, 2, \ldots$ because in the next subsection we use various $\tilde{f}_j$. The $f_j$ obtained in this way forms a dense subset of $E_1 \mathscr{H}$ when $\tilde{f}_j$ is varied. The reason for this is based on the following theorem.

**Theorem 5.1**
For the wave function $f_0(\mathbf{p})$ of a one particle state $f_0 = E_1 Q_0 \Omega$, $|f_0(\mathbf{p})|^2$ is real holomorphic for any local observable $Q_0$.

**Outline of the proof** By restricting the Fourier transform of the JLD representation (4.18) to $(p, p) = m^2, p^0 > 0$, we obtain

$$\begin{aligned} (Q_0 \Omega, E_1 \mathrm{e}^{\mathrm{i}(P,x)} Q_0 \Omega) = \int \mathrm{d}\xi^0 \int \mathrm{d}\boldsymbol{\xi} \rho(m, \xi) \{ & 2\varphi(\xi^0)(\partial/\partial x^0) \Delta_m^+(\xi - x) \\ & - \varphi'(\xi^0) \Delta_m^+(\xi - x) \}. \end{aligned} \tag{5.18}$$

Here $\rho(m, \xi)$ is the integral of $\rho(\mathrm{d}\kappa, \xi)$ over the interval $m - \varepsilon < \kappa < m + \varepsilon$ and does not depend on $\varepsilon$. Since the support of $\rho(m, \xi)\varphi(\xi^0)$ is compact, its Fourier transform

$$\widehat{\rho\varphi}(p) = (2\pi)^{-3} \int \mathrm{e}^{-\mathrm{i}(p,\xi)} \rho(m, \xi)\varphi(\xi^0)\, \mathrm{d}^4\xi$$

is an integral function and the same is true for $\rho\varphi'$. Hence, the Fourier transform of (5.18) with respect to the vector variable $\mathbf{x}$ at $x^0 = 0$ given by

$$|f_0(p)|^2 = 2p^0 \widehat{\rho\varphi}(p) + i(\rho\varphi')^\frown(p) \quad (\text{where } p^0 = (\mathbf{p}^2 + m^2)^{1/2})$$

is real holomorphic. □

Next we show the quasi-locality of $Q_j$. In the expression

$$[Q_j, Q_k(x)] = \iint f_j(\xi) f_k(\eta) [Q_0(\xi), Q_0(\eta + x)]\, \mathrm{d}^4\xi\, \mathrm{d}^4\eta$$

the commutator in the integrand vanishes when $D + \xi$ and $D + \eta + x$ are spacelike due to the covariance and locality of $Q_0 \in \mathfrak{A}(D)$, and hence we obtain

$$\|[Q_j, Q_k(x)]\| \leq 2\|Q_0\|^2 \iint_{\langle x\rangle} |f_j(\xi) f_k(\eta)|\, \mathrm{d}^4\xi\, \mathrm{d}^4\eta,$$

where the integration is performed over $\xi$ and $\eta$ satisfying $\xi - \eta \notin (D - D + x)'$ and this domain of integration is indicated by $\langle x \rangle$. For spacelike $x$, $D - D + \lambda x$ tends to spacelike infinity in the limit $\lambda \to \infty$. Hence, for an appropriate $d > 0$ the $\lambda d$ neighbourhood of the origin is spacelike to $D - D + \lambda x$, i.e. it is in $(D - D + \lambda x)'$. Therefore the domain of integration $\langle \lambda x \rangle$ is contained in the union of the domain $|\xi| \geq \lambda d/2$ and the domain $|\eta| \geq \lambda d/2$ (if $|\xi| < \lambda d/2$ and $|\eta| < \lambda d/2$, then $|\xi - \eta| < \lambda d$) and we obtain the estimate

$$\|[Q_j, Q_k(x)]\| \leq \|Q_0\|^2 (\|f_j\|'_\lambda \|f_k\| + \|f_j\| \|f_k\|'_\lambda),$$

$$\|f\| \equiv \int |f(x)|\, \mathrm{d}^4x, \qquad \|f\|'_\lambda \equiv \int_{|x| > \lambda d/2} |f(x)|\, \mathrm{d}^4x.$$

Since $f = f_j, f_k$ are the Fourier transforms of the $C^\infty$ functions $\tilde{f}_j, \tilde{f}_k$ with compact support,

$$\lim_{\lambda \to \infty} |\lambda|^N \|f\|'_\lambda = 0$$

holds and we obtain the following quasi-locality.

**Theorem 5.2**
If $x$ is spacelike, then for any natural number $N$ the following holds.

$$\lim_{\lambda\to\infty} |\lambda|^N \|[Q_j, Q_k(\lambda x)]\| = 0. \tag{5.19}$$

This limit is uniform over $x \in K$, if $K$ is compact and spacelike.

**Remark 1** In Theorem 5.1, $f(\mathbf{p})$ can be shown to be an integral function on the three-dimensional complex sphere $\{p;\ (p,p) = m^2\}$.

**Remark 2** By the same reason as in eq. (5.16), we obtain for

$$(\Omega, Q_\alpha E_1 Q_\beta(x)\Omega) - (\Omega, Q_\beta(x) E_1 Q_\alpha \Omega) \tag{5.20}$$

an expression of the form (5.18) where $\Delta_m^+$ on the r.h.s. is replaced by $\Delta_m(x) = \Delta_m^+(x) - \Delta_m^+(-x)$. From the two facts that $\Delta_m(x) = 0$ for spacelike $x$ and that we can make the support of $\varphi_0$ arbitrarily small, we conclude that, for those $x \in (D_1 - D_2)'$ for which $[Q_\alpha, Q_\beta(x)]$ vanishes, (5.20) also vanishes. This property is called asymptotic locality [2].

*(b) Asymptotic behaviour of the relativistic wave function*
Starting with a $C^\infty$ function $\hat{g}(\mathbf{p})$ with compact support, we define

$$g(x) = (2\pi)^{-3} \int \mathrm{e}^{-\mathrm{i}(p,x)} \hat{g}(\mathbf{p})\, \mathrm{d}^3\mathbf{p} \tag{5.21}$$

and discuss its asymptotic behaviour for $x^0 \to \infty$.

**Theorem 5.3**
(i) Denote the energy-momentum corresponding to the velocity $\mathbf{v}$ as

$$p_\mathbf{v}^0 = m(1 - \mathbf{v}\cdot\mathbf{v})^{-1/2}, \quad \mathbf{p}_\mathbf{v} = m\mathbf{v}(1 - \mathbf{v}\cdot\mathbf{v})^{-1/2}.$$

The following estimate holds for $t \to \infty$ ($|\mathbf{v}| < 1$)

$$\begin{aligned} g(t, t\mathbf{v}) = {} & [2\pi|t|(1 - \mathbf{v}\cdot\mathbf{v})^{1/2}]^{-3/2} (\sqrt{m}/2)\hat{g}(\mathbf{p}_\mathbf{v}) \\ & \times \exp[-\mathrm{i}\{p_\mathbf{v}^0 t - \mathbf{p}_\mathbf{v}\cdot t\mathbf{v} + (3\pi/4)t/|t|\}] + \mathcal{O}(t^{-5/2}). \end{aligned} \tag{5.22}$$

(ii) There exist constants $A_1$, $A_2$ determined by $\hat{g}$ and a constant $C_N$ determined by $\hat{g}$ and a natural number $N$ such that the following estimate holds:

$$\int |g(t, \mathbf{x})|\, \mathrm{d}^3\mathbf{x} \le A_1(1 + |t|)^{3/2}, \tag{5.23}$$

$$\sup |g(t, \mathbf{x})| \le A_2 |t|^{-3/2}, \tag{5.24}$$

$$\sup\{(1 + |\mathbf{v}|)^N |g(t, t\mathbf{v})|;\ \mathbf{p}_\mathbf{v} \notin \sigma\} \le C_N |t|^{-N}, \tag{5.25}$$

where $\sigma$ is the support of the $C^\infty$ function $\hat{g}(\mathbf{p})$ and if $|\mathbf{v}| \geq 1$ then we write $\mathbf{p}_\mathbf{v} \notin \sigma$ just as a notation.

**Remark 1** If we consider $-t$ and $-x$ instead of $t$ and $x$, then the exponential function $e^{-i(p,x)}$ in eq. (5.21) is changed to $e^{i(p,x)}$ and we obtain a similar result. They are solutions of the Klein–Gordon equation

$$(\Box_x + m^2)g(x) = 0, \qquad \Box_x = (\partial/\partial x^0)^2 - \sum_{i=1}^{3} (\partial/\partial x^i)^2.$$

**Remark 2** It is possible to interpret the estimate of this theorem in terms of a picture that a particle with momentum $\mathbf{p}_\mathbf{v}$ is moving with velocity $\mathbf{v}$. Since a particle at time $t$ with different velocity is spreading with distance proportional to $t$, the density of a particle with velocity $\mathbf{v}$ is proportional to $t^{-3}$. Since the square of the absolute value of a wave function $f$ represents the density, $f$ should be proportional to $t^{-3/2}$. The expression $|t|^{-3/2}\hat{g}(\mathbf{p}_\mathbf{v})$ in eq. (5.22) exactly represents this circumstance. The other positive factors can be computed from the normalization and the main part of the phase is the phase factor $\exp\{-i(p_\mathbf{v}, x)\}$ of the wave function where $x = (t, t\mathbf{v})$. (5.24) represents the uniformity of this estimate over $\mathbf{v}$, (5.25) represents the fact that there are no particles running in other directions than the $\mathbf{v}$ corresponding to the support of $\hat{g}$. The latter fact makes the effective volume of $\mathbf{x}$ integration in (5.23) proportional to $t^3$ and hence the estimate $t^{-3/2}t^3 = t^{3/2}$ in (5.23) follows from (5.24).

**Outline of the proof** As an example we sketch the proof of eq. (5.25). First, setting

$$G(\lambda) = (\mathrm{d}/\mathrm{d}\lambda) \int_{\beta(\mathbf{p}) \leq \lambda} (2\pi)^{-3}\hat{g}(\mathbf{p})\, \mathrm{d}^3\mathbf{p}/(2p^0), \tag{5.26}$$

$$\beta(\mathbf{p}) = p^0 - \mathbf{p} \cdot \mathbf{v}$$

we get

$$g(t, t\mathbf{v}) = \int_{-\infty}^{\infty} e^{-i\lambda t} G(\lambda)\, \mathrm{d}\lambda.$$

If $G(\lambda)$ is in the $C^\infty$ class and $G^{(n)}(\lambda)$ $(n = 0, 1, \ldots)$ tends to 0 faster than $|\lambda|^{-\alpha}$, $(\alpha > 1)$ for $\lambda \to \infty$, then by partial integration we obtain the following estimate:

$$g(t, t\mathbf{v}) = (it)^{-N} \int_{-\infty}^{\infty} e^{-i\lambda t} G^{(N)}(\lambda)\, \mathrm{d}\lambda$$

$$|g(t, t\mathbf{v})| \leq |t|^{-N} \int_{-\infty}^{\infty} |G^{(N)}(\lambda)|\, \mathrm{d}\lambda. \tag{5.27}$$

Now we compute $G(\lambda)$.

First, if $\mathbf{p} \in \sigma$ and $\mathbf{p}_v \notin \sigma$, then $\mathbf{p} \neq \mathbf{p}_v$ and

$$(\partial\beta/\partial p^i) = (p^i/p^0) - v^i \qquad (i = 1, 2, 3)$$

do not vanish at the same time. Thus, setting

$$\rho^i(\mathbf{p}) = J(\mathbf{p})^{-2}\partial\beta/\partial p^i, \qquad J(\mathbf{p}) = \left[\sum_{i=1}^{3}(\partial\beta/\partial p^i)^2\right]^{1/2}$$

we have $J(\mathbf{p}) \neq 0$ and we obtain

$$G(\lambda) = (2\pi)^{-3}\int_{\beta(\mathbf{p})=\lambda} \hat{g}(\mathbf{p})J(\mathbf{p})^{-1}\mathrm{d}^2\sigma(\mathbf{p}), \tag{5.28}$$

$$G^{(n)}(\lambda) = (2\pi)^{-3}\int_{\beta=\lambda} J(\mathbf{p})^{-1}L^n\hat{g}(\mathbf{p})\,\mathrm{d}^2\sigma(\mathbf{p}), \tag{5.29}$$

$$Lf(\mathbf{p}) = \sum_{i=1}^{3}(\partial/\partial p^i)[\rho^i(\mathbf{p})f(\mathbf{p})].$$

Here $\mathrm{d}^2\sigma$ is the surface element of the plane $\{\mathbf{p};\ \beta(\mathbf{p}) = \lambda\}$.

(For $J(\mathbf{p}) \neq 0$ the formula for $G(\lambda)$ can be obtained even for the case of $(\partial\beta/\partial p^i) = 0$ for some $i$ as follows:

$$\begin{aligned} G(\lambda) &= (2\pi)^{-3}\sum_{j=1}^{3}(\mathrm{d}/\mathrm{d}\lambda)\int_{\beta\le\lambda}\hat{g}\{J^{-2}(\partial\beta/\partial p^j)^2\}\,\mathrm{d}^3\mathbf{p} \\ &= (2\pi)^{-3}\sum\int\!\!\int \hat{g}|p^j|\,\mathrm{d}p^k\mathrm{d}p^l \qquad (\{j,k,l\} = \{1,2,3\}) \\ &= (2\pi)^{-3}\int \hat{g}J^{-1}\,\mathrm{d}\sigma. \end{aligned}$$

The $p^j$ in the second line is a solution of $\beta(\mathbf{p}) = \lambda$ as a function of $(p^k, p^l)$ and, if there are several solutions, we have to add up the contributions from all solutions. The third line can be obtained from $\mathrm{d}p^k\mathrm{d}p^l = J^{-1}|\partial\beta/\partial p^j|\,\mathrm{d}\sigma$.

Since the solution $p^j$ of $\beta(\mathbf{p}) = \lambda$ satisfies $\partial p^j/\partial\lambda = (\partial\beta/\partial p^i)^{-1}$, we obtain for $G'(\lambda)$

$$\begin{aligned} G'(\lambda) &= \sum\int(\partial\beta/\partial p^j)^{-1}(\partial/\partial p^j)\{FJ^{-2}|\partial\beta/\partial p^j|\}\,\mathrm{d}p^k\mathrm{d}p^l \\ &= \sum\int|\partial\beta/\partial p^j|^{-1}(\partial/\partial p^j)\{FJ^{-2}\partial\beta/\partial p^j\}\,\mathrm{d}p^k\mathrm{d}p^l \\ &= \int\sum(\partial/\partial p^j)(F\rho^j)J^{-1}\mathrm{d}^2\sigma = \int(LF)J^{-1}\mathrm{d}^2\sigma. \end{aligned}$$

where the abbreviation $F = (2\pi)^{-3}\hat{g}$ is used. Repeating this procedure the formula for $G^{(n)}(\lambda)$ is obtained.)

Since $J(\mathbf{p})^{-1}|\mathbf{v}|$ converges to 1 uniformly in $\mathbf{p}$ for $\mathbf{v} \to \infty$, $|\mathbf{v}|^n L^n$ is a polynomial in $(\partial/\partial p^i)$ with coefficients uniformly bounded in $\mathbf{p} \in \sigma$. Therefore, due to (5.27) we obtain for $|\mathbf{v}| > 1$ the following estimate:

$$|L^n \hat{g}(\mathbf{p})| \leq (1 + |\mathbf{v}|)^{-n} \sum C(D)|D\hat{g}(\mathbf{p})|.$$

Here the sum runs over all differential operators $D$ of order $n$ or smaller. Consequently, by (5.27)

$$\begin{aligned} |g(t, t\mathbf{v}) &\leq |t|^{-n} \int |G^{(n)}(\lambda)| \, \mathrm{d}\lambda \\ &\leq |t|^{-n} \int (2\pi)^{-3} |L^n \hat{g}(\mathbf{p})| \, \mathrm{d}^3\mathbf{p} \\ &\leq C|t|^{-n}(1 + |\mathbf{v}|)^{-n}. \end{aligned}$$

Next, consider the case $|\mathbf{v}| \leq 1$. Since $\hat{g}(\mathbf{p})$ and its derivatives of any order vanish on the boundary of $\sigma$, the integral

$$\int |L^n \hat{g}(\mathbf{p})| \, \mathrm{d}^3\mathbf{p}$$

is continuous for those $\mathbf{v}$ such that $\mathbf{p}_\mathbf{v}$ is in the closure of the complement of $\sigma$. Therefore, using (5.27) we get

$$\begin{aligned} |g(x)| &\leq |t|^{-N} (2\pi)^{-3} \int |L^N \hat{g}(\mathbf{p})| \, \mathrm{d}^3\mathbf{p} \\ &\leq C'|t|^{-N}. \end{aligned}$$

Combining the above two estimates we obtain (5.25). □

*(c) One particle creation operators at time t*

Using the $C^\infty$ function $\hat{g}_j$ with compact support and the $Q_j$ of Section 5.3(a), we define

$$Q_j(t, g_j) = \int Q_j(x) g_j(x) \, \mathrm{d}^3\mathbf{x} \quad (x^0 = t). \tag{5.30}$$

Here $g_j(x)$ is defined in terms of $\hat{g}_j$ in (5.21).

One property of this operator is that due to $Q_j(x)\Omega = e^{i(p,x)} f_j$ it creates a one particle state from the vacuum:

$$Q_j(t, g_j)\Omega = \hat{g}_j(\mathbf{P}) f_j \in E_1 \mathscr{H}. \tag{5.31}$$

The corresponding wave function is given by $\hat{g}_j(\mathbf{p}) f_j(\mathbf{p})$.

$Q_j$ defined by eq. (5.16) is differentiable in the following sense. By

$$Q_j(x) \equiv T(x)Q_jT(x)^*$$
$$= \int Q_0(y+x)f_j(y)\,\mathrm{d}^4y = \int Q_0(y)f_j(y-x)\,\mathrm{d}^4y,$$

$Q_j(x)$ is differentiable with respect to $x$:

$$\left(\frac{\partial}{\partial x^k}\right)Q_j(x) = \int Q_0(y)(\partial/\partial x^k)f_j(y-x)\,\mathrm{d}^4y$$
$$= \int Q_0(y)(-\partial/\partial y^k)f_j(y-x)\,\mathrm{d}^4y.$$

In particular we use the following notation for the time derivative:

$$\dot{Q}_j = (\partial/\partial x^0)Q_j(x)|_{x=0} = \int Q_0(y)(-\partial/\partial y^0)f_j(y)\,\mathrm{d}^4y.$$

Since the $t$ dependence of $Q_j(t, g_i)$ in eq. (5.30) originates from $x^0 = t$ in two places, $Q_j(x)$ and $g_j(x)$, we can write

$$(\mathrm{d}/\mathrm{d}t)Q_j(t, g_j) = \dot{Q}_j(t, g_j) + Q_j(t, \dot{g}_j), \tag{5.32}$$

where $\dot{g}_j$ denotes $(\partial/\partial x^0)g_j$. By (5.31) we get

$$(\mathrm{d}/\mathrm{d}t)Q_j(t, g_j)\Omega = 0. \tag{5.33}$$

In (5.32), $\dot{Q}_j$ satisfies the quasi-locality of Theorem 5.2 and $\dot{g}_j$ is given by the formula (5.21) for $g_j$, with $\hat{g}_j$ being replaced by $\mathrm{i}p^0\hat{g}_j$. Thus the asymptotic behaviour of Theorem 5.3 holds.

*(d) Construction of asymptotic states*

We can now state the main theorem of this section. We construct the state vectors, each of which describes a many particle state asymptotically at remote future ($t = \infty$) and at remote past ($t = -\infty$), as the limit of

$$\Psi_t \equiv Q_1(t, g_1)\cdots Q_n(t, g_n)\Omega. \tag{5.34}$$

**Theorem 5.4**

Suppose that the support of $\hat{g}_j$ for different $j$ is mutually disjoint.

(i) $\Psi_t$ converges strongly for $t \to \pm\infty$, i.e. there exits a vector $\Psi_\pm[Q_1g_1\ldots Q_ng_n]$ in $\mathscr{H}_\omega$ satisfying

$$\|\Psi_+[Q_1g_1\ldots Q_ng_n] - \Psi_t\| \to 0 \quad (t \to +\infty),$$
$$\|\Psi_-[Q_1g_1\ldots Q_ng_n] - \Psi_t\| \to 0 \quad (t \to -\infty).$$

(ii) The inner product between the limit vectors above is given by the following formula:

$$\begin{aligned}(\Psi_+[Q'_1 g'_1 \dots Q'_m g'_m], \Psi_+[Q_1 g_1 \dots Q_n g_n] \\ = (\Psi_-[Q'_1 g'_1 \dots Q'_m g'_m], \Psi_-[Q_1 g_1 \dots Q_n g_n]) \\ = \delta_{mn} \sum_{\nu} \prod_{j} (h'_j, h_{\nu(j)}).\end{aligned} \tag{5.35}$$

Here the summation runs over all permutations $\nu$ of the indices $j = 1, \dots, n$. $h_j$ is the wave function

$$h_j(\mathbf{p}) = \hat{g}_j(\mathbf{p}) f_j(\mathbf{p}) \tag{5.36}$$

of a one particle state given by (5.31) and is determined by $g_j$ and $Q_j$. Similarly for $h'_j$.

(iii) $\Psi_\pm[Q_1 g_1 \dots Q_n g_n]$ are determined by the wave functions $h_1, \dots, h_n$ of (5.36) alone and hence can be denoted by

$$\Psi^{\text{out}}[h_1 \dots h_n] \equiv \Psi_+[Q_1 g_1 \dots Q_n g_n], \tag{5.37}$$

$$\Psi^{\text{in}}[h_1 \dots h_n] \equiv \Psi_-[Q_1 g_1 \dots Q_n g_n]. \tag{5.38}$$

They are symmetric in $h_1, \dots, h_n$.

(iv) With respect to $\mathscr{P}_+^\uparrow$ transformations, the following formulae hold:

$$U(a, \Lambda)\Psi^{\text{out}}[h_1 \dots h_n] = \Psi^{\text{out}}[(a, \Lambda)h_1 \dots (a, \Lambda)h_n]$$

$$U(a, \Lambda)\Psi^{\text{in}}[h_1 \dots h_n] = \Psi^{\text{in}}[(a, \Lambda)h_1 \dots (a, \Lambda)h_n]$$

where

$$[(a, \Lambda)h](\mathbf{p}) = e^{i(a,p)} h(\mathbf{\Lambda}^{-1}\mathbf{p}), \quad p^0 = (\mathbf{p} \cdot \mathbf{p} + m^2)^{1/2}$$

and $\mathbf{\Lambda}^{-1}\mathbf{p}$ denotes the spatial part of the vector $\Lambda^{-1}p$.

(v) If we interpret $\Psi^{\text{out}}$ and $\Psi^{\text{in}}$ in definitions (5.37) and (5.38) as maps from the vectors

$$\Phi_+(h_1 \dots h_n) \qquad (h_j \in \mathscr{H}_1)$$

in the Fock space $F_+(\mathscr{H}_1)$ introduced in (3.58) into the space $\mathscr{H}_\omega$, the closed linear maps $\Psi^{\text{out}}$, $\Psi^{\text{in}}$ from $F_+(\mathscr{H}_1)$ to $\mathscr{H}_\omega$ are uniquely determined by linearity and closure and become isometries.

(vi) Denote the restriction of $U(a, \Lambda)$ to the one particle subspace $\mathscr{H}_1$ by $U_1(a, \Lambda)$ and consider the representation $\Gamma(U_1(a, \Lambda))$ of $\mathscr{P}_+^\uparrow$ on $F_+(\mathscr{H}_1)$ in terms of the $\Gamma(A)$ introduced in Section 3.5. Then the following equations hold:

$$\Psi^{\text{out}}\Gamma(U_1(a,\Lambda)) = U(a,\Lambda)\Psi^{\text{out}}, \tag{5.39}$$

$$\Psi^{\text{in}}(U_1(a,\Lambda)) = U(a,\Lambda)\Psi^{\text{in}}. \tag{5.40}$$

**Outline of proof** (i) If the variation of $\Psi_t$ becomes small rapidly for $t \to \pm\infty$, we can show the convergence of $\Psi_t$. So let us first estimate the time derivative of $\Psi_t$:

$$\frac{\mathrm{d}\Psi_t}{\mathrm{d}t} = \sum_{k=1}^{n} Q_1(t,g_1)\cdots\{D_t Q_k(t,g_k)\}\cdots Q_n(t,g_n)\Omega.$$

Here we abbreviate $\mathrm{d}/\mathrm{d}t = D_t$. Since $A \equiv D_t Q_k(t,g_k)$ is zero on $\Omega$ due to (5.33), we can shift $A$ to the right, passing through operators $Q_l(t,g_l)$ one by one, using the formula $AQ = [A,Q] + QA$, and get

$$A\cdots Q_n(t,g_n)\Omega = \sum_{l>k}\cdots[A, Q_l(t,g_l)]\cdots Q_n(t,g_n)\Omega.$$

First, applying (5.23) to (5.30) we obtain

$$\begin{aligned} \|Q_j(t,g_j)\| &\leq \int \|Q_j(x)\|\,|g_j(x)|\,\mathrm{d}^3\mathbf{x} \qquad (x^0 = t) \\ &\leq \|Q_j\| A_{1j}(1+|t|)^{3/2}. \end{aligned} \tag{5.41}$$

Furthermore, using (5.32) we get

$$[A, Q_l(t,g_l)] = [\dot{Q}_k(t,g_k), Q_l(t,g_l)] + [Q_k(t,\dot{g}_k), Q_l(t,g_l)].$$

Setting $x^0 = y^0 = t$ we get

$$\begin{aligned} &\|[\dot{Q}_k(t,g_k), Q_l(t,g_l)]\| \\ &\quad \leq \int \|[\dot{Q}_k(t,\mathbf{x}), Q_l(t,\mathbf{y})]\|\,|g_k(x)g_l(y)|\,\mathrm{d}\mathbf{x}\,\mathrm{d}\mathbf{y}. \end{aligned} \tag{5.42}$$

Denote the support of $\hat{g}_j$ by $\sigma_j$ and the set of all $\mathbf{v}$ for which $\mathbf{p}_\mathbf{v} \in \sigma_j$ by $V_j$. Then $V_k$ and $V_l$ are compact and, by the assumption on the support of $\hat{g}_j$ they are disjoint and separated by a finite (i.e. non-zero) distance. Thus by (5.25) and (5.23) we get for the integral with the restriction $x \notin tV_k$ (denoted by $\notin$ as a subscript of the integration symbol) the following estimate:

$$\|\dot{Q}_k\|\,\|Q_l\| \int_{\notin} |g_k(x)|\,\mathrm{d}\mathbf{x} \int |g_l(y)|\,\mathrm{d}\mathbf{y}$$
$$\leq \|\dot{Q}_k\|\,\|Q_l\| A_{1l}(1+|t|)^{3/2} C_{Nk}|t|^{3-N} \int (1+|\mathbf{v}|)^{-N}\,\mathrm{d}\mathbf{v}$$
$$\leq \alpha_{Nkl}|t|^{(9/2)-N}. \tag{5.43}$$

A similar estimation holds also for the integral with the restriction $y \notin tV_l$. By using Theorem 5.2 and eq. (5.24) we obtain for the remaining part

$$\int_{tV_k} \mathrm{d}\mathbf{x} \int_{tV_l} \mathrm{d}\mathbf{y} \|[\dot{Q}_k(t,\mathbf{x}), Q_l(t,\mathbf{y})]\|\, |g_k(x) g_l(y)|$$
$$\leq |V_k|\,|V_l|\,|t|^{6-3-N} A_{2k} A_{2l} C'_{Nkl}. \tag{5.44}$$

where, due to the circumstance that $V_k - V_l$ is compact and does not contain zero, we have used Theorem 5.2 in the following form:

$$\|[\dot{Q}_k(t,\mathbf{x}), Q_l(t,\mathbf{y})]\| = \|[\dot{Q}_k, Q_l(0,\mathbf{y}-\mathbf{x})]\|$$
$$= \|[\dot{Q}_k, Q_l(0,t\mathbf{v})]\| \leq C'_{Nkl}|t|^{-N} \qquad (\mathbf{v} \in V_k - V_l).$$

From (5.43) and (5.44) we obtain the estimate of (5.42) and combining it with (5.41) (after redefining the arbitrary natural number $N$), we get

$$\|\mathrm{d}\Psi_t/\mathrm{d}t\| \leq G_N|t|^{-N} \tag{5.45}$$

for any $N$ and an appropriate constant $G_N$. Therefore,

$$\Psi_T = \Psi_0 + \int_0^T (\mathrm{d}\Psi_t/\mathrm{d}t)\,\mathrm{d}t \to \Psi_0 + \int_0^\infty (\mathrm{d}\Psi_t/\mathrm{d}t)\,\mathrm{d}t \qquad (T \to \infty)$$

is absolutely convergent. The proof for $T \to -\infty$ is the same and we have shown (i).

(ii) We will apply the cluster decomposition (4.29) in terms of truncated functions to

$$(\Psi'_t, \Psi_t) = \omega(Q'_m(t,g'_m)^* \cdots Q'_1(t,g'_1)^* Q_1(t,g_1) \cdots Q_n(t,g_n)), \tag{5.46}$$

where $\Psi_t$ is given by (5.34) and

$$\Psi'_t = Q'_1(t,g'_1) \cdots Q'_m(t,g'_m)\Omega.$$

Since the l.h.s. of (5.46) converges to the first and second line of (5.35) for $t \to \pm\infty$, respectively, it is enough to show that the r.h.s. of (5.46) converges to the third line of

(5.35). In the sum (4.29) of the cluster decomposition, the term

$$\prod_k \omega^T(Q'_k(t,g'_k)^* Q_{\nu(k)}(t,g_{\nu(k)})) \tag{5.47}$$

exists only for $m = n$. Since

$$\omega(Q_l(t,g_l)) = (\Omega, h_l) = 0$$

the $\omega^T$ in (5.47) can be replaced by $\omega$. By

$$\omega(Q'_k(t,g'_k)^* Q_l(t,g_l)) = (Q'_k(t,g'_k)\Omega,\ Q_l(t,g_l)\Omega) = (h'_k, h_l)$$

(5.47) is independent of $t$ and gives the third line of (5.35). Consequently, it is enough to show that in the cluster decomposition of the r.h.s. of (5.46) all terms except (5.47) vanish in the limit. Thus let us consider

$$\omega^T(Q_a(t,g_a)\cdots Q_b(t,g_b)) = \int\cdots\int \omega^T(Q_a(t,\mathbf{x}_a)\cdots Q_b(t,\mathbf{x}_b)) \\ \times g_a(t,\mathbf{x}_a)\cdots g_b(t,\mathbf{x}_b)\,\mathrm{d}^3\mathbf{x}_a\cdots\mathrm{d}^3\mathbf{x}_b. \tag{5.48}$$

Let us consider the case where the number of $Q_a \ldots Q_b$ is $n > 2$. First, we estimate all $g$ other than the last $g_b$ by (5.24) as

$$|g_a(t,\mathbf{x})| \le A_2|t|^{-3/2},$$

obtaining the factor $(A_2)^{n-1}|t|^{-3(n-1)/2}$. Next, by the translation invariance of $\omega$ we have

$$\omega^T(Q_a(t,\mathbf{x}_a)\cdots Q_b(t,\mathbf{x}_b)) = \omega^T(Q_a(0,\mathbf{x}_a-\mathbf{x}_b)\cdots Q_b(0,0)). \tag{5.49}$$

Due to Corollary 4.12 the r.h.s. is absolutely integrable with respect to the $(n-1)$ vector variables $\mathbf{x}_a, \ldots$ excluding $\mathbf{x}_b$. The integral of the absolute value will be denoted by $W$. It does not depend on $\mathbf{x}_b$. Finally estimating the $\mathbf{x}_b$ integration of $|g_b(t,\mathbf{x}_b)|$ by (5.23), the absolute value of eq. (5.48) is bounded by

$$WA_1A_2^{n-1}\{(1+|t|)|t|^{-(n-1)}\}^{3/2}.$$

Consequently, it converges to zero as $t \to \infty$ for $n > 2$ and, for $n = 2$ it is uniformly bounded.

The above estimate goes through even if all or a part of $Q(t,g)$ is $Q(t,g)^*$. We now consider the case $n = 2$. For $a \ne b$ the support of $\hat{g}_a$ and that of $\hat{g}_b$ are disjoint. Let their distance be $\varepsilon > 0$. In the estimate of (5.48) (since $n = 2$, there is no part corresponding to "...") we separate the integration region into the following three pieces:

($\alpha$) $\mathbf{x}_a$ such that $\mathbf{p}_\mathbf{v}$ for $\mathbf{v} = \mathbf{x}_a/t$ is not contained in the support of $\hat{g}_a$ ($\mathbf{x}_b$ is arbitrary).
($\beta$) $\mathbf{x}_b$ such that $\mathbf{p}_\mathbf{v}$ for $\mathbf{v} = \mathbf{x}_b/t$ is not contained in the support of $\hat{g}_b$ ($\mathbf{x}_a$ is arbitrary).
($\gamma$) Cases other than ($\alpha$) and ($\beta$) above. In this case, $\mathbf{p}_\mathbf{u}$ for $\mathbf{u} = \mathbf{x}_a/t$ is contained in the support of $\hat{g}_a$, and $\mathbf{p}_\mathbf{v}$ for $\mathbf{v} = \mathbf{x}_b/t$ is contained in the support of $\hat{g}_b$. Hence there exists a constant $\delta > 0$ such that $|\mathbf{u} - \mathbf{v}| \geq \delta$ corresponding to the separation between the supports of $\hat{g}_a$ and of $\hat{g}_b$.

In region ($\alpha$) we bind $\omega^T$ by a constant, use (5.25) for the $\mathbf{x}_a$ integration of $|g_a|$ and (5.23) for the $\mathbf{x}_b$ integration of $|g_b|$. In region ($\beta$) we make the same estimate interchanging $a$ and $b$. For ($\gamma$), $|g_a|$ is estimated for example by (5.24). Using the estimate of an exponential decrease for $|\omega^T|$ in Corollary 4.10, we carry out the $\mathbf{x}_a$ integration and obtain an estimate proportional to

$$(\delta t)^{1/2} \exp(-\delta t).$$

If we estimate the $\mathbf{x}_b$ integration of the remaining $|g_b|$ by (5.23), and combine it with the estimate of $|g_a|$ we obtain a constant bound, and hence we find that in all three regions the expression tends to zero exponentially for $t \to \infty$.

From the above result we have shown that all terms except (5.47) tend to zero, and Theorem 5.4 (ii) is proven.

(iii) Suppose that we are given two sets $Q_1 g_1 \ldots Q_n g_n$ and $Q'_1 g'_1 \ldots Q'_n g'_n$ such that

$$h_j(\mathbf{p}) = \hat{g}_j(\mathbf{p}) f_j(\mathbf{p}) = \hat{g}'_j(\mathbf{p}) f'_j(\mathbf{p}) \qquad (j = 1, \ldots, n)$$

where $f'_j$ is the wave function of $Q'_j\Omega$. By (ii) we get

$$\begin{aligned}&(\Psi_\pm(Q_1 g_1 \ldots Q_n g_n), \Psi_\pm(Q'_1 g'_1 \ldots Q'_n g'_n))\\ &\quad = \|\Psi_\pm(Q_1 g_1 \ldots Q_n g_n)\|^2 = \|\Psi_\pm(Q'_1 g'_1 \ldots Q'_n g'_n)\|^2 = \sum_\nu \prod_j (h_j, h_{\nu(j)}).\end{aligned}$$

Hence

$$\|\Psi_\pm(Q_1 g_1 \ldots Q_n g_n) - \Psi_\pm(Q'_1 g'_1 \ldots Q'_n g'_n)\|^2 = 0$$

holds and we see that $\Psi_\pm(Q_1 g_1 \ldots Q_n g_n)$ is determined by $h_1 \ldots h_n$ alone. Since the above inner product does not depend on the ordering of $h_1 \ldots h_n$ we know at the same time that it is also symmetric in $h_1 \ldots h_n$.

(iv) First let us give a proof for the case of a three-dimensional rotation $R$. For $\Psi_\pm$ of eq. (5.34) we get

$$U(a, R)\Psi_\pm = Q_1^s(t, g_1^s) \cdots Q_n^s(t, g_n^s)\Omega \qquad (s = (a, R)),$$

where

$$Q_j^s(t, g_j^s) = U(a, R)Q_j(t, g_j)U(a, R)^*$$
$$= \int Q_j^s(t, R\mathbf{x})g(t, \mathbf{x})\,\mathrm{d}^3\mathbf{x} = \int Q_j^s(t, \mathbf{x})g(t, R^{-1}\mathbf{x})\,\mathrm{d}^3\mathbf{x}$$

$$Q_j^s \equiv U(a, R)Q_jU(a, R)^*, \quad g_j^s(t, \mathbf{x}) \equiv g_j(t, R^{-1}\mathbf{x}).$$

(Note that $U(a, \Lambda)U(x, \mathbf{1})U(a, \Lambda)^* = U(\Lambda x, \mathbf{1})$.)

The one particle state corresponding to $Q_j^s$ is given by

$$Q_j^s\Omega = U(a, R)Q_j\Omega = U(a, R)f_j \;(\approx \mathrm{e}^{\mathrm{i}(p,a)}f_j(R^{-1}\mathbf{p})).$$

Therefore the $h_j^s$ corresponding to $Q_j^sg_j^s$ is given by

$$h_j^s(\mathbf{p}) = f_j^s(\mathbf{p})g_j^s(\mathbf{p}) = \mathrm{e}^{\mathrm{i}(p,a)}h_j(R^{-1}\mathbf{p}) = (U(a, R)h_j)(\mathbf{p}).$$

Consequently, (iv) is proven for the case of $\Lambda = R$.

Next we give a proof for the case that $a = 0$ and $\Lambda$ is a pure Lorentz transformation, for example $\Lambda_v^1$ along the $x$ axis with velocity $v$. By the same calculation as above we obtain

$$U(0, \Lambda_v^1)\Psi_t = Q_1'(t, g_1')_v \cdots Q_n'(t, g_n')_v\Omega,$$
$$Q_j' = U(0, \Lambda_v^1)Q_jU(0, \Lambda_v^1)^*,$$
$$g_j'(x) = g_j((\Lambda_v^1)^{-1}x),$$
$$Q(t, f)_v \equiv \int Q(\Lambda_v^1 x)f_j(\Lambda_v^1 x)\,\mathrm{d}^3\mathbf{x}|_{x^0=t}. \tag{5.50}$$

Note that by eq. (5.50) we introduced a new symbol $Q(t, f)_v$ since $\Lambda_v^1$ changes the time component of $x$. This is the essential difference from the case of $\Lambda = R$.

For general $Q_jg_j$ we define

$$\Psi_t^v \equiv Q_1(t, g_1)_v \cdots Q_n(t, g_n)_v\Omega. \tag{5.51}$$

If we can prove that $\Psi_t^v$ converges for $t \to \infty$ and the limit is independent of $v$, then similarly as before we obtain

$$\lim U(0, \Lambda_v^1)\Psi_t = \Psi_\pm[Q_1'g_1' \cdots Q_n'g_n'],$$

$$h_j'(\mathbf{p}) = (U(0, \Lambda_v^1)h_j)(\mathbf{p}),$$

and (iv) is proven.

To show that (5.51) has a limit independent of $v$, it is sufficient to show the estimate

$$\| \mathrm{d}\Psi_t^v / \mathrm{d}v \| \leq A|t|^{-3}$$

for a constant $A$ independent of $v$. Because of

$$(\mathrm{d}/\mathrm{d}v) Q_j(t, g_j)_v \Omega = 0$$

it can be shown in the similar way as the estimate of $\mathrm{d}\Psi_t / \mathrm{d}t$.

Since $\mathscr{P}_+^\uparrow$ can be generated from $(a, R)$ and $\Lambda_v^!$ discussed above, (iv) is now established for general $(a, \Lambda)$.

(v) Due to (ii) the maps $\Psi^{\text{out}}$ and $\Psi^{\text{in}}$ preserve the inner product as maps from $F_+(\mathscr{H}_1)$ into $\mathscr{H}_\omega$. Therefore, by first extending these maps to linear maps (because vanishing of linear combinations, i.e. linear relations, are preserved as can be seen by computing their norm) and then to the closure, we obtain isometric operators which map the subspace of $F_+(\mathscr{H}_1)$ spanned by $\Phi_+(h_1 \ldots h_n)$ into $\mathscr{H}_\omega$.

By an appropriate construction of $Q$ and a choice of $\hat{g}$, we can obtain any $C^\infty$ function with compact support for each $h_j$. Hence (v) will be shown if we prove that the closed linear hull of all $\Phi_+(h_1 \ldots h_n)$ with such $h_j$'s of non-overlapping support coincides with $F_+(\mathscr{H}_1)$.

Thus we show that we can approximate $h_1 \otimes \cdots \otimes h_n$ with arbitrary $h_j$ by a linear combination of the same expression in which all $h_j$ are of $C^\infty$ class with mutually non-overlapping compact supports. First, since we can approximate an arbitrary $h_j$ by those $h'_j$ which have compact support, we assume that each $h_j$ has a compact support. Next we divide $\mathbf{R}^3$ into cubes $c_\alpha$, which are sufficiently small and all of the same size, and divide each $h_j$ into a sum of $h_{j\alpha}$, where $h_{j\alpha}$ is the restriction of $h_j$ to $c_\alpha$. If the cubes are taken sufficiently small, then for all $j$, $\alpha$ we can make $\|h_{j\alpha}\|/\|h_j\|$ smaller than any given $\varepsilon > 0$. Then $h_1 \otimes \cdots \otimes h_n$ can be written as a sum of the following terms:

$$h_{1\alpha_1} \otimes \cdots \otimes h_{n\alpha_n} \quad \text{(orthogonal to each other).} \tag{5.52}$$

If we can prove that for any pair $i \neq j$ the sum of the norm squared of all those terms, for which either $\alpha_i$ coincides with $\alpha_j$ or $\alpha_i$ and $\alpha_j$ are neighbours is small, then we can approximate $h_1 \otimes \cdots \otimes h_n$ by a sum of vectors satisfying the desired conditions. (The approximation can be made before symmetrization, the projection operator for symmetrization can be applied afterwards. Thus we can assume that the above vectors are mutually orthogonal for different $(\alpha_1, \ldots, \alpha_n)$s.)

Fix one $\alpha_1$. Then the $\alpha_2$ which is dropped in the above described approximation are 27 cubes including the neighbouring cubes. Therefore, measured in norm squared, they are smaller than $\|h_2\|^2$ by a factor $27\varepsilon^2$. Adding (5.52) over all possibilities of $\alpha_1, \ldots, \alpha_n$ with such an $\alpha_2$, the norm squared of the sum does not exceed $27\varepsilon^2 \|h_1\|^2 \cdots \|h_n\|^2$. Similarly, if we fix $\alpha_1$ and $\alpha_2$, then the $\alpha_3$ to be excluded by the approximation are at most $27 \times 2$ cubes including the neighbouring cubes of $\alpha_1$ and $\alpha_2$. Thus, summing (5.52) over all possibilities of $\alpha_1, \ldots, \alpha_n$ with such an $\alpha_3$, the norm squared of the sum does not exceed $54\varepsilon^2\|h_1\|^2\cdots\|h_n\|^2$. If we perform similar estimates for $\alpha_3, \ldots$ the norm squared of the sum of the terms which are dropped in the approximation (those $(\alpha_1, \ldots, \alpha_n)$ for which $\alpha_i$ and $\alpha_j$ are either identical or adjacent at least for one pair $i \neq j$) is bounded by

$$27\varepsilon^2(1+2+\cdots+(n-1)) = (27/2)n(n-1)\varepsilon^2$$

times $\|h_1\|^2\cdots\|h_n\|^2$. Therefore, taking the cubes sufficiently small and thus making $\varepsilon$ small, we can make this sum arbitrarily small.

Finally, if we approximate $h_{j\alpha}$ by a $C^\infty$ function obtained by smoothing a little near the boundary of the cubes, we can approximate any $h_1 \otimes \cdots \otimes h_n$ by the tensor product of $C^\infty$ functions $h'_1, \ldots, h'_n$ with disjoint compact support (hence, of positive distance), and therefore the proof of (v) is completed.

(vi) is just a rewriting of (iv).

## 5.4 The counter interpretation of asymptotic states—verification of an appropriate asymptotic behaviour

In this section we show that the vectors $\Psi^{\text{out}}(h_1\ldots h_n)$ and $\Psi^{\text{in}}(h_1\ldots h_n)$ introduced in the previous section exhibit the desired behaviour at $t = \pm\infty$. As explained already in Remark 2 of Theorem 5.3, any $h_j(\mathbf{p})$ can be interpreted as describing asymptotically the probability amplitude of a particle moving with velocity $\mathbf{v} = \mathbf{p}/p^0$ (using the unit system where the speed of light is 1, $p^0 = (\mathbf{p}\cdot\mathbf{p}+m^2)^{1/2}$). Therefore we want to show that $\Psi^{\text{in}}$ and $\Psi^{\text{out}}$ can also be interpreted as states of particles moving with velocity $\mathbf{v}_j = \mathbf{p}_j/p_j^0$ with a probability amplitude $h_1(\mathbf{p}_1)\cdots h_n(\mathbf{p}_n)$.

To find out whether there is a particle moving with velocity $\mathbf{v}$, let us use the way of thinking of a counter experiment. As an operator $Q$ representing the counter we use a quasi-local operator with the following properties:

$$Q\Omega = 0, \qquad E_1QE_1 \neq 0, \qquad Q^* = Q. \tag{5.53}$$

The first equation is the requirement that this operator counts 0 for the vacuum and the second equation is the requirement that it gives the response for a one particle state. A quasi-local operator means an operator given by (5.16) and the property of quasi-locality given in Theorem 5.2 is applicable. For the physical interpretation below it is only necessary that $Q$ is almost localized in a certain finite region (if we neglect in eq. (5.16) the tail of $F_j$ which rapidly tends to 0 at remote distances), and it is not important where the operator is exactly localized and of what size this localization is. The reason for this is that if we measure the passing of a particle at two space–time points the location of which is far from each other, then even if the counter has a finite size we can determine the velocity of the particle with a precision inverse proportional to the spatial distance and time difference of the two points. By a similar reasoning also the part of the tail of $F_j$ which we neglect becomes smaller and thus in the limit the effect of the neglected part disappears.

Concretely, we fix one quasi-local observable $Q$ satisfying (5.53) as the observable corresponding to a counter, and we investigate the behaviour of

$$\varphi(Q(t, t\mathbf{v}_1)\cdots Q(t, t\mathbf{v}_n)) \tag{5.54}$$

at $t \to \pm\infty$ of the state $\varphi$, the asymptotic behaviour of which we are interested in, by varying the number $n$ of the counters as well as the velocities $\mathbf{v}_1, \ldots, \mathbf{v}_n$. $(t, t\mathbf{v})$ represents the orbit of a point moving with velocity $\mathbf{v}$ if we move the time $t$. (We have chosen the starting point to be 0, since with the same reasoning as described above, the influence of the difference of the starting points can be neglected in the limit $t \to \infty$.) Therefore, by observing expectation values for changing $t$, $Q(t, t\mathbf{v})$ can be interpreted as the observable which measures a quantity proportional to the probability of existence of an object (particle) moving with velocity $\mathbf{v}$ in the limit $t \to \pm\infty$.

Since a particle with a fixed velocity distribution spreads over distances proportional to time $t$, we can expect the probability density as tending to zero proportional to $t^{-3}$. Therefore to get a meaningful quantity proportional to the velocity distribution of a particle in the limit $t \to \infty$, the expectation value of $t^3 Q(t, t\mathbf{v})$ is an appropriate quantity to be considered.

With this background, the following theorem holds for eq. (5.54), which is the main result of this section [3].

**Theorem 5.5**
Let $h_1, \ldots, h_k$ be one particle states represented by the $C^\infty$ wave functions $h_1(\mathbf{p}), \ldots, h_k(\mathbf{p})$ with mutually disjoint, compact support and consider the vector

$$\Psi = \Psi^{\mathrm{out}}(h_1 \otimes \cdots \otimes h_k) \tag{5.55}$$

in $\mathscr{H}$ given in Theorem 5.4(v). Consider the particle-counter observables

$$R_j(t) = Q_j(t, t\mathbf{v}_j) \qquad (j = 1, \ldots, n),$$

given in terms of quasi-local observables $Q_1, \ldots, Q_n$ satisfying condition (5.52) and distinct velocity vectors $\mathbf{v}_1, \ldots, \mathbf{v}_n$. Then in the limit $t \to +\infty$ the following equations hold

$$\lim t^{3n}(\Psi, R_1(t)\cdots R_n(t)\Psi)$$

$$= \begin{cases} 0 \quad (k < n), & (5.56\text{a}) \\ \prod_{j=1}^{n} \{\Gamma_j(\mathbf{p}_{\mathbf{v}_j})|h_{\alpha_j}(\mathbf{p}_{\mathbf{v}_j})|^2\} \quad \prod_{j=n+1}^{k} \|h_{\alpha_j}\|^2, & (5.56\text{b}) \end{cases}$$

where $\mathbf{p}_\mathbf{v}$ denotes the momentum corresponding to the velocity $\mathbf{v}$ defined in Theorem 5.3(i) and $\alpha_1, \ldots, \alpha_n$ is a permutation of $1, \ldots, n$.

The second expression $b$ holds for the case $k \geq n$ and for the permutation $\alpha$ satisfying

$$\mathbf{p}_{v_j} \in (\text{support of } h_{\alpha_j}) \qquad (j = 1, \ldots, n),$$

and vanishes if there is no such $\alpha$. The quantity $\Gamma_j$ representing the measured expectation value of the particle-counter observable $Q_j$ is given by

$$\Gamma_j(\mathbf{p}) = 2m^{-2}(\pi p^0)^3 Q_j(\mathbf{p}, \mathbf{p}), \tag{5.57}$$

where $Q_j(\mathbf{p}, \mathbf{q})$ is a $C^\infty$ function of $\mathbf{p}$ and $\mathbf{q}$ defined by

$$(h, Q_j h') = \int Q_j(\mathbf{p}, \mathbf{q}) h(\mathbf{p})^* h'(\mathbf{q}) \mathrm{d}\mu(\mathbf{p}) \mathrm{d}\mu(\mathbf{q}), \tag{5.58}$$

$$(\mathrm{d}\mu(\mathbf{p}) = (2p^0)^{-1} \mathrm{d}^3\mathbf{p}, \qquad \text{for arbitrary } h, h' \in \mathscr{H}_1).$$

The same conclusion holds when $\Psi^{\text{out}}$ is replaced by $\Psi^{\text{in}}$ in (5.55) and the limit $t \to -\infty$ is taken in (5.56).

**Interpretation of theorem** In the case $n = k = 1$, the eq. (5.56) for $h \in \mathscr{H}_1$ becomes

$$\lim t^3(h, Q(t, t\mathbf{v})h) = \Gamma(\mathbf{p_v})|h(\mathbf{p_v})|^2 \tag{5.59}$$

and this supports the interpretation of $\Gamma(\mathbf{p_v})$ as giving the measured expectation value of a particle with velocity $\mathbf{v}$. On the basis of this interpretation, the eq. (5.56) for general $n$ and $k$ shows that the asymptotic behaviour of $\Psi$ in eq. (5.55) for $t \to +\infty$ can be interpreted as $n$ particles moving with velocity $\mathbf{v}_1, \ldots, \mathbf{v}_n$ with a probability density $|h_1(\mathbf{p}_{\mathbf{v}_1})|^2 \cdots |h_n(\mathbf{p}_{\mathbf{v}_n})|^2$. (Theorem 5.4(ii) provides a complete information on the quantum mechanical phase.)

**Outline of proof of theorem** First let us show that for any $C^\infty$ function $h_j$ with compact support, there exist a quasi-local observable $Q_j$ and a function $\hat{g}_j$ satisfying (5.36). The basic reason for this is that by Theorem 5.1 the wave function $f_0(\mathbf{p})$ given by (5.15) for the local observable $Q_0$ is real analytic. Therefore, $f_0$ does not become zero in a neighbourhood of some point $\mathbf{p}$. Furthermore, by considering the Lorentz transformed quantity $U(\Lambda, a)Q_0 U(a, \Lambda)^*$ instead of $Q_0$, we can move such a point $\mathbf{p}$ to any point $\mathbf{p}' = \Lambda\mathbf{p}$. Therefore, if the support of $h_j$ is sufficiently small, we can find a $Q_0$ such that $f_0 \neq 0$ on the support of $h_j$. We can now choose a $C^\infty$ function $\tilde{f}_j(p)$ with compact support disjoint from the support of $P^\mu$ on $(\mathbf{1} - E_1)\mathscr{H}$, which coincides with $h_j(\mathbf{p})/f_0(\mathbf{p})$ for

$$\mathbf{p} \in (\text{support of } h_j), \quad p^0 = (\mathbf{p}\cdot\mathbf{p} + m^2)^{1/2}$$

and then $f_j = h_j$ holds for $Q_j$ defined in (5.16) with such a pair of $Q_0$ and $\tilde{f}_j$. We can choose a $C^\infty$ function $g_j$ with disjoint support which is 1 on the support of $h_j$. (Since by assumption the supports of the functions $h_j$ are mutually disjoint compact sets, their mutual distances are not zero.)

By the above discussion, we can approximate $\Psi$ in eq. (5.55) by a vector of the form of (5.34) at each time $t$. Furthermore, by eq. (5.45) it can be approximated to the order of $|t|^{-N+1}$ for any $N$, and hence this approximation is valid even after multiplying $t^{3n}$. Thus, substituting (5.34) into $\Psi$ we can estimate the resulting expression similarly as in the proof of Theorem 5.4(i). Note that by condition (5.53) all factors where $\Omega$ appears on the immediate right or left of $Q_j(t, t\mathbf{v}_j)$ vanish. So, if we can estimate the case $k = n = 1$, then the rest is exactly the same estimate as Theorem 5.4(i).

If we use eq. (5.58), the estimate of $n = k = 1$ case is an asymptotic estimate of a concrete integral, and can be performed in the same way as for Theorem 5.3. For this we use the property that $Q_j(\mathbf{p}, \mathbf{q})$ is a $C^\infty$ function, which is proved as follows. Using the quasi-local $Q$ and $Q'$ we represent the one particle states $h$ and $h'$ as

$$h = Q\Omega, \quad h' = Q'\Omega.$$

Then we have

$$\begin{aligned}(\Omega, Q^*(0,\mathbf{x})Q_jQ(0,\mathbf{y})\Omega)\\ = \iint Q_j(\mathbf{p},\mathbf{q})h(\mathbf{p})^*h'(\mathbf{q})e^{i(\mathbf{p}\cdot\mathbf{x}-\mathbf{q}\cdot\mathbf{y})}\,d\mu(\mathbf{p})\,d\mu(\mathbf{q}).\end{aligned}$$

By the condition (5.53) requiring $Q_j\Omega = Q_j^*\Omega = 0$, the l.h.s. is a truncated expectation value and by Corollary 4.12 of the previous chapter it converges to zero faster than any power of $\mathbf{x}\cdot\mathbf{x} + \mathbf{y}\cdot\mathbf{y}$ in the limit $\mathbf{x}\cdot\mathbf{x} + \mathbf{y}\cdot\mathbf{y} \to \infty$. Therefore its Fourier transform

$$Q_j(\mathbf{p},\mathbf{q})h(\mathbf{p})^*h'(\mathbf{q})(4p^0q^0)^{-1}$$

is a $C^\infty$ function. Since $h$ and $h'$ are $C^\infty$ functions and can be chosen such that they do not vanish in the neighbourhood of any given points $\mathbf{p}$ and $\mathbf{q}$ (which we have already shown), $Q_j(\mathbf{p}, \mathbf{q})$ is also of $C^\infty$ class. □

**Remark 1** If for any given $\mathbf{p}$ there is no quasi-local operator $Q$ satisfying (5.52) and for which $\Gamma(\mathbf{p}) \neq 0$ in a neighbourhood of $\mathbf{p}$, then the above theorem loses its value and its interpretation becomes meaningless. The existence of such a $Q$ can be checked as follows. First from axiom (4) of local observables and the irreducibility which holds for the pure vacuum state, the existence of a local observable $Q_0$ satisfying $E_1Q_0E_1 \neq 0$ follows.

Since $p-q$ is either spacelike or zero for $(p,p)=(q,q)=m^2$, we can choose a $C^\infty$ function $\tilde{f}_j$ with a compact support consisting of spacelike vectors only such that the $Q_j$ in eq. (5.16) satisfy

$$Q_j\Omega = 0, \qquad E_1 Q_j E_1 \neq 0.$$

Then $Q = Q_j^* Q_j$ satisfies (5.53). For example we get

$$E_1 Q E_1 = E_1 Q_j^* E_1 Q_j E_1 + E_1 Q_j^*(\mathbf{1} - E_1) Q_j E_1 \geq E_1 Q_j^* E_1 Q_j E_1 \neq 0.$$

Let us denote the projection operator onto $|\mathbf{p}| < M$ in $E_1\mathscr{H}_\varphi$ by $E^M$. Since $Q(\mathbf{p},\mathbf{q})$ is of $C^\infty$ class and $Q \geq 0$, we have $Q(\mathbf{p},\mathbf{p}) \geq 0$ and

$$\int_{|\mathbf{p}|<M} Q(\mathbf{p},\mathbf{p})\, \mathrm{d}\mu(\mathbf{p}) = \operatorname{Tr} E^M Q E^M > 0$$

for a sufficiently large $M$. Therefore, the $C^\infty$ function $Q(\mathbf{p},\mathbf{p})$ does not vanish at least in a neighbourhood of some vector $\mathbf{q}$. As in the previous case we can shift $\mathbf{q}$ to an arbitrary vector $\mathbf{p}$ by a Lorentz transformation (we consider $U(0,\Lambda)QU^*(0,\Lambda)$). Since we have found a $Q > 0$ for which $Q(\mathbf{p},\mathbf{p}) \neq 0$ in a neighbourhood of an arbitrarily given $\mathbf{p}$, we conclude by taking a finite sum of such $Q$s for various $\mathbf{p}$s that there exists a quasi-local observable $Q$ satisfying $Q(\mathbf{p},\mathbf{p}) \neq 0$ on an arbitrarily given compact set.

**Remark 2** Let the images of the creation and annihilation operators $(a^*,h)$ and $(h,a)$ defined on the Fock space $F(\mathscr{H})$ in Section 3.5 by the isometric map $\Psi^{\text{out}}$ (the image operators acting on $\mathscr{H}$) be denoted by

$$\begin{aligned}(a^*_{\text{out}}, h) &= \Psi^{\text{out}}(a^*,h)(\Psi^{\text{out}})^*,\\ (h, a_{\text{out}}) &= \Psi^{\text{out}}(h,a)(\Psi^{\text{out}})^*.\end{aligned} \tag{5.60}$$

Furthermore, for a linear operator $C$ on $\mathscr{H}_1$, we define $(a^*_{\text{out}}, Ca_{\text{out}})$ by

$$(a^*_{\text{out}}, Ca_{\text{out}})\Psi^{\text{out}}(h_1 \otimes \cdots \otimes h_n) = \sum_j \Psi^{\text{out}}(h_1 \otimes \cdots \otimes Ch_j \otimes \cdots \otimes h_n). \tag{5.61}$$

If we do not require a condition like (5.53) for the quasi-local operator $Q_j$, then the asymptotic behaviour of eq. (5.54) for $t \to +\infty$ is obtained by substituting the following expression for $R_j$:

$$\begin{aligned}R_j =& (\Omega, R_j\Omega)\mathbf{1} + (a^*_{\text{out}}, E_1 R_j \Omega) + (E_1 R_j^* \Omega, a_{\text{out}})\\ &+ (a^*_{\text{out}}, E_1 R_j E_1 a_{\text{out}}) + \mathcal{O}(|t|^{-N}).\end{aligned} \tag{5.62}$$

The first term on the r.h.s. does not depend on $t$. The second and third terms are proportional to $t^{-3/2}$ according to Theorem 5.3 of the previous section and played an essential role in Theorem 5.4. Due to Theorem 5.5 the fourth term is proportional to

$t^{-3}$ and it is the part which plays an essential role in Theorem 5.5. As long as we are considering an asymptotic outgoing $n$ particle state constructed from one particle states $h_1, \ldots, h_n$ with disjoint support, the remainder converges to zero faster than any power of $t$. By changing "out" into "in" exactly the same result holds.

### 5.5 Asymptotic conditions and reduction formula for the $S$-matrix

In the discussions of the previous sections we established the construction of the scattering states of particles and their interpretation. In this section we discuss how one can represent the $S$-matrix of the scattering. The aim is to derive the so-called *reduction formula* which has been obtained by colaboration of Lehmann, Symanzik, and Zimmermann in the 1950s. Actually this *LSZ theory* is one of the first papers on mathematical foundation of quantum field theory [4].

The basis for obtaining this reduction formula is the so-called *LSZ asymptotic condition* which was used as a fundamental premise (assumption) in LSZ theory. Its content is the asymptotic estimate of eq. (5.62) for $t \to \infty$ by the first three terms apart from some mathematical details. By the discussions of the previous sections we can easily prove the following form of the LSZ condition [5].

**Theorem 5.6**
Let $\Psi$ be an outgoing state given by eq. (5.55) with $h_j$ satisfying the same condition as in Theorem 5.5. Let $Q$ be a quasi-local operator constructed from the local operator $Q_0$ in the form of (5.16) and denote the one particle wave functions representing $E_1 Q\Omega$ and $E_1 Q^* \Omega$ by $g_Q(\mathbf{p})$ and $g_{Q^*}(\mathbf{p})$, respectively. Using $C^\infty$ functions $\hat{g}_\pm(\mathbf{p})$ with compact support, we set as follows:

$$h_+(\mathbf{p}) = g_Q(\mathbf{p})\hat{g}+(\mathbf{p})/(2p^0),$$
$$h_-(\mathbf{p}) = g_{Q^*}(\mathbf{p})\hat{g}_-(\mathbf{p})/(2p^0),$$

$$g(x) = (2\pi)^{-3} \int \{\mathrm{e}^{-\mathrm{i}(p,x)}\hat{g}_+(\mathbf{p}) + \mathrm{e}^{\mathrm{i}(p,x)}\hat{g}_-(\mathbf{p})\}\, \mathrm{d}^3\mathbf{p}/(2p^0)$$
$$(p^0 = (\mathbf{p}\cdot\mathbf{p} + m^2)^{1/2}) \tag{5.63}$$

$$Q(t,g) = \int Q(x)g(x)\,\mathrm{d}^3\mathbf{x} \qquad (x^0 = t).$$

We require that either of the following conditions hold:

($\alpha$) $\quad E_1 Q\Omega = Q\Omega, \quad \hat{g}_- = 0$

($\beta 1$) $\quad Q\Omega = 0, \quad \hat{g}_+ = 0.$

Furthermore, either the support of $h_\pm$ is disjoint from the support of any $h_j$ or, for the case of ($\beta$1), we assume the following condition:

$$(\beta 2) \qquad E_1 Q E_1 = 0; \quad \text{if } \Phi \perp \Omega, \text{ then } E_1 Q^* \Phi = 0.$$

Under these conditions

$$\lim t^N \| \{Q(t,g) - (a^*_{\text{out}}, h_+) - (h_-, a_{\text{out}})\} \Psi \| = 0 \tag{5.64}$$

in the limit $t \to +\infty$ for any natural number $N$. (In the case of ($\alpha$), $h_- = 0$. In the case of ($\beta$1), $h_+ = 0$. If the support of $h_-$ is disjoint from the support of all $h_j$, then $(h_-, a_{\text{out}})\Psi = 0$.)

The same result holds if we exchange "out" with "in" and $t \to +\infty$ with $t \to -\infty$. (Also "out" has to be replaced by "in" for $\Psi$.)

The $Q$ satisfying condition ($\alpha$) is obtained by (5.16). The $Q$ satisfying conditions ($\beta$1) and ($\beta$2) is obtained by choosing the support of $\tilde{f}_j$ in eq. (5.16) to be in a neighbourhood of

$$m_- = \{p;\ (p,p) = m^2, \ \ p^0 < 0\}.$$

If the support of $\tilde{f}_j$ (restricted to $m_\pm$) has an overlap with the support of $\hat{g}_\pm$, the corresponding $h_+$ and $h_-$ become non-zero $C^\infty$ functions.

**Remark 1** Denote the projection operator to the image of $\Psi^{\text{out}}$ by

$$P_{\text{out}} = \Psi^{\text{out}} (\Psi^{\text{out}})^*.$$

Then

$$w - \lim P_{\text{out}} Q(t,g)\Psi = \{(a^*_{\text{out}}, h_+) + (h_-, a_{\text{out}})\}\Psi \tag{5.65}$$

holds in the limit $t \to +\infty$ as the weak limit under the condition $(\Omega, Q\Omega) = 0$ alone without conditions ($\alpha$) or ($\beta$). This equation with $P_{\text{out}} = \mathbf{1}$ is usually called the LSZ condition. However, even without the assumption $P_{\text{out}} = \mathbf{1}$ (completeness of the scattering states), the above theorem holds and the reduction formula will be derived utilizing this form of the theorem.

**Remark 2** Under the condition ($\alpha$) or ($\beta$1) and without $P_{\text{out}}$ in eq. (5.65), the convergence of (5.65) in the sense of the strong limit can be shown with exactly the same method as used in the proof of the theorem.

**Outline of the proof of the theorem** Although the estimate of (5.23) for $g$ is described for the case $\hat{g}_- = 0$, the case $\hat{g}_+ = 0$ is obtained from $\hat{g}_- = 0$ by considering the complex conjugate $\overline{g(x)}$ and we obtain the following estimate:

$$\|Q(t,g)\| \le \|Q\| \int |g(t,\mathbf{x})|\, d^3\mathbf{x} < (\text{constant})(1 + |t|)^{3/2}.$$

Therefore by (5.45)

$$\lim |t|^N \|Q(t,g)\Psi - Q(t,g)\Psi_t\| = 0$$

holds with $\Psi_t$ being given by (5.34).

First, consider the case where the support of $h_\pm$ is disjoint from the support of any $h_j$. If $(\alpha)$ holds, then

$$\lim t^N \|Q(t,g)\Psi_t - (a^*_{\text{out}}, h_+)\Psi\|$$

by Theorem 5.4 and the estimate of (5.45) and, combining this with the condition $h_- = 0$ we obtain (5.64). If condition $(\beta 1)$ is valid under the same condition for the support of $h_\pm$, then applying the same estimate as in (5.45) to $\|Q(t,g)\Psi_t\|$ itself and, using $Q(t,g)\Omega = 0$ (due to $Q\Omega = 0$) instead of $(\mathrm{d}/\mathrm{d}t)Q(t,g)\Omega = 0$, we can prove

$$\lim t^{2N} \|Q(t,g)\Psi_t\|^2 = 0.$$

Combining this with $h_+ = 0$ and $(a_{\text{out}}, h_-)\Psi = 0$, we have proved (5.64).

Next let us consider the case where conditions $(\beta 1)$ and $(\beta 2)$ hold. Since we can approximate the limit vector by

$$\lim t^N \left\| (h_-, a_{\text{out}})\Psi - \sum_j \omega(Q(t,g)Q_j(t,g_j)) \prod_{k \neq j} Q_k(t,g_k)\Omega \right\| = 0,$$

it is enough to show that

$$\lim t^{2N} \left\| Q(t,g)\Psi_t - \sum_j \omega(Q(t,g)Q_j(t,g_j)) \prod_{k \neq j} Q_k(t,g_k)\Omega \right\|^2 = 0.$$

Using the cluster decomposition, all terms without the following factors either vanish or become products of two point functions and cancel each other by Corollary 4.12, Theorem 5.3 and condition $(\beta 1)$:

$$\begin{gathered} \omega^T(Q_j(t,g_j)^* Q(t,g) Q_j(t,g)), \\ \omega^T(Q_j(t,g_j)^* Q(t,g)^* Q_j(t,g)), \\ \omega^T(Q_j(t,g_j)^* Q(t,g)^* Q(t,g) Q_j(t,g_j)). \end{gathered} \tag{5.66}$$

Among the above factors the first two vanish due to the first equation in condition $(\beta 2)$ because $Q_j(t,g_j)\Omega$ is a one particle state. For the same reason the last factor becomes zero due to the second equation in $(\beta 2)$ because the truncation of the vacuum expectation value automatically subtracts the vacuum contribution to the intermediate state between $Q(t,g)^*$ and $Q(t,g)$. □

As an application we can obtain the following lemma which gives the fundamental formula for the derivation of the reduction formula.

**Lemma 5.7**
Given the two families of $C^\infty$ functions $h'_i$ $(i = 1, \ldots, k)$ and $h''_j$ $(j = 1, \ldots, l)$ with mutually disjoint support within each group (the support of the $h'_i$ may overlap with the support of the $h''_j$), consider the function $g(x)$ defined by eq. (5.21) in terms of a $C^\infty$ function $\hat{g}(\mathbf{p})$ with compact support, and a quasi-local operator $Q$ defined in the form of (5.16) so as to satisfy the following conditions (which can be achieved by choosing the support of $\tilde{f}$ in a neighbourhood of $m_+$):

$$(\alpha) \qquad E_1 Q\Omega = Q\Omega$$

$$(\beta') \qquad \begin{cases} Q^*\Omega = 0 \\ \text{If } \Phi \perp \Omega, \text{ then } E_1 Q\Phi = 0 \text{ (necessarily, } E_1 Q E_1 = 0). \end{cases}$$

Let $C_1$ and $C_2$ be any bounded linear operators and let $F(C_1, C_2, Q, x)$ be an operator-valued function, twice continuously differentiable with respect to $x$, the degree of growth of its norm being at most of the order of a polynomial of $|x|$ for $x \to \infty$, satisfying

$$F(C_1, C_2, Q, x) = \begin{cases} Q(x)C_1 & (x^0 > T) \\ C_2 Q(x) & (x^0 < -T) \end{cases} \tag{5.67}$$

for a positive number $T$. Then, for vectors

$$\Psi_1 = \Psi^{\text{out}}[h'_1 \ldots h'_k], \qquad \Psi_2 = \Psi^{\text{in}}[h''_1 \ldots h''_l],$$

the following formulae hold:

$$((a^*_{\text{out}}, h)\Psi_1, C_1\Psi_2) - (\Psi_1, C_2(h, a_{\text{in}})\Psi_2) = \mathrm{i}\int g(x)^* K_{mx}(\Psi_1, F(C_1, C_2, Q^*, x)\Psi_2)\,\mathrm{d}^4x, \tag{5.68}$$

$$(\Psi_1, C_2(a^*_{\text{in}}, h)\Psi_2) - ((h, a_{\text{out}})\Psi_1, C_1\Psi_2) = \mathrm{i}\int g(x) K_{mx}(\Psi_1, F(C_1, C_2, Q, x)\Psi_2)\,\mathrm{d}^4x, \tag{5.69}$$

where $h(\mathbf{p}) = g_Q(\mathbf{p})\hat{g}(\mathbf{p})$ and $K_{mx}$ is the partial differential operator

$$K_{mx} = (\partial/\partial x^0)^2 - \sum_j (\partial/\partial x^j)^2 + m^2 \ \ (= \Box_x + m^2). \tag{5.70}$$

In the $\mathrm{d}^4x$ integration, the $\mathrm{d}x^0$ integration is to be performed after the $\mathrm{d}^3x$ integration.

**Proof of lemma** We use the following notation (where $\partial_0 = (\partial/\partial t)$):

$$Q(t, \overleftrightarrow{\partial_0}\, g) = \int \{Q(t, \mathbf{x})\partial_0 g(t, \mathbf{x}) - (\partial_0 Q(t, \mathbf{x}))g(t, \mathbf{x})\}\, \mathrm{d}^3\mathbf{x}. \tag{5.71}$$

Since $\partial_0 g = \dot{g}$ and $\partial_0 Q(t, x) = \dot{Q}(t, x)$ given in eq. (5.32) satisfy the properties assumed for $g$ and $Q$, this expression can be treated in the same way as $Q(t, g)$, and the following relations hold by Theorem 5.6:

$$\lim \|\mathrm{i}Q(t, \overleftrightarrow{\partial_0}\, g)\Psi_1 - (a^*_{\mathrm{out}}, h)\Psi_1\| = 0 \tag{5.72}$$

in the limit $t \to +\infty$, and

$$\lim \|(-\mathrm{i})Q(t, \overleftrightarrow{\partial_0} g)^*\Psi_2 - (h, a_{\mathrm{in}})\Psi_2\| = 0 \tag{5.73}$$

in the limit $t \to -\infty$, respectively, where condition ($\beta'$) for $Q$ becomes conditions ($\beta 1$) and ($\beta 2$) for $Q^*$. Due to $\overleftrightarrow{\partial_0}$ a factor $\pm\mathrm{i}$ appears on the l.h.s. of both equations above and the definition of $h$ is changed by a factor $(2p^0)$ compared to before. Since $g(x)$ in eq. (5.21) satisfies $K_{mx}g(x) = 0$, we get by partial integration

$$\int_{T''}^{T'} \mathrm{d}x^0 \partial_0 \int \{F(C_1, C_2, Q^*, x)\, \overleftrightarrow{\partial_0}\, g(x)^*\}\, \mathrm{d}^3\mathbf{x}$$

$$= \int_{T''}^{T'} \{F(C_1, C_2, Q^*, x)\partial_0^2 g(x)^* - g(x)^* \partial_0^2 F(C_1, C_2, Q^*, x)\}\, \mathrm{d}^4 x.$$

Rewriting $\partial_0^2 g$ in the first term on the r.h.s. as $(\Delta_x - m^2)g$ ($\Delta_x$ is the Laplace operator and the replacement is justified due to $K_{mx}g = 0$) and operating $\Delta_x$ on $F$ by partial integration (for $|x| \to \infty$ the fast decrease of $g$ dominates over the polynomial increase of $\|F\|$ and hence the surface integral does not give a contribution), we see that this is equal to the following expression:

$$= -\int_{T''}^{T'} g(x)^* K_{mx} F(C_1, C_2, Q^*, x)\, \mathrm{d}^4 x$$

Due to (5.67) the l.h.s. is equal to

$$= Q(T', \overleftrightarrow{\partial_0}\, g)^* C_1 - C_2 Q(T'', \overleftrightarrow{\partial_0}\, g)^*$$

for $T' > T$, $T'' < -T$. Taking the limits $T' \to +\infty$ and $T'' \to -\infty$ and using (5.72) and (5.73), we obtain (5.68) where both sides are multiplied by i. We obtain (5.69) by the same computation using $Q$ and $g$ instead of $Q^*$ and $g^*$. □

Before going into the reduction formula we introduce for the $S$-matrix elements an object corresponding to truncated expectation values. Namely, we define the *connected part* $S_c$ of the $S$-matrix by the following formula.

$$S_c(h'_1 \ldots h'_k; h''_1 \ldots h''_l) \equiv \sum_m (-1)^{m-1}(m-1)! \sum_{\nu=1}^{m} S(I'_\nu; I''_\nu), \qquad (5.74)$$

where $\{I_\nu; \nu = 1, \ldots, m\}$ is a partition of the set of indices $\{1, \ldots, k\}$ into $m$ subsets $I'_\nu$ ($I'_\nu \neq$ *empty*), and $\{I''_\nu; \nu = 1, \ldots, m\}$ is a partition of the set of indices $\{1, \ldots, l\}$ into $m$ parts. Summation runs over all partitions into $m$ parts and then over all $m$. The $I'_\nu$ appearing in $S(I'_\nu; I''_\nu)$ represents $h_j$ written out for all indices $j$ belonging to $I'_\nu$ (in the natural order, although the ordering is irrelevant due to the permutation symmetry in the present case), and similarly for $I''_\nu$. By solving (5.74) for $S$, we obtain the following expansion similarly to eq. (4.29):

$$S(h'_1 \ldots h'_k; h''_1 \ldots h''_l) = \sum \prod_\nu S_c(I'_\nu; I''_\nu). \qquad (5.75)$$

The summation runs over all partitions of the pair of index sets $\{1, \ldots, k; 1, \ldots, l\}$ into pairs $I'_\nu, I''_\nu$ of subsets. $S_c(I'_\nu; I''_\nu)$ is an abbreviated notation in the same way as $S(I'_\nu; I''_\nu)$. For the case that one of $I'_\nu$ and $I''_\nu$ contains just one element and the other contains more than one, we have $S(I'_\nu; I''_\nu) = S_c(I'_\nu; I''_\nu) = 0$. The empty set is not allowed for $I'_\nu$ and $I''_\nu$. If both $I'_\nu$ and $I''_\nu$ consist of only a single index, then

$$S_c(h'; h'') = S(h'; h'') = (h', h''). \qquad (5.76)$$

Another concept necessary for the description of the reduction formula is that of the *time ordered product* and the $\tau$ *function* [6].

Given a positive valued $C^\infty$ function $\varphi^n_P$ in $n$ variables for each permutation $P$ of the indices $1, \ldots, n$, all derivatives of $\varphi^n_P$ being bounded and the following conditions being satisfied for $\delta > 0$:

(a) If $t_{P(j+1)} - t_{P(j)} > \delta$ holds for at least one $j$, then

$$\varphi^n_P(t_1 \ldots t_n) = 0.$$

(b) $\sum_P \varphi^n_P(t_1 \ldots t_n) = 1$.
(c) Let $I_1 = \{1, \ldots, n_1\}, \ldots, I_r = \{n_{r-1}+1, \ldots, n_r\}$ be any partition of the set $\{1, \ldots, n\}$ into $r$ parts and $P$ be any permutation of $1, \ldots, n$. For any set of variables $t_1, \ldots, t_n$ satisfying $t_{P(i)} - t_{P(j)} > \delta$ for $k > l$, $j \in I_k$, and $i \in I_l$, and for any permutations $Q_k$ of the indices belonging to $I_k$ for $k = 1, \ldots, r$, the following equation holds:

$$\varphi^n_{PQ}(t_1 \ldots t_n) = \prod_{k=1}^{r} \varphi^{m_k}_{Q_k}(t_{P(n_{k-1}+1)} \ldots t_{P(n_k)})$$

where $Q = \Pi Q_k$, and $m_k = n_k - n_{k-1}$, $n_0 = 0$.

Then

$$T^{\varphi}(Q_1(x_1)\dots Q_n(x_n)) = \sum_P \varphi_P^n(x_1^0\dots x_n^0)Q_{P(1)}(x_{P(1)})\cdots Q_{P(n)}(x_{P(n)}) \qquad (5.77)$$

is called a *time $\varphi$ ordered product* and

$$\tau^T(x_1\dots x_n) = \omega^T(T^{\varphi}(Q_1(x_1)\cdots Q_n(x_n)))$$

is called a *truncated $\tau$ function* or the *connected part of a $\tau$ function.* If $\omega^T$ is replaced by $\omega$ on the r.h.s., it is usually called a *$\tau$ function.*

To show that this definition is meaningful, we show the existence of $\varphi_P^n$ satisfying the above three conditions. First, if we allow discontinuous functions, the following functions $\theta_P^n$ satisfy the three conditions. If

$$t_{P(1)} \geq t_{P(2)} \geq \cdots \geq t_{P(n)} \qquad (5.78)$$

does not hold, then $\theta_P^n(t_1\dots t_n) = 0$ by definition. If (5.78) is satisfied and the equality sign holds for $k_1$ successive times, $k_2$ successive times, etc. (the inequality holding otherwise with at least one inequality sign between them), then we define

$$\theta_P^n(t_1\dots t_n) = (k_1!k_2!\cdots)^{-1}. \qquad (5.79)$$

In particular, if the inequality holds everywhere in (5.78), then $\theta_P^n = 1$. (Since equalities hold with measure 0, the values when equalities occur are not important.)

Usually, the so-called time ordered product is obtained by using this $\theta$ for $\varphi$.

In order to change $\theta_P^n$ to $C^\infty$ functions, we use a $C^\infty$ function $g$ with its support in $[-\delta/2, \delta/2]$ satisfying $\int g(t)\,\mathrm{d}t = 1$. If we set

$$\varphi_P^n(t_1\dots t_n) = \int \theta_P^n(t_1+u_1\dots t_n+u_n)\prod_{j=1}^n g(u_j)\mathrm{d}u_j$$

then, since $\theta_P^n$ satisfies (a), (b), and (c) with $\delta = 0$, $\varphi_P^n$ satisfies the same conditions for the chosen $\delta$. Now we are ready to state the reduction formula.

**Theorem 5.8**
Let $h_1',\dots,h_k'$ and $h_1'',\dots,h_l''$ be a pair of families of $C^\infty$ functions with mutually disjoint compact support (the support of $h_i'$ and $h_j''$ may overlap), let $Q_1',\dots,Q_k'$ and $Q_1'',\dots,Q_l''$ be of the form as defined in (5.16) satisfying conditions ($\alpha$) and ($\beta'$) for $Q$ in Lemma 5.7, and let $\tilde{g}_1',\dots,\tilde{g}_k'$ and $\tilde{g}_1'',\dots,\tilde{g}_l''$ be

$C^\infty$ functions with compact support satisfying the following equations:

$$h'_j(\mathbf{p}) = \tilde{g}'_j(\mathbf{p}) g_{Q'_j}(\mathbf{p}), \quad h''_j(\mathbf{p}) = \tilde{g}''_j(\mathbf{p}) g_{Q''_j}(\mathbf{p})$$

Then except for the case $k = l = 1$ the following formula holds:

$$\begin{aligned} S_c(h'_1 \ldots h'_k; h''_1 \ldots h''_l) = i^{k+l} \int & g'_1(x'_1)^* \cdots g'_k(x'_k)^* \\ & \times g''_1(x''_1) \cdots g''_l(x''_l) K_{mx'_1} \cdots K_{mx''_k} \\ & \times \omega^T(T^\varphi(Q'_k(x'_k)^* \ldots Q'_1(x'_1)^* Q''_1(x''_1) \ldots Q''_l(x''_l))) \\ & \times \mathrm{d}^4x'_1 \cdots \mathrm{d}^4x'_k \mathrm{d}^4x''_1 \cdots \mathrm{d}^4x''_l. \end{aligned} \tag{5.80}$$

For the case $k = l = 1$, the r.h.s. vanishes. Here $\omega^T$ is the *truncated vacuum expectation value* defined in (4.28) and is to be calculated for each term in eq. (5.77). (The integration over spacelike vectors is performed first. After that the time integration converges faster than any inverse power of the variables.)

**Outline of the proof** The proof has a computational aspect and an analytical aspect such as convergence. To start with, we first explain how to push through the calculation to obtain eq. (5.80) without worrying about how strong and how uniform the convergence is. Afterwards we will justify the calculation.

The computation starts from the r.h.s. of (5.80), leading to the l.h.s. However, we first consider the r.h.s., where $\omega^T$ is replaced by $\omega$. In order to apply Lemma 5.7 for the integration of the $x''_l$, we check whether $T^\varphi(\ldots)$ on the r.h.s. of (5.80) has the properties of $F$ in eq. (5.67) as a function of $x''_l$.

We denote the time components of $x'_i$, $x''_j$ as $t'_i$, $t''_j$. Then $T^\varphi$ is the sum over $P$ of those terms with the coefficient

$$\varphi_P^{k+l}(t'_k \cdots t'_1, t''_1 \cdots t''_l). \tag{5.81}$$

When we set $t''_l$ negative and large, (5.81) is non-zero only for those permutations $P$ which do not move the last variable $t''_l$ by condition (a) for $\varphi$ and we obtain

$$T^\varphi(\cdots) = T^\varphi(\cdots Q''_{l-1}(x''_{l-1})) Q''_l(x''_l)$$

On the other hand if $t''_l$ tends to $+\infty$, then (5.81) is nonzero only for those permutations $P$ which move $t''_l$ to the first position and we get

$$T^\varphi(\cdots) = Q''_l(x''_l) T^\varphi(\cdots Q''_{l-1}(x''_{l-1}))$$

(In both cases, due to property (c) of $\varphi$, the number of variables in $T^\varphi(\cdots)$ on the r.h.s. is reduced by one.) Thus (5.67) is satisfied. Furthermore, since $\Psi_1$ is $\Omega$ in the present case, the second term on the l.h.s. of eq. (5.69) vanishes. Thus we obtain the equation

$$\mathrm{i}\int g_l''(x_l'')K_{mx_l''}\omega(T^{\varphi}(\cdots))\,\mathrm{d}^4x_l'' = (\Omega, T^{\varphi}(\cdots)\Psi^{\mathrm{in}}[h_l''])$$

In the same way by rewriting the integration over $x_{l-1}'', x_{l-2}'', \ldots$ one after another, we obtain

$$\begin{aligned}\mathrm{i}^l\int g_1''(x_1'')\cdots g_l''(x_l'')K_{mx_1''}\cdots K_{mx_l''}\omega(T^{\varphi}(\cdots))\mathrm{d}^4x_1''\cdots\mathrm{d}^4x_l''\\ = (\Omega, T^{\varphi}(Q_k'(x_k')^*\cdots Q_1'(x_1')^*)\Psi^{\mathrm{in}}[h_1''\cdots h_l'']).\end{aligned} \tag{5.82}$$

Next, we perform a similar computation by using (5.68) for the $x_k'$ integration. Abbreviating the above expression as $(\Omega, T^{\varphi}(\cdots)\Psi^{\mathrm{in}}[\cdots])$, we get

$$\begin{aligned}\mathrm{i}\int g_k(x_k')^*K_{mx_k'}(\Omega, T^{\varphi}(\cdots)\Psi^{\mathrm{in}}[\cdots])\,\mathrm{d}^4x_k'\\ = (\Psi^{\mathrm{out}}[h_k'], T^{\varphi}(Q_{k-1}'(x_{k-1}')\cdots)\Psi^{\mathrm{in}}[\cdots])\\ - (\Omega, T^{\varphi}(Q_{k-1}'(x_{k-1}')\cdots)(h_k', a_{\mathrm{in}})\Psi^{\mathrm{in}}[\cdots]).\end{aligned} \tag{5.83}$$

The second term on the r.h.s. is

$$\sum_j(\Omega, T^{\varphi}(\cdots)\Psi^{\mathrm{in}}[h_1''\cdots \hat{j}\cdots h_l''])S(h_k', h_j'').$$

(See eq. (3.60) for symbols.) We continue the same calculation successively for the $x_{k-1}'$ integration, the $x_{k-2}'$ integration, etc. The term in the final stage corresponding to the first term in the above equation yields

$$(\Psi^{\mathrm{out}}[h_1'\cdots h_k'], \Psi^{\mathrm{in}}[h_1''\cdots h_l'']) = S(h_1'\cdots h_k', h_1''\cdots h_l''). \tag{5.84}$$

In addition, we obtain terms corresponding to the second term in the above equation. The treatment of these terms is a combinatorial problem and is done as follows.

First let us discuss the cases where $k$ and $l$ are small. For $k=0$, $T^{\varphi}(\cdots)$ is 1 and (5.82) vanishes for $l\neq 0$. On the other hand, if $l=0$, $\Psi^{\mathrm{in}}$ remains equal to $\Omega$ and hence the second term of (5.83) vanishes. For $k\neq 0$, (5.84) $=0$ and consequently the entire (5.80) vanishes. For $l=k=0$ it is defined to be 1.

For the case $k=1$, $T^{\varphi}(\cdots)$ becomes simply $Q_1'(x_1')^*$ and hence eq. (5.82) becomes

$$(Q_1'(x_1')\Omega, \Psi^{\mathrm{in}}[\cdots])$$

Since $Q_1'(x_1')\Omega$ is a one particle state, it satisfies

$$(\Box_x + m^2)Q_1'(x)\Omega = 0$$

So eq. (5.80) is zero independently of $l$.

For the case $l = 1$, $\Psi^{\text{in}}[\cdots]$ is a one particle state and hence $(h'_k, a_{\text{in}})\Psi^{\text{in}}[\cdots]$ is proportional to $\Omega$. For $k \neq 1$ it is zero for the same reason as in the above case of $l = 0$. When proceeding with the further computation of the first term of (5.83), the second term of (5.83) vanishes each time for the same reason, and (5.84), which remains at the last stage, also vanishes due to $l = 1$. Therefore, eq. (5.80) is zero for $k \neq 1$. For the remaining case $k = l = 1$ the expression is zero as is already shown.

Summarizing, eq. (5.80) is 1 for the case $k = l = 0$ and zero otherwise unless $k \geq 2$ and $l \geq 2$.

Now in order to prove (5.80), let us prove the following formula for the $S$-matrix element:

$$S(h'_1 \cdots h'_k, h''_1 \cdots h''_l) = \sum T(\{1, \cdots, k\}\backslash I'; \{1, \cdots, l\}\backslash I'')\mathbf{1}(I'; I''). \tag{5.85}$$

Here $I'$ is a subset of $\{1, \ldots, k\}$ and $\{1, \ldots, k\}\backslash I'$ is the remainder after $I'$ is removed. Similarly for the subset $I''$ of $\{1, \ldots, l\}$. The sum of the r.h.s. runs over all pairs $I'$ and $I''$ with the same number of elements. $\mathbf{1}(\cdots)$ is defined by the following equation.

$$\mathbf{1}(I'; I'') = \sum_P \prod_j S(h'_j; h''_{P(j)}), \tag{5.86}$$

where the product on the r.h.s. is over all $j$ in $I'$, the sum runs over all bijections (1:1 correspondence) $P$ from $I'$ onto $I''$, and $S(h', h'') = (h', h'')$. Furthermore,

$$T(h'_1 \cdots h'_k; h''_1 \cdots h''_l) \equiv T(1 \cdots k; 1 \cdots l), \tag{5.87}$$

represents the expression on the r.h.s. of (5.80), where the truncated vacuum expectation value $\omega^T$ is replaced by the vacuum expectation value $\omega$, and a similar notation $T(J', J'')$ is used also for the subsets $J'$, $J''$ of $\{1, \ldots, k\}$ and $\{1, \ldots, l\}$. We take $T = 1$ for $J' = J'' =$ empty, and $T = 0$ otherwise for $|J'| < 2$ as well as for $|J''| < 2$. (Here we denote the number of elements of $J$ by $|J|$.)

The proof of (5.85) can be given by repeated application of the calculation of (5.83). Formally we prove the following formula by using mathematical induction for $k$ and $n$:

$$\begin{aligned}
&\mathrm{i}^n \int g'_1(x'_1)^* \cdots g'_n(x'_n)^* K_{mx'_1} \cdots K_{mx'_n} \\
&\quad \times (\Psi^{\text{out}}[h'_{n+1} \cdots h'_k], T^{\varphi}(Q'_n(x'_n)^* \cdots Q'_1(x'_1)^*)\Psi^{\text{in}}[h''_1 \cdots h''_l]) \\
&\quad = \sum T(\{1, \cdots, k\}\backslash I'; \{1, \cdots, l\}\backslash I'')\mathbf{1}(I'; I'').
\end{aligned} \tag{5.88}$$

Here $I'$ and $I''$ are subsets of $\{n+1, \ldots, k\}$ and $\{1, \ldots, l\}$ respectively, with the same number of elements, and the sum on the r.h.s. is performed over all such pairs $I'$, $I''$. The case $I' = I'' =$ empty set is to be included and the case $n = 0$ is eq. (5.85) to be proven.

First, in order to perform the mathematical induction for $k$ we consider $k = 1$. In this case $n = 0$ or 1. From the result of the above discussion, both sides become

$(h'_1, h''_1)$ for $k = l = 1, n = 0$, and vanish otherwise. Hence eq. (5.88) holds for $k = 1$. Then we assume that (5.88) is satisfied for $k < k'$ for a certain $k'$ and prove that (5.88) holds for $k = k'$.

We now use mathematical induction on $n$. For $n = k'$ the set $I'$ can only be the empty set and the r.h.s. of (5.88) becomes $T(\cdots)$. Since $\Psi^{\text{out}}[\cdots]$ on the l.h.s. becomes $\Omega$,

$$(\Psi^{\text{out}}[\cdots], T^{\varphi}(\cdots)\Psi^{\text{in}}[\cdots])$$

is equal to (5.82). Therefore, by the definition of $T(\cdots)$ given in (5.87) and by the computation leading to (5.82), eq. (5.88) holds.

Now take an $n'$ ($k' > n' \geq 0$) and assume that (5.88) holds for $n > n'$. Then we prove that (5.88) holds also for $n = n'$.

Applying formula (5.68) of Lemma 5.7 to the l.h.s. of eq. (5.88) with $n = n' + 1$, we obtain the following expression:

$$\begin{aligned} &\mathrm{i}^{n'} \int g'_1(x'_1)^* \cdots g'_{n'}(x'_{n'})^* K_{mx'_1} \cdots K_{mx'_{n'}} \\ &\times \{(\Psi^{\text{out}}[h'_{n'+1} \cdots h'_{k'}], T^{\varphi}(Q'_{n'}(x'_{n'})^* \cdots Q'_1(x'_1)^*)\Psi^{\text{in}}[h''_1 \cdots h''_l]) \\ &- \sum_j (\Psi^{\text{out}}[h'_{n'+2} \cdots h'_{k'}], T^{\varphi}(Q'_{n'})(x'_{n'})^* \cdots Q'_1(x'_1)^*) \\ &\times \Psi^{\text{in}}[h''_1 \cdots \hat{j} \cdots h''_l])S(h'_{n'+1}; h''_j)\}. \end{aligned} \tag{5.89}$$

The expression

$$(\Psi^{\text{out}}[\cdots], T^{\varphi}(\cdots)\Psi^{\text{in}}[\cdots])$$

in each term of the sum over $j = 1, \ldots, l$ on the r.h.s. corresponds to the case $k = k' - 1$ (the sum of the number of $h'_j$ inside $\Psi^{\text{out}}$ and the number of $Q'_j(x'_j)^*$ inside $T^{\varphi}$ is $k' - 1$) and hence, by the inductive assumption with respect to $k$, we can use eq. (5.88). Consequently, inside (5.89) we have

$$\begin{aligned} \sum_j \cdots &= \sum_j \sum T(\{1 \cdots \widehat{n'+1} \cdots k'\} \backslash I'; \{1 \cdots \hat{j} \cdots l\} \backslash I'')\mathbf{1}(I'; I'')S(h'_{n'+1}; h''_j) \\ &= \sum T(\{1 \cdots k'\} \backslash \bar{I}'; \{1 \cdots l\} \backslash \bar{I}''\})\mathbf{1}(\bar{I}'; \bar{I}''), \end{aligned} \tag{5.90}$$

where $I'$ and $I''$ are subsets of $\{n' + 2, \ldots, k'\}$ and $\{1, \ldots, \hat{j}, \ldots, l\}$, respectively, with the same number of elements, $\bar{I}'$ and $\bar{I}''$ are subsets of $\{n' + 1, \ldots, k'\}$ and $\{1, \ldots, l\}$ with the same number of elements, and we have the condition on $\bar{I}'$ that it contains always $n' + 1$. The double sum over $j$ and over a pair $I', I''$ becomes just a single sum over the pair $\bar{I}', \bar{I}''$.

Since (5.88) holds for $n = n' + 1$ by the inductive assumption, all terms of the expression (5.89) together are equal to the following expression.

$$\sum T(\{1, \ldots, k'\}\backslash I'; \{1, \ldots, l\}\backslash I'')\mathbf{1}(I'; I''). \tag{5.91}$$

Here $I'$ and $I''$ are subsets of $\{n'+2, \ldots, k\}$ and $\{1, \ldots, l\}$, respectively, with the same number of elements. Shifting the terms of (5.90) from the l.h.s. to the r.h.s. (i.e. (5.91)) of this equality, we see that the first term in eq. (5.89) remaining on the l.h.s. is equal to the following expression:

$$(5.90) + (5.91) = \sum T(\{1, \ldots, k'\}\backslash I'; \{1, \ldots, l\}\backslash I'')\mathbf{1}(I'; I''). \tag{5.92}$$

This time $I'$ and $I''$ are subsets of $\{n'+1, \ldots, k\}$ and $\{1, \ldots, l\}$ with the same number of elements, and those $I'$ which include $n'+1$ come from (5.90), and those $I'$ which do not include $n'+1$ come from (5.91).

Since this final form of the equality is just (5.88) for $n = n'$, we have proved (5.88) by mathematical induction, and thus also (5.85).

Next we derive (5.80) from (5.85). For that purpose, we first show that (5.80) satisfies (5.75). We substitute the expression given by the r.h.s. of (5.80) into the corresponding $S_c$ on the r.h.s. of (5.75) and calculate the r.h.s. of (5.75), where the sum runs over all partitions of $\{1, \ldots, k\}$ and $\{1, \ldots, l\}$. The case $I'_\nu = I''_\nu =$ empty set is excluded and $S_c(I'_\nu; I''_\nu)$ vanishes if $|I'_\nu| < 2$ or $|I''_\nu| < 2$, except that for the case $|I'_\nu| = |I''_\nu| = 1$, $S_c(I'_\nu; I''_\nu) = (h'_i, h''_j)$ if $I'_\nu = \{i\}$ and $I''_\nu = \{j\}$. We substitute (5.80) when $|I'_\nu|$ and $|I''_\nu|$ are both larger than 1. Taking the sum over those subdivisions $\{J'_\mu; I'_\nu\}$ and $\{J''_\mu, I''_\nu\}$ of the division

$$\{1, \ldots, k\} = J' \cup I', \quad \{1, \ldots, l\} = J'' \cup I'', \quad |I'| = |I''|$$

such that the $I'_\nu$ and $I''_\nu$ included in $I'$ and $I''$ satisfy $|I'_\nu| = |I''_\nu| = 1$, and the $J'_\mu$ and $J''_\mu$ included in $J'$ and $J''$ satisfy $|J'_\mu| \geq 2$, $|J''_\mu| \geq 2$, we obtain

$$\Big\{\sum \prod_\mu T^T(J'_\mu; J''_\mu)\Big\}\mathbf{1}(I'; I''), \tag{5.93}$$

where $T^T(I', I'')$ is the r.h.s. of (5.80) for $\{h'_i, i \in I'\}$ and $\{h''_j, j \in I''\}$.

In order to compare with (5.93), we represent $T(J'; J'')$ of (5.92) in terms of the expression obtained by substituting $\omega$ into $\omega^T$ on the r.h.s. of (5.80), represent $T^\varphi(\cdots)$ by the r.h.s. of (5.77) and then expand $\omega(\cdots)$ in each term as a sum of products of $\omega^T(\cdots)$ using (4.29). Let us define the restriction of a permutation $P$ to a subset $I$ which is not necessarily invariant as a set under $P$, as the permutation of $i \in I$ inside $I$ according to the order of $I(i)$. Then for any given partition $\Pi_\mu \omega^T(J'_\mu, J''_\mu)$ in the expansion of $\omega$, summing $\varphi_P$ over those $P$ such that their restriction $P_\mu$ to each $\{J'_\mu, J''_\mu\}$ is the same for all $\mu$, we obtain $\Pi_\mu \varphi_{P_\mu}$ for $\varphi_P$ concretely constructed via

(5.79). Then we get the expansion

$$\omega(T^{\varphi}(J', J'')) = \sum \prod_{\mu} \omega^{T}(T^{\varphi}(J'_{\mu}, J''_{\mu})) \tag{5.94}$$

and hence we obtain the expansion

$$T(J'; J'') = \sum \prod_{\mu} T^{T}(J'_{\mu}; J''_{\mu})$$

where all possible combinations appear for $J'_{\mu}$ and $J''_{\mu}$ except for the case $J'_{\mu} = J''_{\mu}$ = empty set. However, for $J'_{\mu} < 2$ and $J''_{\mu} < 2$ we have already shown that all $T(J'_{\mu}, J''_{\mu})$ vanish and hence all $T^{T}(J'_{\mu}, mJ''_{\mu})$ as constructed from them in (4.28) also vanish. Thus (5.93) is equal to $T(J; J')\mathbf{1}(I'; I'')$ and, on substituting the r.h.s. of eq. (5.80) into $S_c$s on the r.h.s. of eq. (5.75), we obtain the r.h.s. of (5.85) and hence the l.h.s. of eq. (5.75). Therefore, we have shown that (5.80) satisfies (5.75).

Since $S_c$ is uniquely determined by solving eq. (5.75), the fact that the r.h.s. of (5.80) satisfies eq. (5.75) means that eq. (5.80) is established. Hence, the computational part of the proof of Theorem 5.8 is completed.

Next let us examine the convergence of the integral. The integral over the space vectors $\mathbf{x}'_i$, $\mathbf{x}''_j$ converges well because the vacuum expectation value is bounded and each $g(x)$ tends to zero faster than any power of $|\mathbf{x}|$ for $\mathbf{x} \to \infty$. As for each time variable, due to Theorem 5.6 and Lemma 5.7 we know for each time variable the convergence faster than any power of the variable. Thus we shall show here the uniform convergence over all time variables.

In order to avoid symbols becoming too complicated, we will perform the discussion on the integral on the r.h.s. of (5.80) using $\omega$ instead of $\omega^{T}$ up to the point where we need $\omega^{T}$. By (4.28), $\omega^{T}$ is a sum of products of $\omega$ and also $T^{\varphi}(\cdots)$ can be treated as in the computation for (5.94). Hence the considerations on $\omega$ can be applied to $\omega^{T}$ as it stands.

Equation (5.80) with $\omega^{T}$ being substituted by $\omega$ can be written in the following form, using the converse calculation of the derivation of Lemma 5.7.

$$\begin{aligned} T(\cdots) ={}& (-i)^{k+l} \int \mathrm{d}(x'_1)^0 \cdots \mathrm{d}(x''_{\nu})^0 (\partial/\partial(x'_1)^0) \cdots (\partial/\partial(x''_{\nu})^0) \\ & \times \left\{ \int \mathrm{d}^3\mathbf{x}'_1 \cdots \mathrm{d}^3\mathbf{x}''_{\nu} \omega(T^{\varphi}(\cdots)) \overleftrightarrow{\partial_0} g'_1(x'_1)^* \cdots \overleftrightarrow{\partial_0} g''_{\nu}(x''_{\nu}) \right\}. \end{aligned} \tag{5.95}$$

Considering $\{\cdots\}$ as a function of the time variables, it is sufficient to know its behaviour when a part of the variables tend to $+\infty$, another part of the variables tend to $-\infty$ and the rest of the variables stay finite. Renaming the variables and the corresponding operators as

$$Q^{(1)}(x_1) \cdots Q^{(n)}(x_n)$$

(each $Q^{(j)}$ is either $Q_a'^*$ or $Q_b''$), we will show that $\{\cdots\}$ converges faster than any inverse power of $\lambda$ in the limit $\lambda \to +\infty$, uniformly over $x_j^0 > \lambda$ for $j \le p$, over $(x_j^0) < -\lambda$ for $j > q$ and over

$$\lambda - \delta > x_j^0 > -\lambda + \delta,$$

for all other $j$. Here $\delta$ is the positive number which appears in the property (a) to be satisfied by $\varphi$. If the variable is in this range, then the $\omega$ of (5.95) has the form

$$(\Psi_1, A\Psi_2), \quad A \equiv T^{\varphi}(Q^{(p+1)}(x_{p+1}) \cdots Q^{(q)}(x_q))$$
$$\Psi_1 = \Psi_1(x_1 \cdots x_p) = T^{\varphi}(Q^{(1)}(x_1) \cdots Q^{(p)}(x_p))^* \Omega$$
$$\Psi_2 = T^{\varphi}(Q^{(q+1)}(x_{q+1}) \cdots Q^{(n)}(x_n))\Omega.$$

We estimate $\Psi_1$, $\Psi_2$, and $A$ separately.

First, by the estimate of $\|Q(t,g)\|$ obtained in the first part of the proof of Theorem 5.6,

$$\left\| \int A \overleftrightarrow{\partial_0} g^{(p+1)} \cdots \overleftrightarrow{\partial_0} g^{(q)} \mathrm{d}^3\mathbf{x}_{p+1} \cdots \mathrm{d}^3\mathbf{x}_q \right\|$$
$$= \|T^{\varphi}(Q^{(p+1)}(x_{p+1}^0, \overleftrightarrow{\partial_0} g_{p+1}) \cdots Q^{(q)}(x_q^0, \overleftrightarrow{\partial_0} g_q))\|$$

is bounded by a polynomial of $\lambda$, where $g_k$ is $g_i'$ or $g_j''$ corresponding to $x_k$.

Since $Q_i'$, $Q_i'^*$, $Q_j''$, $Q_j''^*$ satisfy the condition on $Q$ in Theorem 5.6, we obtain the following equation in the limit $\lambda \to \infty$ by successively applying Theorem 5.6 and using the estimate for $\|Q(t, \partial_0 g)\|$:

$$\lim \lambda^N \|\{Q^{(1)}(x_1^0, \overleftrightarrow{\partial_0} g_1)^* \cdots Q^{(p)}(x_p^0, \overleftrightarrow{\partial_0} g_p)^* - \alpha_1^* \cdots \alpha_p^*\}\Omega\| = 0$$
$$\lim \lambda^N \|\{Q^{(q+1)}(x_{q+1}^0, \overleftrightarrow{\partial_0} g_{q+1}) \cdots Q^{(n)}(x_n^0, \overleftrightarrow{\partial_0} g_n) - \beta_{q+1} \cdots \beta_n\}\Omega\| = 0$$

where, depending on whether $Q^{(k)}$ is $Q_a'^*$ or $Q_b''$, we have

$$\alpha_k = -\mathrm{i}(a_{\text{out}}^*, h_a') \quad \text{or} \quad \mathrm{i}(h_b'', a_{\text{out}})$$
$$\beta_k = \mathrm{i}(h_a', a_{\text{in}}) \quad \text{or} \quad -\mathrm{i}(a_{\text{in}}^*, h_b'')$$

Thus, within the approximation up to the order of $|\lambda|^{-N}$, we can make the following substitution:

$$\int \Psi_1 \overleftrightarrow{\partial_0} g_1 \cdots \overleftrightarrow{\partial_0} g_p \mathrm{d}\mathbf{x}_1 \cdots \mathrm{d}\mathbf{x}_p \sim \sum_P B_{P(p)} \cdots B_{P(1)} \varphi_P(x_1^0 \cdots x_p^0)\Omega, \tag{5.96}$$

$$\int \Psi_2 \overleftrightarrow{\partial_0} g_{q+1} \cdots \overleftrightarrow{\partial_0} g_n \mathrm{d}\mathbf{x}_{q+1} \cdots \mathrm{d}\mathbf{x}_n \sim \sum_P C_{P(q+1)} \cdots C_{P(n)} \varphi_P(x_{q+1}^0 \cdots x_n^0)\Omega, \tag{5.97}$$

where depending on whether $Q^{(k)}$ is either $Q_a'^*$ or $Q_b''$

$$\begin{aligned} B_k &= -\mathrm{i}(a^*_{\mathrm{out}}, h'_a) - (a^*_{\mathrm{out}}, h'_a/(2p^0_a))(\partial/\partial x^0_k) \quad \text{or} \\ &\quad \mathrm{i}(h''_b, a_{\mathrm{out}}) - (h''_b/(2p^0_b), a_{\mathrm{out}})(\partial/\partial x^0_k) \\ C_k &= \mathrm{i}(h'_a, a_{\mathrm{in}}) - (h'_a/(2p^0_a), a_{\mathrm{in}})(\partial/\partial x^0_k) \quad \text{or} \\ &\quad -\mathrm{i}(a^*_{\mathrm{in}}, h''_b) - (h''_b/(2p^0_b), a_{\mathrm{in}})(\partial/\partial x^0_k) \end{aligned}$$

In these formulae $(\partial/\partial x^0_k)$ acts on the function $\varphi_p$.

If the $B_k$ mutually commute in eq. (5.96), the product $B_{P(p)} \ldots B_{P(1)}$ does not depend on the permutation $P$. Therefore by $\sum_p \varphi_p = 1$, eq. (5.96) is independent of the time variable and is already the limit vector corresponding to $x^0_1, \ldots, x^0_p \to +\infty$. The $B_k$ are actually not commutative, but since the commutator of two $B$s is proportional to the unit operator $\mathbf{1}$, the $B_k$ become mutually commutative in truncated vacuum expectation values of more than two operators. Therefore if $n > 2$ we may think that eq. (5.96) does not depend on the time variable in $\omega^T$.

Since we may apply exactly the same discussion to (5.97), we conclude that for $n > 2$ the integration over the time rapidly converges uniformly (faster than any power of the time variables). The case $n = 2$ is evident. □

**Remark 1** In Theorem 5.8 we imposed a strong condition on $Q$. The condition that $E_1 Q\Omega = Q\Omega$ or $Q^*\Omega = 0$ is not satisfied by local observables. In order make the theorem applicable to local observables, the following expression can be used in place of $Q(t, \overleftrightarrow{\partial}_0 g)$.

$$\begin{aligned} Q(f_t) &= \int f_t(x) Q(x)\,\mathrm{d}^4 x, \\ f_t(x) &= (2\pi)^{-4} \int \tilde{f}(p) \mathrm{e}^{-\mathrm{i}(p,x)+\mathrm{i}(p^0-\omega(\mathbf{p}))t}\,\mathrm{d}^4 p \end{aligned} \tag{5.98}$$

Here $\omega(\mathbf{p}) = (\mathbf{p}\cdot\mathbf{p} + m^2)^{1/2}$ and $\tilde{f}$ is a $C^\infty$ function with a compact support satisfying the following conditions:

$$\begin{aligned} &\text{support of } \tilde{f} \subset V_+ = \{p;\ (p,p) > 0, p^0 > 0\} \\ &(\text{support of } \tilde{f}) \cap (\text{support of } P^\mu) \subset \{p;\ (p,p) = m^2, p^0 > 0\}. \end{aligned}$$

Actually, if we set $Q' = Q(f_0)$ and if $\hat{g}(\mathbf{p})$ is a $C^\infty$ function with a compact support taking the value 1 if $(\omega(\mathbf{p}), \mathbf{p})$ is in the support of $\tilde{f}$, then we obtain the equality $Q(f_t) = Q'(t, \overleftrightarrow{\partial}_0 g)$ and $Q'$ satisfies conditions $(\alpha)$ and $(\beta')$ of Lemma 5.7. In this case the corresponding one particle state is given by

$$h(\mathbf{p}) = g_Q(\mathbf{p}) \tilde{f}(\omega(\mathbf{p}), \mathbf{p})$$

To obtain the reduction formula for this $Q$, we use a method slightly different from Theorem 5.8. It is based on the following formula. Denoting by $\hat{f}_t(x)$ the function which we obtain when $\tilde{f}(p)(p^0+\omega(\mathbf{p}))^{-1}$ is substituted into $\tilde{f}(p)$ in (5.98), we obtain

$$K_{mx}\hat{f}_t = \mathrm{i}(\partial/\partial t)f_t$$

Using this $\hat{f}_t$, we obtain

$$\begin{aligned}
&((a^*_{\text{out}}, h)\Psi_1, C_1\Psi_2) - (\Psi_1, C_2(h, a_{\text{in}})\Psi_2) \\
&\quad = \mathrm{i}\int \mathrm{d}t \int \tilde{f}_t(x)^* \, \mathrm{d}^4x K_{mx}(\Psi_1, F(C_1, C_2, Q^*, x)\Psi_2) \\
&(\Psi_1, C_2(a^*_{\text{in}}, h)\Psi_2) - ((h, a_{\text{out}})\Psi_1, C_1\Psi_2) \\
&\quad = \mathrm{i}\int \mathrm{d}t \int \tilde{f}_t(x) \, \mathrm{d}^4x K_{mx}(\Psi_1, F(C_1, C_2, Q, x)\Psi_2)
\end{aligned}$$

Here $\Psi_1$, $\Psi_2$, and $F$ are the same quantities as in Lemma 5.7. The discussion concerning the convergence in the limit $t \to \infty$ being also the same, the reduction formula for quasi-local operators also holds for local operators.

**Remark 2** For application, an expression where $K_{mx}$ is brought inside $T^\varphi$ in the reduction formula (5.80) is useful. Its main part is the truncated vacuum expectation value of the time ordered product of the current operators

$$J'_i(x'_i) = K_{mx'_i}Q'_i(x), \quad J''_j(x''_j) = K_{mx''_j}Q''_j(x''_j).$$

Including also the other parts we obtain the following formula

$$\begin{aligned}
&\left(\prod_j K_{mx_j}\right)\omega^T(T^\varphi[Q_1(x_1)\cdots Q_n(x_n)]) \\
&\quad = \omega^T(T^\varphi[\{J_1(x_1) + \overleftarrow{\partial_0}\,(\overleftarrow{\partial_0} + 2\,\overrightarrow{\partial_0})Q_1(x_1)\}\cdots\{J_n(x_n) + \overleftarrow{\partial_0}\,\{\overleftarrow{\partial_0} + 2\,\overrightarrow{\partial_0}\}Q_n(x_n)])
\end{aligned}$$

In this expression, those terms which include at least one $\overleftrightarrow{\partial_0}$ are generally called *equal time commutators.*

**Remark 3** We can also derive the reduction formula by using the non-continuous function $\theta$ (see (5.79)). In this case, the following lemma holds instead of Lemma 5.7 and is used for the derivation.

**Lemma 5.9**
Let $\Psi_1$, $\Psi_2$, $Q$, $\hat{g}$ and $h$ be the same quantities as given in Lemma 5.7. Define $F(C_1, C_2, Q', x)$ as

$$F(C_1, C_2, Q', x) = \begin{cases} Q(x)C_1 & (x^0 > 0) \\ C_2Q(x) & (x^0 < 0) \end{cases}$$

Then the following equations hold

$$((a^*_{\text{out}}, h)\Psi_1, C_1\Psi_2) - (\Psi_1, C_2(h, a_{\text{in}})\Psi_2)$$
$$= \mathrm{i}\int g(x)^*(\Psi_1, F(C_1, C_2, J^*, x)\Psi_2)\,\mathrm{d}^4x$$
$$+ \mathrm{i}\int g(x)^*\,\overleftrightarrow{\partial_0}(\Psi_1, [Q(x)C_1 - C_2Q(x)]\Psi_2)\,\mathrm{d}^3\mathbf{x}|_{x^0=0}$$
$$(\Psi_1, C_2(a^*_{\text{in}}, h)\Psi_2) - ((h, a_{\text{out}}\Psi_1, C_1\Psi_2)$$
$$= \mathrm{i}\int g(x)(\Psi_1, F(C_1, C_2, J, x)\Psi_2)\,\mathrm{d}^4x$$
$$+ \mathrm{i}\int g(x)\,\overleftrightarrow{\partial_0}(\Psi_1, [Q(x)C_1 - C_2Q(x)]\Psi_2)\,\mathrm{d}^3\mathbf{x}\,|_{x^0=0}$$

where $J(x) = K_{mx}Q(x)$.

The proof of this Lemma is obtained by applying the proof of Lemma 5.7 separately to $\int_{-\infty}^0 \mathrm{d}x^0$ and $\int_0^\infty \mathrm{d}x^0$.

By repeated use of this lemma, we can prove the following formulae similarly as in Theorem 5.8:

$$S_c(1\cdots k; 1\cdots l) = \mathrm{i}^{k+1}\int \mathrm{d}^4x'_1\cdots\mathrm{d}^4x''_l g'_1(x'_1)^*\cdots g'_k(x'_k)^*$$
$$\times g''_1(x''_1)\cdots g''_l(x''_l)\tau(1\cdots k; 1\cdots l) \tag{5.99}$$

$$\tau(1\cdots k; 1\cdots l) = \sum \omega^T(T^\theta[J(I_1)\cdots J(I_\nu)]). \tag{5.100}$$

Here $I_j$ is a pair of subsets of $1,\ldots,k$ and $1,\ldots,l$, and a pair of empty sets is excluded. $\{I_1,\ldots,I_\nu\}$ is a partition of $\{(1,\ldots,k),(1,\ldots,l)\}$. Summation is taken over all partitions. $J(I)$ is defined by following equation:

If $I = \{i_1,\ldots,i_a; j_1,\ldots,j_b\}$, for $i_1 < \cdots < i_a$, $j_1 < \cdots < j_b$, then

$$J(I) = L''(j_b)\cdots L''(j_1)L'(i_a)\cdots L'(i_2)J(x'_{i_1})^*$$

$$L'(i)A = [(\partial/\partial x'^0_i) \leftrightarrow Q'_i(x'_i)^*, A]\delta(x'^0_i - x'^0_{i_1}))$$

$$L''(j)A = [(\partial/\partial x''^0_j) \leftrightarrow Q''_j(x''_i), A]\delta(x''^0_j - x'^0_{i_1}))$$

The time derivative acts on $Q$ and $g$, but not on $\delta$. If the part from $(1,\ldots,k)$ is not in $I$ (empty), (the case of $a = 0$), we use $J(x''_{j_1})$ and $x''^0_{j_1}$ instead of $J(x'_{i_1})^*$ and $x'^0_{i_1}$.

Finally we describe the momentum representation of the reduction formula. This representation is used in application. Similarly as in (5.13) we define the distribution $S_c(1 \dots k; 1 \dots l)$ as follows:

$$S_c(h'_1 \cdots h'_k; h''_1 \cdots h''_l) = \int S_c(\mathbf{p}'_1 \cdots \mathbf{p}'_k; \mathbf{p}''_1 \cdots \mathbf{p}''_l) \times h'_1(\mathbf{p}'_1)^* \cdots h'_k(\mathbf{p}'_k)^* h''_1(\mathbf{p}''_1) \cdots h''_l(\mathbf{p}''_l)\, \mathrm{d}\mu(\mathbf{p}'_1) \cdots \mathrm{d}\mu(\mathbf{p}''_k). \tag{5.101}$$

The following notation is also defined in the sense of a distribution by the r.h.s.:

$$\omega(\tilde{T}^{\varphi}(p_1 \cdots p_n; Q_1 \cdots Q_n)\Omega) = (2\pi)^{-4n} \int \mathrm{d}^4x_1 \cdots \mathrm{d}^4x_n \times \exp \mathrm{i}[(p_1, x_1) + \cdots + (p_n, x_n)]\omega(T^{\varphi}(Q_1(x_1) \cdots Q_n(x_n))) \tag{5.102}$$

$\omega^T$ is defined similarly.

**Corollary 5.10**
Consider the following expression:

$$\tilde{\tau}_{\varphi}(p'_1 \cdots p'_k; p''_1 \cdots p''_l) \equiv \prod_i [(p'_i, p'_i) - m^2] \prod_j [(p''_j, p''_j) - m^2] \times \omega^T(\tilde{T}^{\varphi}(p'_k \cdots p'_1 p''_1 \cdots p''_l; Q'^*_k \cdots Q'^*_1 Q''_1 \cdots Q''_l))$$

As independent variables, take $\mathbf{p}'_1, \dots, \mathbf{p}'_k, \mathbf{p}''_1, \dots, \mathbf{p}''_l$ and

$$E'_i \equiv p'^0_i - \omega(\mathbf{p}'_i), \quad E''_j \equiv p''^0_j - \omega(\mathbf{p}''_j) \quad (i = 1, \dots, k, j = 1, \dots, l)$$

For fixed values of $E'_1 \dots E'_k, E''_1 \dots E''_l$ in a neighbourhood of the origin, $\tilde{\tau}_{\varphi}$ is a distribution of $\mathbf{p}'_1, \dots, \mathbf{p}'_k, \mathbf{p}''_1, \dots, \mathbf{p}''_l$ in a neighbourhood of mutually different fixed values of $\mathbf{p}'_1, \dots, \mathbf{p}'_k$ and mutually different fixed values of $\mathbf{p}''_1, \dots, \mathbf{p}''_l$ and it is a $C^{\infty}$ function of $E'_1, \dots, E'_k, E''_1, \dots, E''_l$ (distribution-valued function). In particular, consider this function where the values of all variables $E$ are set to be zero and denote it by $\tilde{\tau}'_{\varphi}$. Then the following formula holds except for $k = l = 1$:

$$S_c(\mathbf{p}'_1 \cdots \mathbf{p}'_l; \mathbf{p}''_1 \cdots \mathbf{p}''_k) = (-2\pi i)^{k+l} \left\{ \prod_i g_{Q'_i}(p'_i)^* \prod_j g_{Q''_j}(p''_j) \right\}^{-1} \times \tilde{\tau}'_{\varphi}(p'_1 \cdots p'_l; -p''_1 \cdots -p''_k). \tag{5.103}$$

The proof follows immediately from Theorem 5.8.

**Remark** $\tilde{\tau}'_{\varphi}$ as well as $S_c$ contain the factor

$$\delta^4(p'_1 + \cdots + p'_k - p''_1 - \cdots - p''_l)$$

which represents the energy-momentum conservation law (where in $S_c$ we put $p'^0_i = \omega(\mathbf{p}'_i), p''^0_j = \omega(\mathbf{p}''_j)$).

Therefore, $\tilde{\tau}'_{\varphi}$ contains the factor

$$\delta(E'_1 + \cdots + E'_k - E''_1 - \cdots - E''_l + \omega(\mathbf{p}'_1) \cdots - \omega(\mathbf{p}''_l))$$

for the $E$ variables. However, after the $\mathbf{p}'_1, \ldots, \mathbf{p}''_l$ integration, it becomes a $C^{\infty}$ function of the $E$ variables.

## 5.6 Analyticity and TCP symmetry

By the reduction formula in the previous section, the connected part $S_c$ of the $S$-matrix can be represented as a distribution of the momenta of incoming and outgoing particles by the Fourier transform (called *Green's Function* below) of the truncated vacuum expectation value (truncated $\tau$ function) of the time ordered product of local operators and hence we can obtain the properties of the $S$-matrix by using the properties of the local observables.

In this section we briefly explain known results about the analyticity (domain of holomorphy) of the Green's functions, the analyticity of the $S$-matrix, and the crossing symmetry and the TCP symmetry of the $S$-matrix. In the derivation of these results, the theory of functions of many complex variables is used extensively. However, only references will be given for details.

*Analyticity of Green's functions*

In the study of the analyticity of the $S$-matrix, the analyticity of the function defined in eq. (5.102) as a distribution of $n$ four-dimensional momenta

$$p = (p_1, \ldots, p_n) \quad p_j = (p_j^0, \mathbf{p}_j) \tag{5.104}$$

becomes fundamental. After extension of $p$ in (5.104) to complex variables, a function $\tilde{r}(p)$ in many complex variables $p$ can be defined such that (5.102) coincides with its boundary values in a sense described later. Since such an $\tilde{r}(p)$ is uniquely determined by the condition that eq. (5.102) coincides with its boundary values, $\tilde{r}(p)$ is called the *analytic continuation* of (5.102). In the following we explain briefly the domain of holomorphy of $\tilde{r}(p)$ and its real points, the boundary values of $\tilde{r}(p)$ and how it can be constructed.

The quantity which has a great influence on the domain of holomorphy of $\tilde{r}(p)$ is the energy–momentum spectrum of the vector

$$Q_j\Omega - (\Omega, Q_j\Omega)\Omega. \tag{5.105}$$

If the scattering states of a particle of mass $m$ form a complete set, (in terms of the notation in Remark 1 of Theorem 5.6, this means $P_{\text{out}} = 1$ or $P_{\text{in}} = 1$), then the energy momentum spectrum of the state (5.105), which is orthogonal to the vacuum, is contained in

$$\{p; (p,p) \geq m^2, p^0 > 0\}. \tag{5.106}$$

As we mentioned briefly in Remark 2 of Theorem 5.8, if we consider the Fourier transform of the expression which is obtained by putting the Klein–Gordon partial differential operator inside the time ordering $T^{\varphi}$, then the term which is obtained by taking the current operator $J$ as the $Q_j$ of the Green's function becomes the main term (where $J(x) = K_{mx}Q(x)$). Inside the domain of holomorphy of this main term obtained from its general properties (as explained below), all other terms turn out to be also holomorphic. Therefore, the study of the domain of holomorphy of $\tilde{r}(p)$, where the $Q_j$ are substituted by $J_j$, becomes important. In this case the spectrum of (5.105) is contained in the set of (5.106) with $m$ replaced by $2m$. The reason is that the $K_{mx}$ in the definition of $J$ is proportional to $(p, p) - m^2$ after Fourier transformation and it annihilates the contribution of the one particle states to (5.105) (namely $E_1 J\Omega = 0$). Below we explain the domain of holomorphy of $\tilde{r}(p)$ for this case.

Due to the translation invariance of $\omega$, $\tilde{r}(p)$ is given by

$$\tilde{r}(p) = (2\pi)^{-4(n-1)} \int \mathrm{d}x_1 \cdots \mathrm{d}x_{n-1} \exp \mathrm{i}[(p_1, x_1) + \cdots + (p_{n-1}, x_{n-1})]$$
$$\times \omega^T(T^{\varphi}(J_1(x_1) \cdots J_{n-1}(x_{n-1})J_n)), \tag{5.107}$$

as a function on the plane

$$\{p; p_1 + \cdots + p_n = 0\} \equiv H. \tag{5.108}$$

In the expression of $S_c(\mathbf{p}_1, \ldots, \mathbf{p}_k; \ -\mathbf{p}_{k+1}, \ldots, -\mathbf{p}_n)$ it appears in the form of a $\delta$ function $\delta^4(\Sigma_j p_j)\tilde{r}(p)$ (with all other terms). Here $\delta^4(\Sigma p_j)$ is the product of Dirac $\delta$ functions of each component of $\Sigma p_j$ and represents the energy-momentum conservation.

Let us introduce a notation necessary for the description of the domain of holomorphy: convex cones $C_i$ and $V_i$ determined by energy momenta of subsystems and the set $S_{ij}$ of all points of contact of two adjacent convex cones $C_i$ and $C_j$.

For a non-trivial subset $I$ of $\{1, \ldots, n\}$ we define

$$p(I) = \sum_{j \in I} p_j. \tag{5.109}$$

(For the complement $I^c$ of $I$ we have $p(I^c) = -p(I)$ on $H$.) In particular, if we look at the energy components

$$H^0 = \{p^0 = (p_1^0 \cdots p_n^0);\ p_1^0 + \cdots + p_n^0 = 0\} \tag{5.110}$$

then the hyperplanes

$$H^0(I) = \{p^0 \in H^0; p^0(I) = \sum_{j \in I} p_j^0 = 0\}$$

with all possible $I$ divide $H^0$ into convex polyhedral cones. Labelling these convex polyhedral cones by an index $i$ and denoting them by $C_i$, we can describe each $C_i$ as

$$C_i = \{p^0 \in H^0; \sigma_i^I p^0(I) \geq 0, \forall I\} \tag{5.111}$$

in terms of the sign $\sigma_i^I = \pm 1$, indicating on which side of the hyperplane $H^0(I)$ the cone $C_i$ is located. We define the corresponding four-dimensional open convex cone as

$$V_i = \{p \in H; \sigma_i^I p(I) \in V_+, \forall I\}, \tag{5.112}$$

where $V_+$ is the future cone of the energy-momentum spectrum which appears in Theorem 4.5 (2). (Note that it is an open set here.)

If for two convex polyhedral cones $C_i$ and $C_j$ $(i \neq j)$ $\sigma_i^I = \sigma_j^I$ holds for $I \neq I_{ij}$ (namely $C_i$ and $C_j$ are on the same side of $H^0(I)$) excluding one $I_{ij}$, then $C_i$ and $C_j$ are adjacent across the hyperplane $H^0(I_{ij})$ in between. In this case

$$S_{ij} = \{p \in H; \sigma_i^I p(I) \in V_+, (\forall I \neq I_{ij}), p(I_{ij}) = 0\}. \tag{5.113}$$

This is a relative open subset of $\bar{V}_i \cap \bar{V}_j$.

**Theorem 5.11**

1. The hyperfunction $\tilde{r}(p)$ defined by (5.107) on the hyperplane $H$ of eq. (5.108) is a boundary value of the function $\tilde{r}(\zeta)$ of many complex variables on the complexification

$$H + iH = \{\zeta = (\zeta_1 \cdots \zeta_n); \zeta_j \in \mathbf{C}^4, \zeta_1 + \cdots + \zeta_n = 0\} \tag{5.114}$$

of $H$, given by

$$\tilde{r}(p) = \lim_{\varepsilon \to +0} \tilde{r}(p + i\varepsilon p). \tag{5.115}$$

2. $\tilde{r}(\zeta)$ is holomorphic in (a neighbourhood of) each point of the following sets

$$\mathscr{T}_i \equiv \{p + iq; q \in V_i\}, \tag{5.116}$$

$$\sum_{ij}(2m) = \{p + iq; q \in S_{ij}, p(I_{ij})^2 < (2m)^2\}, \tag{5.117}$$

where $p(I)^2$ represents $(p(I), p(I))$.

3. $\tilde{r}(\zeta)$ is holomorphic at the following real points:

$$\{p \in H; p(I)^2 < (2m)^2 \quad (\forall I)\}. \tag{5.118}$$

**Remark 1** For a holomorphic function in many complex variables there appears a phenomenon such that all functions which are holomorphic in a certain connected domain $D$ are also holomorphic in a larger domain $\tilde{D}$ ($\supset D$). The maximal such $\tilde{D}$ is called the *envelope of holomorphy* of $D$ [7]. (For one variable, $\tilde{D} = D$). If we also find out for $\tilde{r}(p)$ the envelope of holomorphy of the union (which is a connected set) of all $V_i$ of (5.112) and all $\Sigma_{ij}$ of (5.116) (for all pairs $(i,j)$ for which $C_i$ and $C_j$ are adjacent), then knowledge about the domain of holomorphy of $\tilde{r}(p)$ is increased. For $n = 2$ the envelope of holomorphy is known. Also for each pair of adjacent $C_i$, $C_j$, the envelope of holomorphy of the triple $V_i$, $V_j$, $S_{ij}$ can be discussed analogously to the case $n = 2$.

**Remark 2** By the last comment given in Remark 1, in a neighbourhood of any given real point $p$ satisfying $p(I_{ij})^2 < (2m)^2$, a holomorphy neighbourhood of $\Sigma_{ij}(2m)$ can be chosen to be of a uniform size inside the region of $\sigma_i^I q(I) \in V_+ (I \neq I_{ij})$. Then we can conclude that $\tilde{r}(p)$ is holomorphic in a neighbourhood of any relative inner point of the intersection $\cap\overline{\Sigma}_{ij}(2m)$ of the closure of some of $\Sigma_{ij}(2m)$. (3) of the theorem is a special case where one takes the intersection over all possible $ij$.

**Remark 3** The boundary value (5.115) means

$$\int \tilde{r}(p)f(p)\,\mathrm{d}p = \lim_{\varepsilon\to+0} \int \tilde{r}(p + \mathrm{i}\varepsilon p)f(p)\,\mathrm{d}p \tag{5.115$'$}$$

for any $C^\infty$-function $f$ with compact support ($\mathrm{d}p = \mathrm{d}^4p_1 \cdots \mathrm{d}^4p_{n-1}$). The l.h.s. denotes $\tilde{r}(f)$ as a distribution.

If $p(I)$ is a timelike vector ($p(I)^2 > 0$) for all $I$, then $p$ belongs to one of $V_i$ and the $p + \mathrm{i}\varepsilon p$ on the r.h.s. is contained in the domain of holomorphy of $\tilde{r}$. In particular, if the support of $f$ is in $V_i$, then

$$\int \tilde{r}(p)f(p)\,\mathrm{d}p = \lim \int \tilde{r}(p + \mathrm{i}\varepsilon q)f(p)\,\mathrm{d}p \quad (q \in V_i) \tag{5.119}$$

holds and the limit is uniform over $q$ in a compact subset of $V_i$. If $p(I)$ is spacelike for $I = I_{ij}$ and timelike for $I \neq I_{ij}$, then $\tilde{r}(p + \mathrm{i}\varepsilon q)$ is holomorphic for $q$ belonging to the union of $V_i$, $V_j$ and a neighbourhood of $\Sigma_{ij}(2m)$, and

$$\tilde{r}(p) = \lim_{\varepsilon\to+0} \tilde{r}(p + i\varepsilon q) \tag{5.120}$$

holds in the sense of (5.119). Also in the more general case where several $p(I)$ are spacelike, the r.h.s. of (5.120) is holomorphic for $q$ in a correspondingly wider domain and (5.120) is satisfied.

**Remark 4** The r.h.s. of (5.119) exists without any restriction on the support of $f$ and it determines the distribution

$$\tilde{r}_i(p) = \lim_{\varepsilon\to+0} \tilde{r}(p + i\varepsilon q) \quad (q \in V_i). \tag{5.121}$$

In the same sense as (5.107), $\tilde{r}_i(p)$ is the Fourier transform of the following function:

$$r_i(x) = \sum_P \varphi^i_P(x^0)\omega(J_{P(1)}(x_{P(1)}) \cdots J_{P(n)}(x_{P(n)})) \qquad (5.122)$$

where $x^0 = (x^0_1, \ldots, x^0_n)$ and $\varphi^i_P$ is obtained from $\varphi_P$ (used to define the time $\varphi$ ordered product $T^\varphi$ in (5.77)) as follows. The Fourier–Laplace transform $\tilde{\varphi}_P(\zeta^0)$ of $\varphi_P$ is a rational function and it is independent of $P$ except for the sign $\pm 1$. The inverse Fourier transform of the boundary value $\tilde{\varphi}^i_P(p^0)$ of $\tilde{\varphi}_P(p^0 + i\varepsilon q^0)$ for $q_0 \in C_i$ as $\varepsilon \to +0$ is $\varphi^i_P(x^0)$. Note that $\omega$ in (5.122) can be replaced by $\omega^T$ without changing $r_i(x)$, i.e. the difference of $\omega$ and $\omega^T$ cancels out after the summation.

The function $r_i(x)$ defined by (5.122) is called a *generalized retarded function* or simply *r-function* (after multiplying $(-i)^n$). In particular for the case where $\varphi_P$ is taken to be the discontinuous function $\theta_P$, the $r_{i_R}$ corresponding to $i \equiv i_R$ satisfying

$$\sigma^I_i = +1 \text{ for } 1 \in I, \quad \sigma^I_i = -1 \text{ for } 1 \notin I,$$

can be represented by the summations of the vacuum expectation value of the multiple commutators over permutations $P$ which do not change 1 ($P(1) = 1$):

$$\begin{aligned} &r_{i_R}(x_1; x_2, \ldots, x_n) \\ &\quad = \sum \theta_P(x^0)\omega([\cdots[[A_1(x_1), A_{P(2)}(x_{P(2)})], A_{P(3)}(x_{P(3)})]\cdots A_{P(n)}(x_{P(n)})]). \quad (5.123)\end{aligned}$$

This is the retarded function used in the LSZ theory (up to the overall constant $(-i)^n$) [8].

From the expression (5.123), $r_{i_R}$ is seen to vanish unless $x^0_1 \geq x^0_j$ for all $j$. The support of a general $r_i(x)$ is likewise restricted to those $x$ where $x^0$ is inside the polar set of $C_i$. Also for adjacent $C_i$ and $C_j$ and for $P(I_{ij})^2 < (2m)^2$, $\tilde{r}_i(p) = \tilde{r}_j(p)$. Furthermore, if the $I'$ and $I''$ and their complements do not have any inclusion relation and, for the adjacent four cones $C_{i(\sigma'\sigma'')}(\sigma' = \pm, \sigma'' = \pm)$

$$\sigma^{I'}_{i(\sigma',\sigma'')} = \sigma'\sigma^{I'}_i, \quad \sigma^{I''}_{i(\sigma',\sigma'')} = \sigma''\sigma^{I''}_i,$$

$$\sigma^{I'}_{i(\sigma',\sigma'')} = \sigma^I_i \quad \text{for } I \neq I', I'' \qquad (i = i(+,+))$$

holds, then

$$r_{i(+,+)} + r_{i(-,-)} = r_{i(+,-)} + r_{i(-,+)}$$

This is called the *Steinmann identity* [9].

**Proof of Theorem** First defining $r_i$ by eq. (5.122), we examine the support of $r_i(x)$ mentioned in Remark 4. Using that result about the support we can show that the Fourier–Laplace transform $\tilde{r}_i(\zeta)$ defines a holomorphic function in $\mathscr{T}_i$. Then, showing that for adjacent $V_i$ and $V_j$ the $\tilde{r}_i(p)$ and $\tilde{r}_j(p)$ coincide for $p(I_{ij})^2 < (2m)^2$ as was mentioned in Remark 4 and applying the edge of the wedge theorem [10], we see that $\tilde{r}_i(\zeta)$ and $\tilde{r}_j(\zeta)$ are the same one analytic function which is also holomorphic on $\Sigma_{ij}(2m)$. The relations between $r_i(x)$ and $\omega^T(T^\varphi(\cdots))$ in eq. (5.107) can be seen from the relations between $\varphi'_P$ and $\varphi_P$ and we obtain (5.115). We omit the details of the proof [11]. □

Note that there are also Steinmann's formulation using multiple commutators as a main tool [12] and Ruelle's formulation using the idea of cohomology [13].

*Lorentz invariance of the S-matrix and TCP symmetry*

The unitary operators $U(a, \Lambda)$ in Theorem 5.4 (iv) and (vi) are obtained from the $\mathscr{P}_+^\uparrow$ invariance of the vacuum state as described in the beginning of Section 5.3, and are common for $\Psi^{\text{out}}$ and $\Psi^{\text{in}}$. Therefore

$$\begin{aligned}
&(\Psi^{\text{out}}[h'_1 \cdots h'_k], \Psi^{\text{in}}[h''_1 \cdots h''_l]) \\
&\quad = (U(a,\Lambda)\Psi^{\text{out}}[\cdots], U(a,\Lambda)\Psi^{\text{in}}[\cdots]) \\
&\quad = (\Psi^{\text{out}}[(a,\Lambda)h'_1 \cdots (a,\Lambda)h'_k], \Psi^{\text{in}}[(a,\Lambda)h''_1 \cdots (a,\Lambda)h''_l]).
\end{aligned}$$

The invariance under translations $(a, \mathbf{1})$ among the above relations gives a factor

$$\delta^4(p'_1 + \cdots + p'_k - p''_1 - \cdots - p''_l)$$

corresponding to the energy-momentum conservation law for the $S$-matrix as we have already mentioned, where $p_i'^0 = \omega(\mathbf{p}'_i)$ and $p_i''^0 = \omega(\mathbf{p}''_i)$. On the other hand, the Lorentz transformation $\Lambda$ gives rise to the *Lorentz invariance of the S-matrix* in the following theorem.

**Theorem 5.12**
For all $\Lambda \in \mathscr{L}_+^\uparrow$ the following equation holds:

$$S(\mathbf{p}'_1 \cdots \mathbf{p}'_k; \mathbf{p}''_1 \cdots \mathbf{p}''_l) = S(\mathbf{\Lambda p}'_1 \cdots \mathbf{\Lambda p}'_k; \mathbf{\Lambda p}''_1 \cdots \mathbf{\Lambda p}''_l)$$

where $\mathbf{\Lambda p}$ is defined in terms of $\mathbf{p}$ and $\mathbf{\Lambda}$ as follows. We write the spatial part of $\Lambda p$ as $\mathbf{\Lambda p}$ where $\Lambda p$ is obtained by application of the Lorentz transformation $\Lambda$ on the four vector $p = (p^0, \mathbf{p})$ with $p^0 = \omega(\mathbf{p})$.

Using the above results the *TCP invariance of the S-matrix* is proved by Epstein [14].

**Theorem 5.13**
The following equation holds:

$$S(\mathbf{p}'_1 \cdots \mathbf{p}'_k; \mathbf{p}''_1 \cdots \mathbf{p}''_l) = S(\mathbf{p}''_1 \cdots \mathbf{p}''_l; \mathbf{p}'_1 \cdots \mathbf{p}'_k). \tag{5.124}$$

**Remark 1** The above equation represents the symmetry under the exchange of incoming and outgoing particles. Actually in a more general case we have to change particles to antiparticles and antiparticles to particles at the same time. Here we consider only one kind of particle. Therefore, automatically particles and antiparticles coincide and thus we obtain the above form of the theorem.

Particles and antiparticles are defined as follows. For a local observable $Q$, assume that the state (denoted by $E_{1m}Q\Omega$) obtained by projecting $Q\Omega$ to the energy-momentum spectrum $p^2 = m^2$ be a particle with mass $m$. Then we obtain the JLD representation of both $\omega(QQ^*)$ and $\omega(Q^*Q)$ in terms of the same measure $\mu$ as defined from $\omega([Q^*, Q(x)])$ by (4.19). Thus, we see that $E_{1m}Q^*\Omega \neq 0$ and this represents the antiparticle state. Therefore, a particle with momentum $\mathbf{p}'_i$ on the r.h.s. of (5.124) is an antiparticle with momentum $\mathbf{p}'_i$ on the l.h.s. and the same is true for $\mathbf{p}''_j$.

**Remark 2** If the scattering states form a complete system, i.e. if $\mathscr{H}^{\text{out}}$ and $\mathscr{H}^{\text{in}}$ coincide with $\mathscr{H}_\omega$, then (5.124) is equivalent to the existence of an involutive anti-unitary operator $\Theta$ ("involutive" means $\Theta^2 = 1$) satisfying (and uniquely defined by) the following equation:

$$\begin{aligned} \Theta\Psi^{\text{out}}[h_1 \cdots h_n] &= \Psi^{\text{in}}[\bar{h}_1 \cdots \bar{h}_n], \\ \Theta\Psi^{\text{in}}[h_1 \cdots h_n] &= \Psi^{\text{out}}[\bar{h}_1 \cdots \bar{h}_n]. \end{aligned} \tag{5.125}$$

Here $\bar{h}$ is the complex conjugate of $h$. Actually, (5.124) can be obtained from

$$(\Theta\Psi, \Theta\Phi) = (\Phi, \Psi)$$

by substituting $\Psi = \Psi^{\text{in}}[\ldots]$ and $\Phi = \Psi^{\text{out}}[\ldots]$.

**Remark 3** The existence of $\Theta$ has been first derived by Jost [15] within the framework of quantum field theory from the basic postulates of the theory as an operator implementing space–time inversion ($x \to -x$) and charge conjugation (particle $\rightleftharpoons$ antiparticle). Since it changes the sign of the time it interchanges "in" and "out", it does not change the momentum $\mathbf{p}$ which is proportional to the velocity $\mathrm{d}x/\mathrm{d}t$ (since $x \to -x$, $t \to -t$), and it interchanges particle and antiparticle, thus leading to (5.125). If we translate the properties of $\Theta$ in quantum field theory to the language of the local observables, we obtain

$$\Theta\mathfrak{A}(D)\Theta = \mathfrak{A}(-D). \tag{5.126}$$

However, it is not known whether the operator $\Theta$ in Remark 2 has this property in general. Note that under the assumption of completeness, the $\Theta$ in Remark 2 satisfies

$$\Theta U(a, \Lambda)\Theta = U(-a, \Lambda), \quad \Theta\Omega = \Omega. \tag{5.127}$$

Since $\Theta$ is an antilinear operator ($\Theta \mathrm{i} = -\mathrm{i}\Theta$), it commutes with the energy momentum $P^\mu$ in spite of (5.127).

Concerning the abbreviation, TCP stands for time reversal (T), charge conjugation (C), and parity (P) and it is sometimes called PCT or CTP or in some other order.

**Remark 4** Space–time inversion $x \to -x$ belongs to the full Lorentz group, but not to the restricted Lorentz group $\mathscr{L}_+^\uparrow$. On the other hand, in the complex Lorentz group obtained by complexification of the Lorentz group $\mathscr{L}_+^\uparrow$, the space–time inversion is in the connected component of the unit element along with $\mathscr{L}_+^\uparrow$.

Therefore Jost derived an operator $\Theta$, which represents the symmetry under space–time inversion (TP), from the $\mathscr{L}_+^\uparrow$ invariance and analyticity of quantum field theory. However, it is accompanied by complex conjugation (in order to change the direction of the time translation $\exp \mathrm{i}tP^0$ and at the same time to keep the energy $P^0$ positive, the change $\mathrm{i} \to -\mathrm{i}$ is unavoidable because it is impossible to make $P^0 \to -P^0$). As a result the symmetry including charge conjugation $C$ is obtained.

**Outline of the proof of Theorem 5.13** Rewrite the pure Lorentz transformation in direction of the $x$-axis

$$
\begin{aligned}
x_0 &\to (\cosh \chi)x_0 + (\sinh \chi)x_1 \\
x_1 &\to (\sinh \chi)x_0 + (\cosh \chi)x_1
\end{aligned}
$$

as $\Lambda_\lambda$ with a complex parameter $\lambda = \exp \chi$ and regard the function $\tilde{r}(\Lambda_\lambda p)$ as a function of $\lambda$ with its value given as a hyperfunction of $p$. Then $\tilde{r}$ becomes holomorphic in $\lambda \in \mathbf{C}/\{0\}$ and $p \in \mathscr{T}_i$ and, in the limit of $\operatorname{Im} p$ tending to 0 within $V_i$, converges to $r_i$ for $\lambda = 1$, and converges to $r_{i'}$ corresponding to $-C_i = C_{i'}$ for $\lambda = -1$. From these relations and the Lorentz invariance of the $S$-matrix, Theorem 5.13 follows. For the details of proof, see the reference of Epstein given above.

*The crossing symmetry and analyticity of the S-matrix*

The analyticity of the $S$-matrix elements $S_c(\mathbf{p}_1, \mathbf{p}_2, -\mathbf{p}_3, -\mathbf{p}_4)$ from two particles to two particles is well studied and the same result as for the case of the Wightman field is obtained. To describe it briefly, we first define suitable independent variables.

There are the following four constraints for the 12 real variables $\mathbf{p}_1, \mathbf{p}_2, \mathbf{p}_3, \mathbf{p}_4$:

$$
\begin{aligned}
&p_1 + p_2 + p_3 + p_4 = 0 \\
&(p_1^0 = \omega(\mathbf{p}_1),\ p_2^0 = \omega(\mathbf{p}_2),\ p_3^0 = -\omega(\mathbf{p}_3),\ p_4^0 = -\omega(\mathbf{p}_4))
\end{aligned} \tag{5.128}
$$

Furthermore, by Theorem 5.12, the $S$-matrix is invariant under the group $\mathscr{L}_+^\uparrow$ with six parameters. Thus, we can treat it as a function of two independent variables. Normally we take the following three Lorentz invariant variables. Due to a linear relation among them there are effectively only two variables:

$$
\begin{aligned}
s &= (p_1 + p_2)^2 = (p_3 + p_4)^2, \\
t &= (p_1 + p_3)^2 = (p_2 + p_4)^2, \\
u &= (p_1 + p_4)^2 = (p_2 + p_3)^2,
\end{aligned} \tag{5.129}
$$

$$
s + t + u = 4m^2. \tag{5.130}
$$

If $p_1$ and $p_2$ are incoming particles, $s$ is a real number satisfying $4m^2 \leq s$, $\sqrt{s}$ representing the total energy (including the mass) in the centre of mass system and, for a fixed $s$, $t$ is a real number satisfying $0 \geq t \geq 4m^2 - s$, $\sqrt{t}$ representing the momentum transfer between the particles in the centre of mass system. The real values of $s, t, u$ in the domain $s \geq 4m^2$, $t \leq 0$, $u \leq 0$ satisfying $s + t + u = 4m^2$ are called *physical points* of the 1–2 system. Interchanging the role of $s, t, u$ we also obtain the physical points of the 1–3 system as well as of the 1–4 system. For example, for a fixed $u \leq 0$ the real values $s$ satisfying $s \geq 4m^2 - u$ are the physical points of the 1–2 system, and the real values $s$ satisfying $s \leq u$ become the physical points of the 1–3 system.

By the investigations of the domain of holomorphy of $\tilde{r}(p)$ [16], it is known that there exists an analytic function $F(s, t)$ of $s$ and $t$ (also of $u$) which is holomorphic in a sufficiently small neighbourhood of each physical point, except for the zeroes of certain analytic functions of $s, t, u$, and the real points of the energy variable ($\mathrm{Im}\, s = 0$ for $s \geq 4m^2$). Then the $S$-matrix is given by a boundary value as

$$\lim_{\varepsilon \to +0} F(s + \mathrm{i}\varepsilon(s - m^2), t). \tag{5.131}$$

For the 1–2 system, take $p_1^0$ and $p_2^0$ positive, $p_3^0$ and $p_4^0$ negative, and compute $s \geq 4m^2$, $0 \geq t \geq 4m^2 - s$ using (5.129). Then (5.131) gives $S_c(\mathbf{p}_1, \mathbf{p}_2, -\mathbf{p}_3, -\mathbf{p}_4)$. For the 1–3 system, take $p_1^0$ and $p_3^0$ positive, $p_2^0$ and $p_4^0$ negative, and compute $t \geq 4m^2$, $0 \geq s \geq 4m^2 - t$ by using (5.129). Then $S_c(\mathbf{p}_1, \mathbf{p}_3, -\mathbf{p}_2, -\mathbf{p}_4)$ is given by the same (5.131). Similarly for the 1–4 system.

Furthermore, $F(s, t)$ is known to be holomorphic in $\mathrm{Im}\, s \neq 0$ and $|s| > R(u)$, if we fix any $u < 0$ [17]. Namely, in the complex plane of $s$, $F(s, t)$ is holomorphic outside a circle with radius $R(u)$ and except for the real axis. Combining this with the holomorphy results in a neighbourhood of the physical points, the $S$-matrix of the 1–2 system and of the 1–3 system are related by analytic continuation. This property is called the *crossing symmetry of the S-matrix*. We can represent this graphically as in Fig. 5.1. In general, the process with incoming particles 1 and 2 and outgoing particles 3 and 4 is related to the process with an incoming pair, particle 1 and antiparticle $\overline{3}$, and an outgoing pair, antiparticle $\overline{2}$ and particle 4. Exactly the same thing can be proved for $F(s, t)$ considered as a function of $s$ with $t < 0$ being fixed.

Furthermore, for a fixed $u < 0$ (or a fixed $t < 0$), it is proved that $F(s, t)$ increases at most polynomially in $s$ as $s$ tends to $+\infty$ [18].

These results give the so-called dispersion relations if we add the holomorphy in a finite region (for example for $s \leq R(t)$ with a fixed $t$) [19].

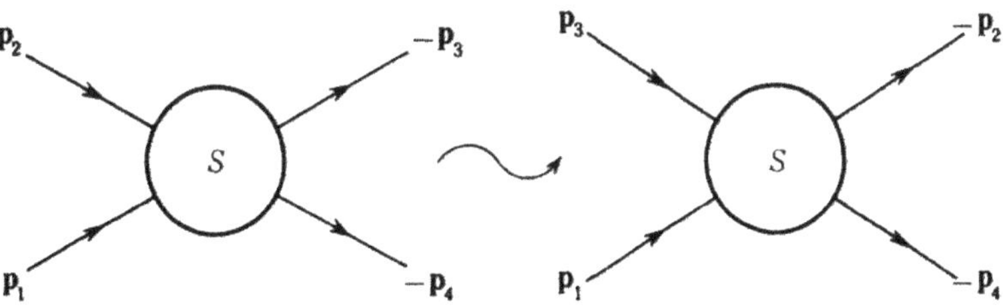


FIG 5.1 Crossing symmetry.

# 6
# SECTOR THEORY

In the previous sections we treated the irreducible representation space associated with the vacuum state. In this chapter we treat other representation spaces necessary to describe particle systems with the vacuum as a background. Such representation spaces are characterized as transportable localized excitations on the vacuum background. If we assume Haag's duality of the vacuum representation, such a localized excitation is described by a localized endomorphism of the observables and its equivalence classes correspond to different representations called *sectors* (eigensubspaces of the superselection rules in each of which unrestricted superposition principle holds). From the discussion about composition and conjugation of localized endomorphisms, we define the statistics (of a sector) which represents the behaviour of more than two equivalent localized excitations under permutation, and we derive the connection of spin and statistics. We also introduce the lattice of local field algebras as operators intertwining different sectors, and the gauge group acting on them. Then the local observable algebras are the gauge invariant part of the local field algebras and sectors are inequivalent representations of the gauge group. The local field algebras can be chosen to satisfy the standard commutation or anticommutation relations between space-like separated regions and we can derive the connection of spin and statistics of sectors. In this way we obtain a deep understanding of the reason why Fermi fields in spatially separated regions anticommute.

## 6.1 Superselection rules and localized excitations

The theory developed in the previous chapter describes the features of a finite number of particles scattered by mutual interaction with the vacuum as a background, where we take the pure vacuum state $\omega$ as a starting point and consider the representation space $\mathscr{H} = \mathscr{H}_\omega$ associated with $\omega$. A premise for this description is the existence of a subspace $\mathscr{H}_1$ of $\mathscr{H}$ formed by one particle states (i.e. one particle states are superposable with the vacuum state).

On the other hand in Section 3.4, stable particles (we consider so-called bound states of several particles also as stable particles) are formulated as irreducible representations of $\tilde{\mathscr{P}}_+^\uparrow$ and we gave a classification in terms of their mass $m$ and spin $j$. In the special case of half-integer spin (for example the spin of an electron is $1/2$), we have a double-valued representation of the relativistic transformation group which is the inhomogeneous Lorentz group $\mathscr{P}_+^\uparrow$, and the univalence superselection rule $u$ introduced in Section 3.3 is $-1$. (Univalence means single-valuedness. $u$ is called by this name because it distinguishes single-valued and double-valued representations. We may also call it the single-valuedness superselection rule). Since $u$ represents a rotation by 360°, it does not change observables or states. For a representation $\pi$ of observables, the representing operator $\pi(A)$ of an observable $A$ commutes with $u$. If the unit vectors $\Phi_1$ and $\Phi_2$ in this representation space satisfy

$$u\Phi_1 = \Phi_1, \quad u\Phi_2 = -\Phi_2, \tag{6.1}$$

then not only the state

$$\omega_j(A) = (\Phi_j, \pi(A)\Phi_j) \qquad (j = 1, 2,\ A \in \mathfrak{A}), \tag{6.2}$$

is invariant under (6.1) which represents a rotation by 360 degrees, but the state

$$\omega(A) = (\Phi, \pi(A)\Phi)$$

given by a unit vector

$$\Phi = \alpha\Phi_1 + \beta\Phi_2 \quad (0 < |\alpha| < 1,\ |\alpha|^2 + |\beta|^2 = 1) \tag{6.3}$$

which is not invariant under $u$, is also invariant under the transformation (6.1), due to $[u, \pi(A)] = 0$ and $u^* u = 1$, and is simply a mixture

$$\omega = |\alpha|^2\omega_1 + |\beta|^2\omega_2 \tag{6.4}$$

of $\omega_1$ and $\omega_2$.

Generally, if two states $\omega_1$ and $\omega_2$, satisfy the condition that in any representation space which contains vectors $\Phi_1$ and $\Phi_2$ representing $\omega_1$ and $\omega_2$, any linear combination (6.3) of $\Phi_1$ and $\Phi_2$ represents only a mixture (6.4), then $\omega_1$ and $\omega_2$ are said to be *not superposable* (*not coherent*) [1]. A selfadjoint (or normal) operator $A$ on the representation space of the observables is called a *superselection rule* if any two states $\varphi_1$ and $\varphi_2$ satisfying the following condition are not superposable. Condition: $\varphi_1$ and $\varphi_2$ are vector states given by the vectors $\Phi_1$ and $\Phi_2$ with disjoint $A$ spectrum. Superselection rules necessarily commute with the representing operators of observables. In the above example, $\omega_1$ and $\omega_2$ are not superposable and $u$ is a superselection rule.

**Theorem 6.1**
A necessary and sufficient condition for two states $\omega_1$ and $\omega_2$ to be not superposable is the disjointness of the associated GNS representations $\pi_{\omega_1}$ and $\pi_{\omega_2}$. (We refer to Section 2.3 for the GNS representation, and to Section 2.2 for disjoint representations.)

**Proof** Given vectors $\Phi_1$ and $\Phi_2$ in a representation space $\mathscr{H}_\pi$ of a representation $\pi$ giving states $\omega_1$ and $\omega_2$ by eq. (6.2). Denote by $E_j$ the projection operator to the subspace

$$\mathscr{H}_j \equiv \text{closure of } \pi(\mathfrak{A})\Phi_j \quad (j = 1, 2) \tag{6.5}$$

generated from each $\Phi_j$. Then $E_j \in \pi(\mathfrak{A})'$. Since we can regard the representations $\pi_1$, $\pi_2$ associated with the states $\omega_1$, $\omega_2$ as the restriction of $\pi$ to the corresponding subspaces $\mathscr{H}_j$, the condition that $\pi_1$ and $\pi_2$ are disjoint is equivalent to

$$E_2\pi(\mathfrak{A})'E_1 = 0. \tag{6.6}$$

(This is because the intertwining map $U$ from $\mathscr{H}_1$ to $\mathscr{H}_2$ can be identified with the operator $E_2UE_1$ in $\pi(\mathfrak{A})'$.)

First, if $\pi_1$ and $\pi_2$ are disjoint then $E_1E_2 = E_2E_1 = 0$. Hence $\pi(A)\Phi_1 \in \mathscr{H}_1$ and $\Phi_2 \in \mathscr{H}_2$ are orthogonal for any $A \in \mathfrak{A}$ and

$$(\Phi_2, \pi(A)\Phi_1) = 0, \quad (\Phi_1, \pi(A)\Phi_2) = (\pi(A^*)\Phi_1, \Phi_2) = 0.$$

Therefore eq. (6.4) is satisfied. Since this is the case for any $\pi$, $\omega_1$, and $\omega_2$ are not superposable.

Next, let us consider the case where $\pi_1$ and $\pi_2$ are not disjoint. Since any element of a von Neumann algebra can be written as a linear combination of four unitary operators in the same algebra, there is at least one unitary operator $U$ in $\pi(\mathfrak{A})'$ satisfying $E_1UE_2 \neq 0$. Consequently, the vectors $\Psi_1$ and $\Psi_2$ satisfying

$$(\Psi_2, U\Psi_1) \neq 0 \qquad (\Psi_1 \in \mathscr{H}_1, \Psi_2 \in \mathscr{H}_2)$$

exist and therefore by eq. (6.5) there exist observables $A_1$ and $A_2$ for which

$$(\pi(A_2)\Phi_2, U\pi(A_1)\Phi_1) \neq 0.$$

In this case, putting $\Phi_1' = U\Phi_1$, we obtain

$$(\Phi_1', \pi(A)\Phi_1') = (U\Phi_1, \pi(A)U\Phi_1) = (\Phi_1, \pi(A)\Phi_1) = \omega_1(A)$$

due to $U \in \pi(\mathfrak{A})'$ and $U^*U = 1$, and hence $\Phi_1'$ also represents $\omega_1$. Furthermore, since

$$\begin{aligned}(\Phi_2, \pi(A_2^*A_1)\Phi_1') &= (\pi_2(A_2)\Phi_2, \pi(A_1)U\Phi_1)\\ &= (\pi_2(A_2)\Phi_2, U\pi(A_1)\Phi_1) \neq 0,\end{aligned} \tag{6.7}$$

the vector state $\omega$ given by $\Phi = \alpha\Phi_1' + \beta\Phi_2$ has the property that

$$\omega(A) - |\alpha|^2\omega_1(A) - |\beta|^2\omega_2(A) = \mathrm{Re}\,\{c(\Phi_2, \pi(A_2^*A_1)\Phi_1')\}$$

with $A = cA_2^*A_1$ does not vanish by (6.7), for example for $c = (\Phi_2, \pi(A_2^*A_1)\Phi_1')^*$, and hence eq. (6.4) does not hold. Therefore, $\omega_1$ and $\omega_2$ are superposable. □

On the other hand, since $\pi(\mathfrak{A})' = \mathbf{C}1$ for an irreducible representation $\pi$, the only superselection rule is given by a constant multiple of the identity operator (trivial case). For any two different vectors $\Phi_1$, $\Phi_2$ of the representation space, the state $\Phi$ given by (6.3) represents a different state for a different value of $\alpha/\beta$. (Such representation spaces are called *coherent*.)

In a representation associated with a Lorentz invariant vacuum state, a single-valued representation of $\mathscr{P}_+^\uparrow$ is obtained due to the $\mathscr{P}_+^\uparrow$ invariance of the vacuum state (see Theorem 2.33), and the univalence superselection rule is $\mathbf{1}$. In particular, in this space there are no vectors representing particles with half-odd integer spin. In order to describe such particles it is necessary to consider a representation disjoint from the vacuum representation along with the latter. This is the purpose of the present section.

Generally there is an infinite number of disjoint representations. We can imagine that, for example, the representation associated with an equilibrium state of a gas of particles uniformly distributed over an infinite space is disjoint from the vacuum representation. The question is which representation among these various possibilities is suitable for treating the states of a finite number of particles in vacuum. In the following we first discuss the selection criteria of such a representation.

In this book we consider representations which satisfy the following two criteria of the Doplicher–Haag–Roberts analysis [2].

The first criterion says that the representation space represents localized excitations of the vacuum. If the excitation is localized in a space–time domain $D$, then in the causal complement $D'$ (see eq. (3.5)) where no effect can propagate from $D$, the representation $\pi$ of this excitation has to be equivalent to the vacuum representation $\pi_\omega$.

$$\pi|\mathfrak{A}(D') \cong \pi_\omega|\mathfrak{A}(D') \quad \text{(unitary equivalence)}. \tag{6.8}$$

Here, we denote by $\mathfrak{A}(D')$ the $C^*$ algebra generated by the union $\cup\mathfrak{A}(D_1)$ over all double cones $D_1$ contained in $D'$. (A definition of $\mathfrak{A}(D')$ is needed since $D'$ is not bounded.)

The second criterion is that this localized excitation is required to be transportable. Namely, eq. (6.8) is required not only for a domain $D$, but also for a translated domain $D + a$ (for arbitrary $a$). (If (6.8) holds for a given $D$, then it holds in any domain containing $D$, which is evident from the monotone property of local observables.)

Our aim is to consider all unitary equivalence classes $[\pi]$ of representations $\pi$ which satisfy the above two criteria, and to clarify their structure. To continue this analysis we make the following two assumptions.

One fundamental assumption is Haag's *causal duality*, which requires in the vacuum representation $\pi_\omega$ the following relations for any double cone $D$. (It is possible to perform our analysis even if the assumption is limited to $D$ above a fixed size. For simplicity we require here the assumption for any $D$.)

$$\pi_\omega(\mathfrak{A}(D))' = \left[\bigcup\{\pi_\omega(\mathfrak{A}(D_1));\ D_1 \subset D'\}\right]''. \tag{6.9}$$

Inside the square bracket on the r.h.s. the union is over all double cones $D_1$ contained in the causal complement $D'$ of $D$ (i.e. all $D_1$ spacelike to $D$), and using the notation of eq. (6.8) the r.h.s. can be written as $\pi_\omega(\mathfrak{A}(D'))''$. The fact that the r.h.s. is contained in the l.h.s. is the locality axiom (3) for local observables, and causal duality requires the stronger condition of the equality. In the cases where the local observables can be constructed from the Wightman field by the method of Section 4.9, the causal duality is also proven.

A secondary, technical assumption is the following property B by Borchers [3].

**Property B** If the open future cones $D$ and $D_1$ have an inclusion relation $\bar{D} \subset D_1$, then for any non-zero projection operator $E$ in $\mathfrak{A}(D)$ there exists an isometry $W$ in $\mathfrak{A}(D_1)$ satisfying $WW^* = E$.

This property holds if $\mathfrak{A}(D)$ is a type-III factor (in this case we can take $W \in \mathfrak{A}(D)$). Even if this is not the case, the property B can be shown by additional use of $\mathscr{P}_+^\uparrow$ invariance and positivity of the energy in cases where the weak additivity of Theorem 4.13 holds.

**Remark 1** If the GNS representation $\pi = \pi_\varphi$ associated with the state $\varphi$ satisfies criterion (6.8) of localized excitations, the state $\varphi$ and the vacuum state $\omega$ are not distinguished at infinite distance in the following sense:

$$\lim_{n\to\infty} \|(\varphi - \omega)|_{\mathfrak{A}_n}\| = 0 \qquad (\mathfrak{A}_n = \mathfrak{A}(D'_n)). \tag{6.10}$$

Here $D_n$ is a monotone increasing sequence of double cones which covers the whole space–time for $n \to \infty$. $D'_n$ is its causal complement and the $C^*$ algebra generated by all $\mathfrak{A}(D_\alpha)$, $D_\alpha \subset D'_n$, is denoted by $\mathfrak{A}(D'_n)$. Conversely we can obtain (6.8) for a sufficiently large $D$ from (6.10), assuming the property B.

**Remark 2** As a quantity (superselection rule) which distinguishes different localized excitations, we can imagine some kind of charge associated to excitations. For example the number of heavy particles (baryon number). However, as for the electric charge, since we can measure it via the electric field at infinite distance by Gauss' law, a localized excitation with such a charge does not satisfy (6.10) and we cannot expect that such an object satisfies the locality criterion of (6.8). Furthermore, the so-called topological charge which is determined by the behaviour at infinite distance also will be excluded by the criterion (6.8). Therefore, a representation satisfying the condition (6.8) does not include all types of particle pictures.

**Remark 3** It is a natural standpoint to consider all excited states created from a vacuum with a finite excitation energy without going into the details of the particle picture. In mathematical terms, the following criterion for representations $\pi$ of local observables has been proposed by Borchers: On the representation space $\mathscr{H}_\pi$ of $\pi$ the criterion requires that there exists a representation $U$ of $\tilde{\mathscr{P}}_+^\uparrow$ which realizes a transformation $gA$ of an observable $A$ by $g \in \mathscr{P}_+^\uparrow$ through the relation

$$U(\tilde{g})\pi(A)U(\tilde{g})^* = \pi(gA) \quad (\tilde{g} \in \tilde{\mathscr{P}}_+^\uparrow)$$

and the spectrum of the generator $P^\mu$ of the translation part is contained in the closed future cone $\bar{V}_+$.

**Remark 4** The following even more specialized condition than the one above is considered to be appropriate for a sector describing particles with a certain charge in a theory without massless particles. This condition has been introduced by Buchholz and Fredenhagen [4] and has been used for derivation of interesting results.

**Condition BF** In a representation $\pi$ of observables, there exists a representation $T(a)$ of the translation group which realizes the translation $(a, \mathbf{1}) \in \mathscr{P}_+^\uparrow$ of observables by

$$T(a)\pi(A)T(a)^* = \pi((a, \mathbf{1})A) \qquad (A \in \mathfrak{A})$$

and the spectrum of its generators $P^\mu$ ($P = (P^\mu)_{\mu=0,1,2,3}$ determined by $T(a) = \mathrm{e}^{\mathrm{i}(P,a)}$) consists of a one particle part $\{p; (p,p) = m^2, p > 0\}$ of mass $m > 0$ and the other part contained in $\{p; (p,p) \geq M^2,\ p^0 > 0\}$ where $M > m$.

In this case, the weak limit

$$\omega(A)\mathbf{1} = \mathrm{w}-\lim_{\lambda\to\infty} \pi((\lambda a, \mathbf{1})A) \quad (A \in \mathfrak{A})$$

exists for any spacelike vector $a$, the functional $\omega$ determined by this equation is a vacuum state, the unitary equivalence (6.8) holds for a spacelike half infinite cone $C$ (specified below) instead of the double cone $D$, and hence the representation $\pi$ can be interpreted as giving an excitation localized in $C$ with vacuum $\omega$ as a background. This has been shown by Buchholz and Fredenhagen, who carried out a similar analysis to DHR for such a situation. Here, $C$ is an infinite convex cone around the central axis of an infinite spacelike ray $\{a + \lambda e;\ \lambda > 0\}$ extending from any vertex point $a$ in the direction of a spacelike vector $e$ (we may also call it a string) with an aperture angle at the vertex small enough so that all tangents are spacelike. It is given as

$$C = \bigcup_{\lambda>0} \{a + \lambda(e + D)\}$$

where $D$ is an open double cone with vertex $\pm e'$, with a positive timelike vector $e'$ such that $e \pm e'$ are spacelike.

The BF analysis treating a transportable excitation localized in $C$ is considered to be capable of also treating excitations with topological charges.

**Convention** From now on, we fix the vacuum $\omega$ and the associated representation space $\mathscr{H} = \mathscr{H}_\omega$ and do not distinguish an observable $A$ from its representation $\pi_\omega(A)$, $\mathfrak{A}$ from $\pi_\omega(\mathfrak{A})$, and $\mathfrak{A}(D)$ from $\pi_\omega(\mathfrak{A}(D))$. Each $\mathfrak{A}(D)$ is taken to be a von Neumann algebra, the vector representing the vacuum is denoted by $\Omega$ and the vacuum representation is represented by the identity map $\iota$. Since we discuss only one vacuum state and its excitation states, we simplify our notation by adopting this convention. (In general, $\pi_\omega$ is not necessarily faithful.) Furthermore in Section 6.5 we also assume the separability of $\mathscr{H}$. To summarize, we assume that the vacuum representation is a faithful irreducible representation on a separable Hilbert space.

## 6.2 Localized endomorphisms and sectors

For a representation $\pi$ satisfying condition (6.8) there exists a unitary operator $V$ which maps the representation space $\mathscr{H}_\pi$ to $\mathscr{H}$ satisfying

$$V\pi(A)V^* = A \quad (A \in \mathfrak{A}(D')). \tag{6.11}$$

Defining for all $A \in \mathfrak{A}$ a homomorphism $\rho$ from $\mathfrak{A}$ to $\mathscr{B}(\mathscr{H})$ by

$$\rho(A) = V\pi(A)V^* \in \mathscr{B}(\mathscr{H}) \tag{6.12}$$

then, the representation $\pi$ on $\mathscr{H}_\pi$ and the representation $\rho \cdot \iota$ on $\mathscr{H}$ are unitarily equivalent as representations of $\mathfrak{A}$. Therefore, it is sufficient to treat

$$\pi_\rho(A) \equiv \rho(A) \quad (A \in \mathfrak{A}) \tag{6.13}$$

instead of $\pi$, in treating the unitary equivalence class $[\pi]$ of a representation $\pi$ which represents a localized excitation. $\pi_\rho$ is the representation defined on the same Hilbert space as the vacuum representation and condition (6.8), being equivalent to (6.11), becomes simply the following condition for $\pi_\rho$:

$$\rho(A) = \pi_\rho(A) = A \quad (A \in \mathfrak{A}(D')). \tag{6.14}$$

In order to characterize such a $\rho$, we introduce the following definition.

**Definition 6.2**
If the endomorphism $\rho$ of $\mathfrak{A}$ satisfies

$$\rho(A) = A$$

($A$ is kept invariant) for any $A \in \pi_\omega(\mathfrak{A}(D'))$, then $\rho$ is said to be *localized* in $D$ and is also called a *localized endomorphism* with support in $D$, where $D$ is always an open double cone and is called a support of $\rho$.

**Theorem 6.3**
The map $\rho$ defined by (6.12) is a localized endomorphism with support in $D$, preserving the operator norm and mapping $\mathbf{1}$ to $\mathbf{1}$.

**Proof** For any $D_0$ and $A \in \mathfrak{A}(D_0)$, take a double cone $D_1$ containing $D$ of eq. (6.14) and $D_0$. Then, first of all by locality, the equation

$$[\pi_\rho(A), \pi_\rho(B)] = \pi_\rho([A, B]) = 0$$

holds for any $B \in \mathfrak{A}(D_1')$ because $D_0$ and $D_1'$ are spacelike. Furthermore, $D_1' \subset D'$ implies $B \in \mathfrak{A}(D_1') \subset \mathfrak{A}(D')$ and we can apply (6.14) to get

$$\pi_\rho(B) = B.$$

Using the causal duality (6.9) we obtain from these two equations

$$\pi_\rho(A) \in \mathfrak{A}(D_1')' = \mathfrak{A}(D_1)'' = \mathfrak{A}(D_1)$$

and we conclude

$$\rho(A) = \pi_\rho(A) \in \mathfrak{A}$$

Since the union of $\mathfrak{A}(D_0)$ is dense in $\mathfrak{A}$ we get

$$\rho(\mathfrak{A}) \subset \mathfrak{A}$$

and thus $\rho$ is an endomorphism.

The preservation of norm, $\|\rho(A)\| = \|A\|$, and $\rho(\mathbf{1}) = \mathbf{1}$ are evident since the definition (6.11) is a unitary transformation. □

**Remark** Each $\rho(\mathfrak{A}(D))$ is isomorphic to $\mathfrak{A}(D)$ via the map $\rho$. Therefore it is a von Neumann algebra on $\mathscr{H}$.

When we discuss transportability and unitary equivalence classes of a representation, it is necessary to describe the unitary equivalence of a representation in terms of relations between the corresponding localized endomorphisms.

**Lemma 6.4**
Given localized endomorphisms $\rho_1$ and $\rho_2$. The unitary equivalence of the corresponding representations

$$\pi_{\rho_1} \cong \pi_{\rho_2} \tag{6.15}$$

is equivalent to the existence of an inner automorphism $\sigma$ satisfying

$$\rho_1 = \sigma\rho_2. \tag{6.16}$$

If the double cone $D$ contains supports of $\rho_1$ and $\rho_2$, then $\sigma$ is an inner automorphism of $\mathfrak{A}$ given by a unitary operator $U$ as

$$\sigma = \mathrm{Ad}\, U, \quad (\mathrm{Ad}\, U)(A) \equiv UAU^* \quad (A \in \mathfrak{A}). \tag{6.17}$$

**Proof** The unitary operator $U$ which represents unitary equivalence (6.15) as

$$\pi_{\rho_1}(A) = U\pi_{\rho_2}(A)U^* \qquad (A \in \mathfrak{A})$$

belongs to $\mathfrak{A}(D')'$ by eq. (6.14), which represents the locality of $\rho_1$ and $\rho_2$. Therefore, by the causal duality it follows that $U$ is in $\mathfrak{A}(D)$. □

If $\rho_1$ is a localized endomorphism with support in $D_1$ and the corresponding representation $\pi_{\rho_1}$ represents a transportable localized excitation, then for any double cone $D_2$ obtained by parallel transport of $D_1$ there exists a localized endomorphism $\rho_2$ with support in $D_2$ satisfying (6.15), and hence (6.16). Such a $\rho_1$ is called *transportable*. We denote the set of all transportable localized endomorphisms with support in $D$ by $\Delta(D)$. If $D_1 \subset D_2$ then $\Delta(D_1) \subset \Delta(D_2)$.

Define the product $\rho_1\rho_2$ of two endomorphisms $\rho_1$ and $\rho_2$ by

$$(\rho_1\rho_2)(A) = \rho_1(\rho_2(A)) \qquad (A \in \mathfrak{A}). \tag{6.18}$$

Then it is evident from eq. (6.14) that $\rho_1\rho_2$ is also an endomorphism and any double cone containing both a support of $\rho_1$ and a support of $\rho_2$ provides a support of $\rho_1\rho_2$. Further, if for any double cone $D_a$ obtained by translation of $D$, there exist endomorphisms $\rho_1'$ and $\rho_2'$ with support in $D_a$ and satisfying

$$\rho_i' = \sigma_i\rho_i, \quad \sigma_i = \mathrm{Ad}\, U_i \qquad (i = 1, 2)$$

then $\rho_1'\rho_2'$ has a support in $D_a$ and

$$\rho_1'\rho_2' = \sigma_1\rho_1\sigma_2\rho_2 = \sigma\rho_1\rho_2 \quad (\sigma = \mathrm{Ad}\, U,\ U = U_1\rho_1(U_2)).$$

Therefore, if $\rho_1$ and $\rho_2$ are transportable, then the product $\rho_1\rho_2$ is also transportable. In particular, we see that $\Delta(D)$ is a semigroup.

Let $\mathscr{I}(D)$ be the set of all localized inner automorphisms Ad $U$, $U \in \mathfrak{A}$ and $\mathscr{I}$ the union of $\mathscr{I}(D)$ for all $D$. Furthermore, let $\Delta$ be the union of $\Delta(D)$ for all $D$. By Theorem 6.3 and Lemma 6.4 we see that the unitary equivalence classes of representations which describe transportable localized excitations are represented by $\mathscr{I}\backslash\Delta$.

**Theorem 6.5**
$\Delta(D)$ and $\Delta$ are semigroups and $\mathscr{I}\backslash\Delta$ is a commutative semigroup. If $\bar{D}_1$ and $\bar{D}_2$ lie spacelike to each other and $\rho_j \in \Delta(D_j)$, $(j = 1, 2)$, then $\rho_1\rho_2 = \rho_2\rho_1$.

**Proof** We have already shown that $\Delta(D)$ is a semigroup. Therefore, $\Delta$ is also a semigroup. If $\sigma_i = \mathrm{Ad}\, U_i$, $(i = 1, 2)$, then

$$(\sigma_1\rho_1)(\sigma_2\rho_2) = \sigma(\rho_1\rho_2), \quad \sigma = \mathrm{Ad}(U_1\rho_1(U_2))$$

holds and hence the product defined by (6.18) preserves the equivalence class. Therefore, $\mathscr{I}\backslash\Delta$ is a semigroup. Finally, let us show that if $\bar{D}_1$ and $\bar{D}_2$ lie spacelike to each other ($\bar{D}_1' \supset \bar{D}_2$) and $\rho_1 \in \Delta(D_1)$, $\rho_2 \in \Delta(D_2)$, then $\rho_1\rho_2 = \rho_2\rho_1$. If this holds, then due to the transportablility of $\rho \in \Delta$, the commutativity of $\mathscr{I}\backslash\Delta$ follows immediately.

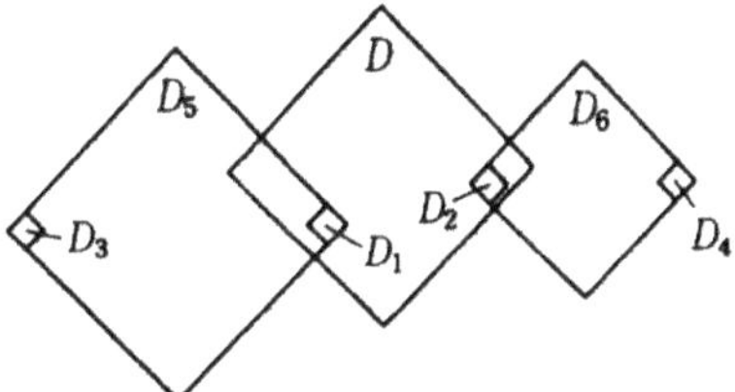


FIG 6.1. Relative positions of domains.

For a given $D$ and $A \in \mathfrak{A}(D)$ we show first that $\rho_1\rho_2$ and $\rho_2\rho_1$ coincide for $A$. For this purpose we transport $D_1$ and $D_2$ to $D_3$ and $D_4$ respectively, in such a way that the following conditions hold (Fig. 6.1).

(i) $D_3$ and $D_4$ both lie spacelike to $D$
(ii) $D_2$ is spacelike to $D_4$, $D_1$ is spacelike to $D_3$
(iii) $D_2$ is spacelike to a double cone $D_5$ which contains $D_1$ and $D_3$
(iv) $D_1$ is spacelike to a double cone $D_6$ which contains $D_2$ and $D_4$
(v) $D_5$ is spacelike to $D_6$.

Due to the transportability of $\rho_1$ and $\rho_2$ there exist $\rho_3 \in \Delta(D_3)$ and $\rho_4 \in \Delta(D_4)$ satisfying

$$\rho_3 = (\mathrm{Ad}\, U_{31})\rho_1 \quad (U_{31} \in \mathfrak{A}(D_5))$$

$$\rho_4 = (\mathrm{Ad}\, U_{42})\rho_2 \quad (U_{42} \in \mathfrak{A}(D_6))$$

Then $A \in \mathfrak{A}(D)$ is invariant due to (i) and hence

$$\rho_3\rho_4(A) = \rho_4\rho_3(A) \; (=A). \tag{6.19}$$

On the other hand by (iii), (iv) and (v) we get

$$\rho_2(U_{31}) = U_{31}, \quad \rho_1(U_{42}) = U_{42}, \quad U_{31}U_{42} = U_{42}U_{31}.$$

Therefore, by (6.19)

$$\begin{aligned}\rho_1\rho_2(A) &= (\mathrm{Ad}\, U_{42}^*U_{31}^*)\rho_3\rho_4(A) \\ &= (\mathrm{Ad}\, U_{31}^*U_{42}^*)\rho_4\rho_3(A) = \rho_2\rho_1(A)\end{aligned}$$

Since the union of $\mathfrak{A}(D)$ over all $D$ is dense in $\mathfrak{A}$, we obtain the commutativity of $\rho_1$ and $\rho_2$. □

We denote the set of all $\rho \in \Delta$ for which $\pi_\rho$ is irreducible ($\pi_\rho(\mathfrak{A})' = \mathbf{C}\mathbf{1}$) by $\Delta^{\mathrm{irr}}$. For $\rho \in \Delta^{\mathrm{irr}}$ we call the set of all vector states of $\pi_\rho$ (which are pure states) a *sector*. $\mathcal{I}\backslash\Delta^{\mathrm{irr}}$ is in 1 : 1 correspondence with the sectors and can be considered as labels of sectors. Therefore it is sometimes called *charge quantum number*.

When considering the irreducible decomposition, we need the situation where the direct sum and the subrepresentations (restriction to invariant subspaces) can freely be taken among the representations representing transportable localized excitations. This is guaranteed by the following theorem.

**Theorem 6.6**
$\mathscr{I}\backslash\Delta$ is closed under direct sum and restriction (taking subrepresentations) of representations.

**Proof** We apply the property B for this proof. First, let the supports of $\rho_1, \rho_2 \in \Delta$ be $D_1$ and $D_2$, respectively. Fix a (big) open double cone $D$ containing $D_1$ and $D_2$ and a (small) double cone $D_0$ such that $\overline{D}_0 \subset D$. Furthermore, we fix a projection operator $E \neq 0, 1$ which belongs to $\mathfrak{A}(D_0)$ and set $P_1 = E$, $P_2 = 1 - E$. By the property B there exist isometries $W_i\,(i = 1, 2)$ in $\mathfrak{A}(D)$ satisfying $P_i = W_i W_i^*$. Define

$$\rho(A) = W_1\rho_1(A)W_1^* + W_2\rho_2(A)W_2^* \quad (A \in \mathfrak{A}). \tag{6.20}$$

Then $\rho(\mathbf{1}) = P_1 + P_2 = \mathbf{1}$ and $W_i^* W_j = \delta_{ij}\mathbf{1}$. Hence

$$\rho(A_1)\rho(A_2) = \rho(A_1 A_2).$$

Therefore $\pi_\rho$ is unitarily equivalent to $\pi_{\rho_1} \oplus \pi_{\rho_2}$, and the restrictions of $\pi_\rho$ to $P_1\mathscr{H}$ and $P_2\mathscr{H}$ are unitarily equivalent to $\pi_{\rho_1}$ and $\pi_{\rho_2}$, respectively. If $A \in \mathfrak{A}(D)'$, then $A$ is invariant under $\rho_i$ $(i = 1, 2)$, because a support of $\rho_i$ is contained in $D_i \subset D$ and hence $A$ commutes with $W_i$ due to $W_i \in \mathfrak{A}(D)$. Therefore, if $A \in \mathfrak{A}(D)'$ then $\rho(A) = A$. Consequently a support of $\rho$ is contained in $D$ and $\rho$ is a localized homomorphism.

Let a double cone $D^a$ be a translate of $D$. Then the translate $D_i^a$ of $D_i$ satisfies $D_i^a \subset D^a$ and there exist localized homomorphisms $\rho_i^a$ with support in $D_i^a$, with the associated cyclic representation unitarily equivalent to $\pi_{\rho_i}$ ($\rho_i \in \Delta$ is transportable.) Construct $\rho^a$ from $\rho_1^a$ and $\rho_2^a$ similarly to the above equation (6.20). This $\rho^a$ has its support in $D^a$ and the corresponding representation is unitarily equivalent to $\pi_\rho$. Therefore, $\rho$ is transportable. Consequently, if $\rho_1, \rho_2 \in \Delta$ then there exists a $\rho \in \Delta$ satisfying

$$\pi_\rho \simeq \pi_{\rho_1} \oplus \pi_{\rho_2}. \tag{6.21}$$

By the requirement (6.21), the unitary equivalence class of such a $\pi_\rho$ is unique and

$$[\rho] \in \mathscr{I} \setminus \Delta$$

is uniquely determined. (Below we denote $[\rho] = [\rho_1] + [\rho_2]$.)

Next let $\rho \in \Delta(D)$ and assume that $E\mathscr{H}$ is an invariant subspace of $\pi_\rho(\mathfrak{A})$ for a nontrivial projection operator $E$. In this case, $E \in \pi_\rho(\mathfrak{A})'$. Since $\rho(A) = A$ for $A \in \mathfrak{A}(D)'$, the following equations hold:

$$\mathfrak{A}(D)' = \pi_\rho(\mathfrak{A}(D)') \subset \pi_\rho(\mathfrak{A})$$
$$E \in \pi_\rho(\mathfrak{A})' \subset \mathfrak{A}(D)'' = \mathfrak{A}(D) \quad \text{(causal duality)}$$

For any fixed double cone $D_3$ which contains $\bar{D}$, there exists an isometry $V \in \mathfrak{A}(D_3)$ satisfying $VV^* = E$ by the property B. Now, put

$$\rho_3(A) = V^* \rho(A) V$$

Then $\rho_3$ is a transportable localized endomorphism with support in $D_3$ and $\pi_{\rho_3}$ is equivalent to the restriction of $\pi_\rho$ to $E\mathcal{H}_\omega$. □

## 6.3 Permutation of excitations and statistics of a sector

In this section we introduce the statistics parameter for describing the behaviour of excitations under permutation and classify the sectors.

Suppose we are given two localized endomorphisms $\rho, \rho'$. An operator $T$ satisfying

$$\rho'(A)T = T\rho(A) \qquad (A \in \mathfrak{A}) \tag{6.22}$$

is an intertwiner between the representations $\pi_\rho$ and $\pi_{\rho'}$. It will be called an intertwiner between $\rho$ and $\rho'$ and the triple $\rho, \rho', T$ will be denoted by

$$\mathbf{T} = (\rho'|T|\rho). \tag{6.23}$$

If a double cone $D$ contains supports of both $\rho$ and $\rho'$, then since

$$\rho'(A) = \rho(A) = A \qquad (A \in \mathfrak{A}(D'))$$

$T$ commutes with $A$ by (6.22) and we get

$$T \in \mathfrak{A}(D')' = \mathfrak{A}(D). \tag{6.24}$$

By Hermitian conjugation and product of operators $T$ we can define conjugation and product of intertwiners via

$$(\rho'|T|\rho)^* = (\rho|T^*|\rho'), \tag{6.25}$$

$$(\rho''|T_2|\rho') \circ (\rho'|T_1|\rho) = (\rho''|T_2 T_1|\rho). \tag{6.26}$$

Furthermore we can define the following *crossed product*

$$(\rho_2'|T_2|\rho_2) \times (\rho_1'|T_1|\rho_1) = (\rho_2'\rho_1'|T_2\rho_2(T_1)|\rho_2\rho_1). \tag{6.27}$$

By a simple calculation we get the following properties:

$$\mathbf{T}_3 \times (\mathbf{T}_2 \times \mathbf{T}_1) = (\mathbf{T}_3 \times \mathbf{T}_2) \times \mathbf{T}_1 \quad \text{(associativity)}, \tag{6.28}$$

$$(\mathbf{T}_2 \times \mathbf{T}_1)^* = \mathbf{T}_1^* \times \mathbf{T}_2^*, \tag{6.29}$$

$$(\mathbf{T}_2' \circ \mathbf{T}_2) \times (\mathbf{T}_1' \circ \mathbf{T}_1) = (\mathbf{T}_2' \times \mathbf{T}_1') \circ (\mathbf{T}_2 \times \mathbf{T}_1). \tag{6.30}$$

If a support of $\rho_1$ is spacelike to a support of $\rho_2$ and a support of $\rho_1'$ is spacelike to a support of $\rho_2'$, then $\mathbf{T}_1 = (\rho_1'|T_1|\rho_1)$ and $\mathbf{T}_2 = (\rho_2'|T_2|\rho_2)$ are said to be *causally disjoint*.

**Lemma 6.7**
If $\mathbf{T}_1$ and $\mathbf{T}_2$ are causally disjoint, then

$$\mathbf{T}_1 \times \mathbf{T}_2 = \mathbf{T}_2 \times \mathbf{T}_1. \tag{6.31}$$

**Proof** (6.31) is equivalent to

$$T_1 \rho_1(T_2) = T_2 \rho_2(T_1). \tag{6.32}$$

Take a double cone $\bar{D}_1$ sufficiently large to contain a support of $\rho_1'$ and a double cone $\bar{D}_2$ sufficiently large to contain a support of $\rho_2'$ which are located spacelike to each other. (The sizes of $\bar{D}_1$ and $\bar{D}_2$ are to be taken larger than the sizes of supports of $\rho_1$ and $\rho_2$.) Since double cone supports of $\rho_1'$ and $\rho_2'$ are mutually spacelike, this is always possible.

Since $\rho_1$ and $\rho_2$ are transportable, we can obtain $\bar{\rho}_1, \bar{\rho}_2$ with their supports in $\bar{D}_1$ and $\bar{D}_2$, respectively, by successively transporting their supports as will be specified below. Let $U_j$ be the intertwining unitary operator of this transport and

$$\bar{\mathbf{T}}_j = \mathbf{T}_j \circ \mathbf{U}_j \quad (j = 1, 2,\ \mathbf{U}_j = (\rho_j | U_j | \bar{\rho}_j))$$

Since $\bar{\mathbf{T}}_j \in \mathfrak{A}(\bar{D}_j)$ by (6.24), we obtain

$$\bar{\rho}_1(\bar{T}_2) = \bar{T}_2, \quad \bar{\rho}_2(\bar{T}_1) = \bar{T}_1, \quad \bar{T}_1\bar{T}_2 = \bar{T}_2\bar{T}_1.$$

Hence eq. (6.32) is satisfied for $\bar{T}_1$ and $\bar{T}_2$ and we get

$$\bar{\mathbf{T}}_1 \times \bar{\mathbf{T}}_2 = \bar{\mathbf{T}}_2 \times \bar{\mathbf{T}}_1. \tag{6.33}$$

The transporting unitary $\mathbf{U}_j$ is constructed by a product of many unitary intertwiners as

$$\mathbf{U}_j = \mathbf{U}_{jn} \circ \cdots \circ \mathbf{U}_{j1}, \qquad \mathbf{U}_{jk} = (\rho_{j(k+1)} | U_{jk} | \rho_{jk}),$$

$$\rho_{j1} = \bar{\rho}_j, \quad \rho_{j(n+1)} = \rho_j, \qquad j = 1, 2.$$

For each $k$, when we transport $\rho_2$ ($\rho_{2k} \neq \rho_{2(k+1)}$) we fix $\rho_1$ ($\rho_{1k} = \rho_{1(k+1)}$), and when we transport $\rho_1$ we fix $\rho_2$. We move $\rho_2$ little by little such that there exists always a double cone $D_k$ which includes supports of $\rho_{2k}$ and $\rho_{2(k+1)}$ and which lies spacelike to a support of $\rho_{1k}$. Since supports of $\bar{\rho}_1$ and $\bar{\rho}_2$ as well as supports of $\rho_1$ and $\rho_2$ are located spacelike to each other, the transport of supports described above, always keeping supports of $\rho_{1k}$ and $\rho_{2k}$ mutually spacelike, is possible.

Putting

$$\bar{\mathbf{T}}_{jk} \equiv \mathbf{T}_j \circ \mathbf{U}_{jn} \circ \cdots \circ \mathbf{U}_{jk}, \quad \bar{\mathbf{T}}_{j(n+1)} \equiv \mathbf{T}_j$$

let us prove the commutativity recursively:

$$\bar{\mathbf{T}}_{1k} \times \bar{\mathbf{T}}_{2k} = \bar{\mathbf{T}}_{2k} \times \bar{\mathbf{T}}_{1k}. \tag{6.34}$$

For $k = 1$ (6.34) holds since it is the same as (6.33). Hence we show that if (6.34) holds for $k$ it also holds for $(k+1)$. For the case when we transport $\rho_2$ at the $k$th step, $\bar{\mathbf{T}}_{1(k+1)} = \bar{\mathbf{T}}_{1k}$ holds and we get

$$\begin{aligned}\bar{\mathbf{T}}_{1(k+1)} \times \bar{\mathbf{T}}_{2(k+1)} &= (\bar{\mathbf{T}}_{1k} \times \bar{\mathbf{T}}_{2k}) \circ (\mathbf{I}_{1k} \times \mathbf{U}^*_{2k}) \\ &= (\bar{\mathbf{T}}_{2k} \times \bar{\mathbf{T}}_{1k}) \circ (\mathbf{U}^*_{2k} \times \mathbf{I}_{1k}) = \bar{\mathbf{T}}_{2(k+1)} \times \bar{\mathbf{T}}_{1(k+1)}\end{aligned}$$

where $\mathbf{I}_{1k} = (\rho_{1k}|\mathbf{1}|\rho_{1k})$. Note that since the double cone $D_k$ which contains supports of $\rho_{2k}$ and $\rho_{2(k+1)}$ lies spacelike to a support of $\rho_{1k}$, $\mathbf{I}_{1k} \times \mathbf{U}^*_{2k} = \mathbf{U}^*_{2k} \times \mathbf{I}_{1k}$ is satisfied by the same reason as (6.33). Furthermore, we have also used (6.34) for the case of $k$ in this computation. In this way we obtain (6.34) for $(k+1)$ if it holds for $k$ and therefore by mathematical induction eq. (6.34) holds for all $k$. In particular, for $k = n+1$ we get (6.31). □

Now we are ready to discuss permutations of $n$ excitations. For $\rho_1, \ldots, \rho_n \in \Delta$ choose $\rho_j^{(0)}$ equivalent to $\rho_j$ such that supports of $\rho_j^{(0)}$ lie spacelike to each other for different indices $j$ and fix the intertwiner $\mathbf{U}_j = (\rho_j^{(0)}|U_j|\rho_j)$ for each $\rho_j$ and $\rho_j^{(0)}$. The permutation $1, \ldots, n \to p(1), \ldots, p(n)$ of the indices $1, \ldots, n$ is denoted by $p$. The unit element of the permutation group $\mathfrak{S}$ is denoted by $e$. Furthermore we write

$$\mathbf{U}_{p(1)} \times \cdots \times \mathbf{U}_{p(n)} = \mathbf{U}(p^{-1}). \tag{6.35}$$

Namely, $\mathbf{U}(p)$ is what we obtain after replacing $p$ by $p^{-1}$ on the l.h.s.. We define

$$\boldsymbol{\varepsilon}_p(\rho_1 \cdots \rho_n) = \mathbf{U}^*(p) \circ \mathbf{U}(e). \tag{6.36}$$

Since supports of $\rho_j^{(0)}$ lie mutually spacelike, we obtain for each $p$, by Theorem 6.5,

$$\rho_{p^{-1}(1)} \cdots \rho_{p^{-1}(n)} = \rho_1 \cdots \rho_n.$$

Thus, the product of eq. (6.36) can be defined and $\boldsymbol{\varepsilon}_p$ is of the following form

$$(\rho_{p^{-1}(1)} \cdots \rho_{p^{-1}(n)}|\varepsilon_p|\rho_1 \cdots \rho_n).$$

This $\boldsymbol{\varepsilon}_p$ has the following property.

**Theorem 6.8**

1. $\boldsymbol{\varepsilon}_p$ depends neither on $\rho_j^{(0)}$ nor on $\mathbf{U}_j$.
2. If supports of $\rho_j$ lie spacelike to each other then
$$\boldsymbol{\varepsilon}_p(\rho_1 \cdots \rho_n) = \mathbf{1}.$$
3. $\boldsymbol{\varepsilon}_q(\rho_{p^{-1}(1)} \ldots \rho_{p^{-1}(n)}) \circ \boldsymbol{\varepsilon}_p(\rho_1 \ldots \rho_n) = \boldsymbol{\varepsilon}_{pq}(\rho_1 \ldots \rho_n), \quad (p, q \in \mathfrak{S}_n).$
4. For the case $m < n$, denote the permutation of $m$ and $m+1$ by $\tau_m$. Then
$$\boldsymbol{\varepsilon}_{\tau_m}(\rho_1 \cdots \rho_n) = \mathbf{1}_{\rho_1} \times \cdots \times \mathbf{1}_{\rho_{m-1}} \times \boldsymbol{\varepsilon}_{\tau_m}(\rho_m \rho_{m+1}) \times \cdots \times \mathbf{1}_{\rho_{m+1}} \times \cdots \times \mathbf{1}_{\rho_n}.$$

**Proof** (1) Let us take $\bar{\rho}_j^{(0)}$, $\bar{\mathbf{U}}_j$ instead of $\rho_j^{(0)}$, $\mathbf{U}_j$. Then the difference can be represented by $\mathbf{V}_j = \bar{\mathbf{U}}_j \circ \mathbf{U}_j^*$. Since supports of $\bar{\rho}_j^{(0)}$ lie mutually spacelike and supports of $\rho_j^{(0)}$ lie mutually spacelike for $\mathbf{V}_j = (\bar{\rho}_j^{(0)}|V_j|\rho_j^{(0)})$, the $\mathbf{V}_j$ are causally disjoint from each other. Therefore, we obtain $\mathbf{V}(p) = \mathbf{V}(e)$ by Lemma 6.7 and we get the following result by eq. (6.30).

$$\bar{\mathbf{U}}(p) \circ \mathbf{U}^*(p) = \bar{\mathbf{U}}(e) \circ \mathbf{U}^*(e), \quad \bar{\mathbf{U}}^*(p) \circ \bar{\mathbf{U}}(e) = \mathbf{U}^*(p) \circ \mathbf{U}(e)$$

Therefore, $\boldsymbol{\varepsilon}_p$ depends on neither $\rho_j^{(0)}$ nor $\mathbf{U}_j$.

(2) If supports of the $\rho_j$ are mutually spacelike, we can take $\rho_j^{(0)} = \rho_j$ and $U_i = \mathbf{1}$ and we obtain $\boldsymbol{\varepsilon}(p) = \mathbf{1}$.

(3) is evident by definition of $\boldsymbol{\varepsilon}_p$.

(4) is obtained by applying eq. (6.30) to the definition of $\boldsymbol{\varepsilon}_p$.

**Remark** Also for the case of $n$ intertwiners $\mathbf{T}_j = (\rho_j'|T_j|\rho_j)$, permutations are defined by

$$\mathbf{T}(p) = \mathbf{T}_{p^{-1}(1)} \times \cdots \times \mathbf{T}_{p^{-1}(n)}.$$

After replacing $\rho_j'$ and $\rho_j$ by endomorphisms with mutually spacelike supports we obtain $\mathbf{T}(p) = \mathbf{T}(e)$ by Lemma 6.7. Therefore, we can represent a permutation of $\mathbf{T}$ by the following formula using $\boldsymbol{\varepsilon}_p$.

$$\mathbf{T}(p) \circ \boldsymbol{\varepsilon}_p(\rho_1 \cdots \rho_n) = \boldsymbol{\varepsilon}_p(\rho_1' \cdots \rho_n') \circ \mathbf{T}(e). \tag{6.37}$$

For the special case $\rho_1 = \cdots = \rho_n = \rho$, we write

$$\boldsymbol{\varepsilon}_p(\rho \cdots \rho) = \boldsymbol{\varepsilon}_\rho^{(n)}(p) = (\rho^n|\boldsymbol{\varepsilon}_\rho^{(n)}|\rho^n) \tag{6.38}$$

and call $\varepsilon_\rho^{(n)}$ the *permutation operator*.

**Corollary 6.9**

1. $p \to \varepsilon_\rho^{(n)}(p)$ is a unitary representation of the permutation group $\mathfrak{S}_n$ and its equivalence class is determined solely by the equivalence class of $\rho$.
2. $\varepsilon_\rho^n(p) \in \rho^n(\mathfrak{A})'$.

**Proof** (1) follows from Theorem 6.8 (3). (2) Since $\varepsilon_\rho^{(n)}$ is the intertwiner from $\rho^n$ to $\rho^n$, this is self-evident by the definition of an intertwiner. □

**Remark** The permutation represented by $\varepsilon_\rho^{(n)}$ can be interpreted as follows. Let us consider $\rho_1, \rho_2, \ldots, \rho_n$ which are equivalent to $\rho$ and have supports lying spacelike to each other. (Above we denoted them by $\rho_j^{(0)}$.) This corresponds to placing $n$ excitations, each represented by $\rho$, at $n$ locations which are located spacelike to each other. Fix the intertwiner $(\rho_j|U_j|\rho)$ between $\rho$ and $\rho_j$. Then the vector $U_j^*\Omega$ in the representation space of $\pi_\rho$ gives rise to the state corresponding to excitations $\rho_i$ in the vacuum:

$$(U_j^*\Omega, \pi_\rho(A)U_j^*\Omega) = (\Omega, \pi_{\rho_j}(A)\Omega), \tag{6.39}$$

and in the representation $\rho^n$ we obtain

$$|p\rangle = U^*_{p^{-1}(1)} \times \cdots \times U^*_{p^{-1}(n)}\Omega, \tag{6.40}$$

which represents the product state of the states (6.39). By Theorem 6.5 the state $\langle p|\pi_{\rho^n}(A)|p\rangle$ given by (6.40) does not depend on the permutation $p$ (on the order of the product of $\rho_1 \dots \rho_n$). However, the vector (6.40) can depend on $p$ and $\varepsilon_\rho^{(n)}$ gives exactly the mutual transformation among the vectors (6.40):

$$\varepsilon_\rho^{(n)}(q)|p\rangle = |qp\rangle. \tag{6.41}$$

Namely, we can interpret it as representing a permutation of $n$ excitations of the same kind (spacelike separated).

In particular, for the case $n = 2$, $\mathfrak{S}_2$ consists of $e$ and the transposition $\tau_1 = (1,2)$ and the operator

$$\varepsilon_\rho = \varepsilon_\rho^{(2)}(\tau_1), \tag{6.42}$$

called the *statistics operator* is of interest. Its property is given by the following corollary.

**Corollary 6.10**

1. $\varepsilon_\rho \in \rho^2(\mathfrak{A})'$.
2. $(\varepsilon_\rho)^2 = \mathbf{1}$.
3. If $\rho' = \sigma_V\rho$ then

$$\varepsilon_{\rho'} = \sigma_W\varepsilon_\rho, \qquad W = V\rho(V).$$

4. Denote $\mathbf{1}_\rho^k \equiv (\rho^k|\mathbf{1}|\rho^k)$, $\boldsymbol{\varepsilon}_\rho = (\rho^2|\varepsilon_\rho|\rho^2)$. Then

$$\boldsymbol{\varepsilon}_\rho^{(n)}(\tau_m) = \mathbf{1}_\rho^{m-1} \times \boldsymbol{\varepsilon}_\rho \times \mathbf{1}_\rho^{n-m-1}$$

where $\tau_m$ is the transposition of $m$ and $m+1$.

**Proof** (1) and (2) are Corollary 6.9 itself. (3) follows by a simple calculation. (4) holds due to Theorem 6.8. □

**Remark** Since $\mathfrak{S}_n$ is generated by transpositions, it follows from (4) that $\varepsilon_\rho^{(n)}(p)$ is determined by $\varepsilon_\rho$.

Above we considered the permutation of excitations. Next we consider the removal (from our sight) of these excitations by transportation to infinite distances.

**Definition 6.11**
If a positive linear map $\phi$ and a localized endomorphism $\rho$ of a $C^*$ algebra $\mathfrak{A}$ satisfy the following relations, then $\phi$ is called a *left inverse* of $\rho$.

$$\phi(A\rho(B)) = \phi(A)B \qquad (A, B \in \mathfrak{A}), \tag{6.43}$$

$$\phi(\rho(A)B) = A\phi(B) \qquad (A, B \in \mathfrak{A}). \tag{6.44}$$

The removal of the excitation $\rho$ by transportation to infinite distances gives a left inverse of $\rho$ due to the following lemma.

**Lemma 6.12**
Suppose we are given a sequence of sufficiently large double cones $D_n$ such that, in the limit $n \to \infty$, $D_n$ is transported to an infinite distance so that it is spacelike to any given double cone. Let $U_n$ be a unitary operator which transports a $\rho$ with support in a certain double cone to an equivalent $\rho_n$ with support in $D_n$ ($\rho_n = \sigma_{U_n}\rho$). Then there exists a subnet $\sigma_{U_{n(\alpha)}}$ of the sequence of maps $\sigma_{U_n}$ which has a limit map $\phi$ in the following sense and $\phi$ is the left inverse of $\rho$:

$$\text{w-}\lim_{\alpha} U_{n(\alpha)} A U^*_{n(\alpha)} = \phi(A) \qquad (A \in \mathfrak{A}). \tag{6.45}$$

**Proof** The existence of a subnet with a limit (6.45) follows from the compactness of the unit sphere of the set of all positive linear maps with respect to the topology determined by the pointwise convergence in the weak operator topology. For $A$ with a bounded support, $D_n$ and the support of $A$ are spacelike for sufficiently large $n$ and satisfy

$$U_n\rho(A)U_n^* = \rho_n(A) = A.$$

So the following equation holds:

$$w - \lim U_{n(\alpha)}\rho(A)BU^*_{n(\alpha)} = A\,\text{w-}\lim_{\alpha} U_{n(\alpha)}BU^*_{n(\alpha)} = A\phi(B).$$

Since the elements $A$ with bounded support are dense in $\mathfrak{A}$, the above equation also holds for general $A$. (6.43) is proved in the same way. □

**Lemma 6.13**
For a given $\rho$, the set of all left inverse maps is a non-empty, compact convex set.

Proof is evident from definition. Non-emptiness is guaranteed by Lemma 6.12. Now we are ready to introduce the *statistical parameter* which gives the classification of sectors.

**Theorem 6.14**
If $\rho \in \Delta$ is irreducible ($\pi_\rho(\mathfrak{A})' = \mathbf{C1}$), and $\phi_\rho$ is its left inverse, then

1. $\phi(\varepsilon_\rho) = \lambda \mathbf{1}$.
2. $\lambda$ is zero or $\pm d^{-1}$, where $d$ is a natural number.
3. The value of $\lambda$ uniquely determines the sector $[\rho]$.

**Proof** (1) Since $\varepsilon_\rho$ commutes with $\rho^2(A)$ (where $A \in \mathfrak{A}$) due to Corollary 6.10(1), we obtain

$$0 = \phi(\varepsilon_\rho \rho^2(A) - \rho^2(A)\varepsilon_\rho) = \phi(\varepsilon_\rho)\rho(A) - \rho(A)\phi(\varepsilon_\rho)$$

from the properties of the left inverse. Therefore, by the assumption that $\rho$ is irreducible, we get

$$\varepsilon_\rho \in \rho(\mathfrak{A})' = \mathbf{C1}, \quad \text{i.e. } \varepsilon_\rho = \lambda \mathbf{1} .$$

(2) Performing the same calculation as above starting from Corollary 6.9(2), we see that $\phi^{n-1}(\varepsilon_\rho^n(p))$ is a constant multiple of the identity operator, i.e. it is of the form

$$\phi^{n-1}(\varepsilon_\rho^n(p)) = \omega_\lambda^n(p)\mathbf{1}. \tag{6.46}$$

Since $\phi$ is a positive linear map which maps identity to identity, $\omega_\lambda^n(p)$ gives a state over the group algebra of $\mathfrak{S}_n$. The value of this state can be obtained by using repeatedly the following formula obtained from Corollary 6.10(4).

$$\phi(\varepsilon_\rho^n(p)) = \begin{cases} \varepsilon_\rho^{n-1}(p') & \text{(for the case } p(1) = 1) \\ \lambda\varepsilon_\rho^{n-1}(p') & \text{(for the case } p(1) \neq 1) \end{cases}$$

where $p'$ is the permutation of $1 \ldots n-1$ obtained by rewriting $t = 2 \ldots n$ to $t-1 = 1 \ldots n-1$, after representing $p$ as a product of cyclic permutations of disjoint subsets of $1 \ldots n$ and removing the letter 1.

As a result, for the product of cyclic permutations of disjoint subsets, $\omega_\lambda^n$ becomes a product of the value of each cyclic permutation and, for the cyclic permutation of $k$ numbers, it gives the value $\lambda^{k-1}$. In particular for the projection operators

$$E_s^n = \sum p/n!, \quad E_a^n = \sum (\operatorname{sgn} p)p/n!$$

to complete symmetry and complete antisymmetry (in the group algebra), the state $\omega_\lambda^n$ takes the following values (sgn is the sign of the permutation):

$$\omega_\lambda^n(E_s^n) = (1+\lambda)(1+2\lambda)\cdots(1+(n-1)\lambda)/n!$$
$$\omega_\lambda^n(E_a^N) = (1-\lambda)(1-2\lambda)\cdots(1-(n-1)\lambda)/n!$$

In order that these values are positive for all $n$, the following must hold:

$$\lambda = 0 \quad \text{or } 1 + \mathrm{d}\lambda = 0 \quad \text{or } 1 - \mathrm{d}\lambda = 0.$$

Here $d$ is a natural number.

(3) Given the left inverses $\phi_1$ and $\phi_2$, set

$$\phi_j(\varepsilon_\rho) = \lambda_j \mathbf{1} \quad (j = 1, 2).$$

By Lemma 6.13, the map $\alpha\phi_1 + (1-\alpha)\phi_2$ is also a left inverse for $0 < \alpha < 1$ and we get

$$(\alpha\phi_1 + (1-\alpha)\phi_2)(\varepsilon_\rho) = (\alpha\lambda_1 + (1-\alpha)\lambda_2)\mathbf{1}$$

By (2) the equation

$$\alpha\lambda_1 + (1-\alpha)\lambda_2 = 0 \quad \text{or} \quad \pm (d_\alpha)^{-1} \quad (d_\alpha \text{ is a natural number})$$

must hold for any $0 < \alpha < 1$. For this to hold, we must have $\lambda_1 = \lambda_2$.

Next, we replace $\rho$ by an element $\sigma_V\rho = \rho'$ in the same equivalence class. If $\phi$ is one of the left inverses of $\rho$, then we can confirm by calculation that $\phi' = \phi\sigma_{V^*}$ is a left inverse of $\rho'$. On the other hand, by Corollary 6.10(3), $\varepsilon_{\rho'} = \sigma_W(\varepsilon_\rho)$ where $W = V\rho(V)$, and we obtain

$$\phi'(\varepsilon_{\rho'}) = \phi(\sigma_{\rho(V)}\varepsilon_\rho) = \sigma_V\phi(\varepsilon_\rho) = \sigma_V(\lambda\mathbf{1}) = \lambda\mathbf{1},$$

namely, we obtain again the same $\lambda$. □

Let $\lambda_\rho$ denote the statistical parameter of a sector $[\rho]$. For a given value of the statistical parameter $\lambda_\rho$ the representation $\varepsilon_\rho^n$ of $\mathfrak{S}_n$ is determined as follows.

**Theorem 6.15**

If a left inverse $\phi$ of $\rho \in \Delta$ satisfies $\phi(\varepsilon_\rho) = \lambda\mathbf{1}$, the Young tableaux associated with irreducible components of the unitary representation $\varepsilon_\rho^n$ of $\mathfrak{S}_n$ are all those satisfying the following conditions:

1. For the case $\lambda = d^{-1}$: the length of columns are $d$ or smaller.
2. For the case $\lambda = -d^{-1}$: the length of rows are $d$ or smaller.
3. For the case $\lambda = 0$: no restriction.

**Outline of proof** The minimal projection operators of the centre of the group algebra of $\mathfrak{S}_n$ are in one-to-one correspondence to the Young tableaux. Thus the problem is to determine which minimal projections do not vanish in a representation $\varepsilon_\rho^n$.

The case (3) $\lambda = 0$. This is the easiest case. If we set $\lambda = 0$ in the formula for $\omega_\lambda^n(p)$ after (6.46), then $\omega_\lambda^n$ is seen to vanish for $p \neq e$ and hence, it is a tracial state, extracting the coefficients of the unit element for any group algebra element. Therefore, it is faithful on the group algebra (i.e. if $\omega_0^n(A^*A) = 0$ then $A = 0$) and hence the representation $\varepsilon_\rho^n$ of any minimal central projection does not vanish by (6.46). Thus, all Young tableaux appear without restriction.

For the case (1) and (2), if $\varepsilon_\rho^n$ is zero for a positive element of the group algebra, then $\omega_\lambda^n$ is also zero due to (6.46). On the other hand the inequality

$$\|\phi(A^*A)\| \geq \lambda^2 \|A^*A\| \tag{6.47}$$

is satisfied for any left inverse $\phi$. Therefore, substituting $A^*A$ into $\varepsilon_\rho^n(\cdot)$, we see that if $\omega_\lambda^n$ vanishes then $\varepsilon_\rho^n$ also vanishes. (For proof of (6.47), see reference [5].)

Thus, it is sufficient to determine the minimal central projection operator $E$ satisfying $\omega_\lambda^n(E) \neq 0$. $\omega_\lambda^n$ is concretely given already and coincides with a faithful tracial state over the following representation.

Let $\mathscr{H}_d$ be a Hilbert space of finite dimension $d$. Consider its $n$th tensor product power

$$\mathscr{H}_d^{\otimes n} = \mathscr{H}_d \otimes \cdots \otimes \mathscr{H}_d \quad \text{(product of } n \text{ factors)}$$

and a representation of $\mathfrak{S}_n$ which acts as permutations of the factors of this tensor product:

$$\pi(p)(\xi_1 \otimes \cdots \otimes \xi_n) = \xi_{p^{-1}(1)} \otimes \cdots \otimes \xi_{p^{-1}(n)}.$$

Furthermore, multiplying the sign $\operatorname{sgn} p$ of each permutation $p$, we introduce the following representation of $\mathfrak{S}_n$:

$$\pi'(p) = (\operatorname{sgn} p)\pi(p)$$

Then for $\lambda = 1/d$ the following equations hold:

$$d^{-n}\operatorname{Tr}(\pi(p)) = \omega_\lambda^n(p), \tag{6.48}$$

$$d^{-n}\operatorname{Tr}(\pi'(p)) = \omega_{-\lambda}^n(p). \tag{6.49}$$

Since there are only $d$ linearly independent vectors in $\mathscr{H}_d$, one cannot antisymmetrize more than $d$ vectors for $\pi(p)$. Thus, for the case (1) $\lambda = d^{-1}$, the length of each column in any relevant Young tableau is $d$ or less. For the case (2) $\lambda = -d^{-1}$, due to the factor $\operatorname{sgn} p$ in $\pi'(p)$, the length of each row of a Young tableau must be $d$ or less. □

Specializing to the case $\lambda = \pm 1$, the following theorem holds.

**Theorem 6.16**
For $\rho \in \Delta$, the following conditions are equivalent:

(a) $\rho$ is an automorphism.
(b) $\rho^2$ is irreducible ($\rho^2(\mathfrak{A}') = \mathbf{C}\mathbf{1}$).
(c) $\varepsilon_\rho = \pm\mathbf{1}$.

**Proof** Since $\mathfrak{A}' = \mathbf{C}\mathbf{1}$ originally, (b) follows from (a). If (b) holds, then due to Corollary 6.10(1) and (2), (c) is satisfied. Finally we assume (c). $\varepsilon_\rho$ is obtained by putting $\rho_1 = \rho_2 = \rho$ and $n = 2$ in (6.36). If we set $\rho_1^{(0)} = \rho$, $U_1 = \mathbf{1}$ for the constituents $(\rho_j^{(0)}|U_j|\rho)$ of $\mathbf{U}$, then we get

$$\varepsilon_\rho = U_2^{-1}\rho(U_2). \tag{6.50}$$

If we choose a support of $\rho_2^{(0)}$ spacelike to the support of $\rho$ and to $D$, we obtain for $A \in \mathfrak{A}(D)$

$$A = \rho_2^{(0)}(A) = U_2\rho(A)U_2^*$$

If $\varepsilon_\rho = \pm 1$, then by (6.50) we get $U_2 = \pm\rho(U_2)$ and $A = \rho(U_2 A U_2^*) \in \rho(\mathfrak{A})$. Thus, $\rho(\mathfrak{A})$ is dense in $\mathfrak{A}$ and therefore $\rho(\mathfrak{A}) = \mathfrak{A}$ holds. Thus, we have shown (a). □

**Summary** If $\rho$ is irreducible, then for the equivalence class $[\rho]$ of an irreducible representation $\pi_\rho$ of $\mathfrak{A}$ we can introduce the statistical parameter $\lambda$ of the sector $[\rho]$, taking either the value 0 or $\pm d^{-1}$, which describes the behaviour under permutation of a number of excitations $[\rho]$ in vacuum. $d(\rho) = d$ is called the *statistical dimension* of the sector $[\rho]$.

The $\omega_\lambda^n(p)$ which describes a permutation is given by (6.48), (6.49) for the case $\lambda \neq 0$ and, since its behaviour is just identical with an $n$ particle system of para-Bosons or para-Fermions of rank $d$, it is called *Bose statistics* for $\lambda > 0$, *para-Bose statistics of order d* for $\lambda = d^{-1}$, *Fermi statistics* for $\lambda < 0$ and *para-Fermi statistics of order d* for $\lambda = -d^{-1}$, respectively. The case of order 1 corresponds to usual Bosons and Fermions, and is exactly the case where $\rho$ is an automorphism by Theorem 6.16. Since $\lambda = 0$ corresponds to $d = \infty$, we call that case *infinite statistics*, and the cases $\lambda \neq 0$ are called *finite statistics*.

If a sector $[\rho]$ satisfies the three conditions that (1) there exists a unitary representation of the translation group which realizes the translation of observables in the representation space $\mathscr{H}_\rho$ of $\pi_\rho$, (2) the spectrum of its energy momentum is contained in the future cone $V_+$ (condition of positive energy) and (3) there exists a sector $[\rho']$ such that $\rho\rho'$ contains the vacuum representation (identity representation) ($[\rho']$ is the *charge conjugate sector* of sector $[\rho]$, see Section 6.4), then it has been shown that $[\rho]$ possesses finite statistics [6]. In particular, under the condition BF, such premises hold and infinite statistics can be excluded.

If $D$ is contained in a support of $\rho$, $\rho(\mathfrak{A}(D))$ is a von Neumann subalgebra of $\mathfrak{A}(D)$ (by causal duality in the vacuum representation). For a subfactor $N$ of a type $II_1$ factor $M$, Jones introduced the Jones index $[M : N]$, which is generalized to a general von Neumann subalgebra as a minimal index. This minimal index and the statistical dimension $d(\rho)$ of $\rho$ have the following relation [7]:

$$d(\rho)^2 = [\mathfrak{A}(D) : \rho(\mathfrak{A}(D))]$$

The above analysis has been described for the case of irreducible $\rho \in \Delta^{\text{irr}}$. For a general $\rho \in \Delta$, there exists a $\phi$ among the left inverses $\phi$ of $\rho$ such that $\phi(\varepsilon_\rho)^2$ is a constant multiple of the identity operator, and this $\phi$ is called a *standard left inverse*. For a standard left inverse $\phi$, $\rho$ is said to have *infinite statistics* if $\phi(\varepsilon_\rho) = 0$, and *finite statistics* if $\phi(\varepsilon_\rho) \neq 0$. A necessary and sufficient condition for $\rho$ to have finite statistics is that it has an irreducible decomposition as a direct sum of a finite number of irreducible $\rho_i \in \Delta^{\text{irr}}$ with finite statistics as follows:

$$\pi_\rho = \sum \pi_{\rho_i}$$

If there is a $\rho \in \Delta$ which is equivalent to a direct sum of an infinite number of irreducible $\rho_i$ with finite statistics, this $\rho$ has infinite statistics, but the structure of a general $\rho$ with infinite statistics is not known.

The discussion and results of this section do not apply to spaces of lower dimensions. More complicated statistics, which are called the braid-group statistics, appear in two-dimensional spaces for the DHR analysis and in three-dimensional spaces for the BF analysis [8].

### 6.4 Charge conjugate sector

The construction of the left inverse of a localized endomorphism $\rho$ is obtained by (6.45) as a weak limit of inner automorphisms $\sigma_{U_n}$ which transport an excitation $\rho$ to spacelike infinity, and thus by the method of removing the excitation $\rho$ by transporting it to infinite distances. In fact, for $U_n^*$ which represents the inverse procedure, the support of $\rho_n$ becomes spacelike to $D$ and $\rho_n(A) = A$ for sufficiently large $n$ for a given $A \in \mathfrak{A}(D)$, and

$$U_n^* A U_n = U_n^* \rho_n(A) U_n = \rho(A)$$

Therefore, also for general $A \in \mathfrak{A}$,

$$\lim_{n\to\infty} \sigma_{U_n^*}(A) = \rho(A) \tag{6.51}$$

is satisfied.

Thus, while the left inverse $\phi$ which we get as a limit of $\sigma_{U_n}$ is a positive map into itself and not an endomorphism, it must have the meaning of removing something corresponding to the excitation $\rho$ from the vacuum state, and after this has been done, an excitation inverse to $\rho$ must remain and is expected to give the charge conjugate excitation in quantum field theory.

In this section, we want to explain how this idea holds for the case of a sector with finite statistics. The set of all localized endomorphisms $\rho$ of $\mathfrak{A}$ satisfying the following two conditions is denoted $\Delta_f^{\text{irr}}$:

(a) $\rho(\mathfrak{A})$ is irreducible.
(b) $\lambda_\rho \neq 0$ (finite statistics).

**Theorem 6.17**
For $\rho \in \Delta_f^{\text{irr}}$ there exists a $\bar{\rho} \in \Delta_f^{\text{irr}}$ such that $\bar{\rho}\rho$ contains the vacuum representation (identity representation) $\iota$, and the sector $[\bar{\rho}]$ is uniquely determined by the sector $[\rho]$. In this case, $\bar{\rho}\rho$ contains the vacuum representation with multiplicity 1 and $\lambda_\rho = \lambda_{\bar{\rho}}$.

The sector $[\bar{\rho}]$ in this theorem is called the *charge conjugate sector* of the sector $[\rho]$. Since $[\bar{\rho}\rho] = [\rho\bar{\rho}]$ by Theorem 6.5, we obviously have $[\bar{\bar{\rho}}] = [\rho]$. Namely, if we take charge conjugation twice, then we obtain the original sector. The relation between charge conjugation and left inverse is as follows.

**Theorem 6.18**
1. The left inverse $\phi$ of $\rho \in \Delta_f^{\text{irr}}$ is unique, and is obtained in terms of $U_n$ in Lemma 6.12 as

$$\operatorname*{w-lim}_{n\to\infty} U_n A U_n^* = \phi(A). \tag{6.52}$$

2. The GNS representation $\pi_\phi$ associated with the state $\omega(\phi(A))$, $A \in \mathfrak{A}$, obtained from the vacuum state $\omega(A) = (\Omega, A\Omega)$ and the left inverse $\phi$, is unitarily equivalent to $\pi_{\bar{\rho}}(A) = \bar{\rho}(A)$.
3. There exists an intertwiner $\mathbf{R} = (\bar{\rho}\rho|R|\iota)$ between $\bar{\rho}\rho$ and $\iota$ such that

$$\bar{\mathbf{R}} \equiv \operatorname{sgn}(\lambda_\rho)\boldsymbol{\varepsilon}(\bar{\rho}, \rho) \circ \mathbf{R} \equiv (\rho\bar{\rho}|\bar{R}|_\ell) \tag{6.53}$$

(sgn $(\lambda_\rho)$ is the sign of $\lambda_\rho$) satisfies

$$\bar{R}^*\rho(R) = 1, \quad R^*\bar{\rho}(\bar{R}) = 1 \tag{6.54}$$

and the left inverse $\phi$ is given by

$$\phi(A) = d(\rho)^{-1}R^*\bar{\rho}(A)R \quad (A \in \mathfrak{A}). \tag{6.55}$$

In order to discuss the particle picture of the charge conjugate sector, we further impose the covariance condition on $\rho$. Among the $\rho$s in $\Delta_f^{\text{irr}}$, we denote the set of those satisfying the following condition by $\Delta_s$.

(c) There exists a continuous unitary representation $g \in \tilde{\mathscr{P}}_+^\uparrow \to U_\rho(g)$ of $\tilde{\mathscr{P}}_+^\uparrow$ on $\mathscr{H}$ satisfying the following equation:

$$U_\rho(g)\rho(A)U_\rho(g)^* = \rho(gA) \qquad (A \in \mathfrak{A}, g \in \tilde{\mathscr{P}}_+^\uparrow), \tag{6.56}$$

where $gA$ represents the transformation of $A$ by an element $g$ in $\mathscr{P}_+^\uparrow$.

**Theorem 6.19**

1. If $\rho \in \Delta_s$ then $\bar{\rho} \in \Delta_s$.
2. The $U_\rho(g)$ satisfying (6.56) is unique.
3. For equivalent $\rho_1, \rho_2 \in \Delta_s$ with intertwiner $\mathbf{R} = (\rho_2|R|\rho_1)$, the operator $R$ is also an intertwiner of $U_{\rho_1}$ and $U_{\rho_2}$.
4. The mass spectrum of $U_\rho$ and that of $U_{\bar{\rho}}$ are quasi-equivalent.

If the sector $[\rho]$ contains one particle states (irreducible representations of $\tilde{\mathscr{P}}_+^\uparrow$) with finite multiplicity then the following result holds.

**Theorem 6.20**

If representations of mass $m$ are contained in the representation $U_\rho$ of $\tilde{\mathscr{P}}_+^\uparrow$ of $\rho \in \Delta_s$ with a finite total multiplicity, then the following holds:

(a) There exists a unitary map $C$ between the mass $m$ subspaces $\mathscr{H}_\rho^m$ and $\mathscr{H}_{\bar{\rho}}^m$ of $U_\rho$ and $U_{\bar{\rho}}$ intertwining $U_\rho$ and $U_{\bar{\rho}}$. (Therefore spin and multiplicity of the particle are the same with $\rho$ and $\bar{\rho}$.)

(b) Between the spin $s$ of the particle and the sign $\operatorname{sgn}\lambda$ of the statistical parameter $\lambda$, the following relation holds (for all particles in $[\rho]$):

$$(-1)^{2s} = \operatorname{sgn}\lambda. \tag{6.57}$$

Equation (6.57) is called the *connection of spin and statistics*. Its proof is known in field theory [9], except for the case of an infinite multiplicity of particles where there exists a counterexample.

The sector of $\rho = \bar{\rho}$ is classified into two types, real and pseudo-real. It is proved for any pseudo-real sector that particle and charge conjugate particle (called anti-particle) must be different. This corresponds to Carruthers' theorem in quantum field theory.

We do not give proofs of theorems in this section due to a limitation on the length of this monograph [10].

## 6.5 Operator algebra system of fields and gauge groups

In this section, we introduce operators like the Dirac field, bridging between vacuum and one particle states, which are not superposable with the vacuum due to superselection rules. When supports of these operators lie in spacelike separated regions we derive normal commutation relations, i.e. *anticommutation relations* among the operators which bridge to para-Fermi statistics sectors and *commutation relations* otherwise. As summarized in Section 6.1, the basic assumptions are a faithful, irreducible vacuum representation on a separable Hilbert space, satisfying causal duality and property B. The mathematical background is the duality theorem by Doplicher and Roberts [11].

The DHR analysis originally started from the consideration of how one can reconstruct the gauge group and the original non-gauge invariant fields from the operator algebra system of local observables generated by gauge invariant operators in a quantum field theory including Fermi fields [12]. The structure of the gauge group and the field operator algebra system to be reconstructed are as follows. Note that the vacuum representation $\pi_0$ is given such that it is a faithful irreducible representation of a $C^*$ algebra $\mathfrak{A}$ generated by a von Neumann algebra $\mathfrak{A}(D)$ of local observables on a separable Hilbert space $\mathscr{H}_0$.

**Definition 6.21**
By a *local operator algebra system of fields* $\mathscr{F}$ with *gauge group* $G$ the following triple $(\pi, G, \mathscr{F})$ satisfying the properties $(\alpha)$ – $(\delta)$ below is meant:

1. A Hilbert space $\mathscr{H}$ with a representation $\pi$ of $\mathfrak{A}$ which contains a $\pi(\mathfrak{A})$ invariant subspace $\mathscr{H}_0$ such that the subrepresentation $\pi_0$ on $\mathscr{H}_0$ is the faithful and irreducible vacuum representation,
2. a compact group $G$ of some unitary operators on $\mathscr{H}$ (compactness in the strong operator topology) which become the identity operator (the trivial representation) on $\mathscr{H}_0$, and
3. a von Neumann algebra $\mathscr{F}(D)$ on $\mathscr{H}$ corresponding to each double cone $D$ and a $C^*$ algebra $\mathscr{F}$ generated by all $\mathscr{F}(D)$.
   Properties:
   ($\alpha$) An automorphism $\alpha_g$ of $\mathscr{F}$ is induced by a unitary transformation $A \in \mathscr{F} \to gAg^* \in \mathscr{F}$ for $g \in G$ and gives an automorphism of $\mathscr{F}(D)$ for each $D$. The set of all elements of $\mathscr{F}$ and $\mathscr{F}(D)$ invariant under all $g \in G$ (fixed point algebras $\mathscr{F}^G$ and $\mathscr{F}(D)^G$, respectively) coincide with $\pi(\mathfrak{A})$ and $\pi(\mathfrak{A}(D))$.
   ($\beta$) The $C^*$ algebra $\mathscr{F}$ of fields is irreducible on $\mathscr{H}$. $(\mathscr{F}' = \mathbf{C}1)$
   ($\gamma$) For any $\mathscr{F}(D)$, $\mathscr{H}_0$ is cyclic. $(\overline{\mathscr{F}(D)\mathscr{H}_0} = \mathscr{H})$
   ($\delta$) If $D_1$ and $D_2$ lie mutually spacelike, then $\mathscr{F}(D_1)' \supset \pi(\mathfrak{A}(D))$.

The relation between the field $\mathscr{F}$, the gauge group $G$ and the observables $\mathfrak{A}$ is explained by ($\alpha$) in the above definition. However, for the commutation relations of fields in two domains which lie spacelike to each other, the condition ($\delta$) is the minimal condition, and the idealized form is the following normal commutation relation.

**Definition 6.22**

A pair consisting of a gauge group $G$ and a local operator algebra system of fields $\mathscr{F}$ are said to *satisfy normal commutation relations*, if there exists a central element $k$ of $G$, of which the square is the identity ($k^2 = e$) and, with respect to the $Z_2$ grading determined by $k$, the $\mathscr{F}$ satisfies the following *graded local commutativity*.

Graded local commutativity: Let $D_1$ and $D_2$ be mutually spacelike, and let

$$F_\sigma \in \mathscr{F}(D_1), \quad F'_\sigma \in \mathscr{F}(D_2), \quad \alpha_k F_\sigma = \sigma F_\sigma, \quad \alpha_k F'_\sigma = \sigma F'_\sigma$$

($\sigma = \pm$). Then the following holds:

$$F_+ F'_+ = F'_+ F_+, \quad F_+ F'_- = F'_- F_+, \quad F_- F'_- = -F'_- F_-. \qquad (6.58)$$

The reason for introducing the gauge group and the fields is to describe all sectors which represent transportable localized excitations by one Hilbert space and an irreducible $C^*$ algebra $\mathscr{F}$ on it, starting from a single vacuum representation. We introduce a terminology which describes the situation that this is achieved.

**Definition 6.23**

A pair consisting of a gauge group and a local operator algebra system of fields is called *complete*, if the representation $\pi$ of a $C^*$ algebra $\mathfrak{A}$ of observables ($\pi(\mathfrak{A}) = \mathscr{F}^G$) contains all irreducible representations $\pi_1$ with finite statistics representing transportable localized excitations ($\pi_1(\mathfrak{A}(D))$ is unitarily equivalent to $\pi_0(\mathfrak{A}(D))$, for all double cones $D$).

Finally as the last definition of this section we define the equivalence of the triples $(\pi, G, \mathscr{F})$.

**Definition 6.24**

A pair of triples $(\pi_i, G_i, \mathscr{F}_i)$, $i = 1, 2$, of a gauge group and a local operator algebra systems of fields are *equivalent* if there exists a unitary operator $W$ from the representation space $\mathscr{H}_1$ of $\pi_1$ to the representation space $\mathscr{H}_2$ of $\pi_2$ such that the following intertwining relations hold:

$$W\pi_1(A) = \pi_2(A)W \qquad (A \in \mathfrak{A})$$

$$WG_1 = G_2 W$$

$$W\mathscr{F}_1(D) = \mathscr{F}_2(D)W \qquad (D \text{ is any double cone}).$$

Since the aim is now clear by the above definitions, we give the result as a theorem [13].

**Theorem 6.25**
If an irreducible faithful vacuum representation $\pi_0$ of a system of local observables $\mathfrak{A}$ on a separable Hilbert space $\mathscr{H}_0$ satisfies causal duality and property B, then there exist a local operator algebra system of fields and a gauge group with the given $(\pi_0, \mathfrak{A}, \mathscr{H}_0)$, satisfying normal commutation relations and which is complete. Its equivalence class is unique.

Therefore, our goal is achieved.

**Remark** If we do not require the normal commutation relations in Definition 6.22, then the choice of the gauge group and the local operator algebra system of fields $\mathscr{F}$ related to the given $\pi_0$ (even when thinking of equivalence classes) is not unique. We can for example construct, from several copies of the same free fields, para-Boson and para-Fermion fields which do not satisfy normal commutation relations. (These fields satisfy neither commutation relations nor anticommutation relations.) Even in such a case, the above theorem implies that, with the local algebra system of observables fixed (which are required to commute at spacelike separated regions due to causality), we can find fields with normal commutation relations. Such a result has been proved in quantum field theory in the case where Wightman fields of two spacelike separated regions satisfy commutation or anticommutation relations by converting given Wightman fields into fields satisfying normal commutation relations via Klein transformation (without changing the observables) [14].

The gauge groups and fields whose existence is demonstrated by the above theorem possess various other properties beyond those mentioned in the above definitions, and these properties are closely related to the statistics analysis of the previous section. The following theorem explains such a situation [13].

**Theorem 6.26**
A gauge group and a local operator algebra system of fields have the following properties:

(a) $\pi(\mathfrak{A})' \cap \mathfrak{F} = \mathbf{C}1$
(b) Given an automorphism $\gamma$ of the $C^*$ algebra $\mathscr{F}$. In order for the automorphism $\gamma$ to coincide with $\alpha_g$ for some $g \in G$, it is necessary and sufficient that all elements of $\mathfrak{A}$ are fixed points under $\gamma$.
(c) $\pi(\mathfrak{A})' = G''$, $\pi = \oplus d(\xi)\pi_\xi$
where the sum extends over the equivalence classes of all irreducible representations $\pi_\xi$ with finite statistics of order $d(\xi)$, satisfying $\pi_\xi(\mathfrak{A}(D)) \cong \pi_0(\mathfrak{A}(D))$ for all double cones $D$.

Furthermore, if fields satisfy normal commutation relations, the following holds.

(d) The grading of $\mathscr{H}$ by $k \in G$ exactly corresponds to the distinction between para-Bose statistics and para-Fermi statistics. Namely, for a vector $\Phi$ belonging to the representation space $\mathscr{H}_\xi$ of an irreducible representation $\pi_\xi$ in the direct sum of the above (c), the following formula holds:

$$k\Phi = (\operatorname{sgn} \lambda_\xi)\Phi$$

(e) A local operator algebra system of fields $\mathscr{F}(D)$ satisfies the following twisted duality. Set

$$\mathscr{F}^t(D) \equiv V\mathscr{F}(D)V^*, \quad V = 2^{-1/2}(1 + ik) \tag{6.59}$$

Then

$$\mathscr{F}^t(D) = \mathscr{F}(D')' \tag{6.60}$$

holds where the definition of $\mathscr{F}(D')$ is similar to the case of $\mathfrak{A}(D')$.

$$\mathscr{F}(D') = \text{the } C^*\text{algebra generated by } \{\mathscr{F}(D_1);\ D_1 \subset D'\}.$$

The multiplicity $d(\xi)$ of $\pi_\xi$ in the above irreducible decomposition of (c) is exactly the dimension of $\mathscr{H}_d$ in the proof of Theorem 6.15, and supports the representation space of the permutation operators in the representation space $\xi^n$. Furthermore, from the above (c), we immediately see that $G$ is commutative if and only if $d(\xi) = 1$ for all sectors $\xi$ (usual Bose and Fermi statistics). In fact, the DHR analysis and the structure of the operator algebra system of fields is much simpler for this case than in the general case.

Proof of the above two theorems will be omitted due to the limitation on length [13].

Using the operator algebra system of fields the existence of which we have shown in this section, we can develop the scattering theory including particles which are not superposable with the vacuum, in the same way as in the case of the $S$-matrix theory of Chapter 5.

## 6.6 Theory of local charges

If in classical field theory a Lagrangian is invariant under a global transformation such as the gauge group in the previous section, then one can define a current density which satisfies a conservation equation, a result known as Noether's theorem. In the corresponding quantum field theory, the integration of the current generates locally a transformation of the field by $G$ and gives a representation of the Lie algebra of $G$. In this section we discuss the local representation of such a $G$ and its Lie algebra.

Let us assume that, for a local algebra system of fields $\mathscr{F}(D)$ introduced in the previous section, there is a unitary representation $g \in G \to U(g)$ of the connected Lie group $G$ satisfying the following conditions:

$$U(g)\Omega = \Omega$$
$$U(g)\mathscr{F}(D)U(g)^* = \mathscr{F}(D) \quad \text{(for all double cones).} \tag{6.61}$$

The gauge group $G$ in the previous section has these properties. Even if there is no gauge group and $\mathscr{F}(D) = \mathfrak{A}(D)$, sometimes there exists such a nontrivial $G$ which, in this case, is called an *internal symmetry*.

We denote automorphisms of $\mathfrak{A}$ induced by $U(g)$ as $\alpha_g$.

$$\alpha_g(A) = U(g)AU(g)^* \quad (A \in \mathfrak{A}). \tag{6.62}$$

**Definition 6.27**

$G$ is called *locally realizable* if a continuous representation of $G$ by unitary operators in $\mathscr{F}(D_2)$

$$g \in G \to V_g \in \mathscr{F}(D_2)$$

exists for any pair of open double cones $D_1$, $D_2$ satisfying $\bar{D}_1 \subset D_2$, and the following two conditions are satisfied:

$$V_g F V_g^* = \alpha_g(F) \quad (F \in \mathscr{F}(D_1)) \tag{6.63}$$

$$\alpha_h(V_g) = V_{hgh^{-1}} \quad (g, h \in G) \tag{6.64}$$

The group of motion $\mathscr{P}_+^\uparrow$ or $\tilde{\mathscr{P}}_+^\uparrow$ of special relativity theory do not preserve $\mathscr{F}(D)$ even as a set. However, if we only consider a neighbourhood of the unit element of $G$, then since the domain $D$ moves only a little, it is possible to formulate a situation similar to the above. This is especially important when considering Lie algebras.

Let $U(\tilde{L})$ be a continuous unitary representation of $\tilde{\mathscr{P}}_+^\uparrow$ satisfying

$$\tilde{L} \in \tilde{\mathscr{P}}_+^\uparrow \to U(\tilde{L}), \quad U(\tilde{L})AU(\tilde{L})^* = \alpha_{\tilde{L}}A \qquad (A \in \mathfrak{A})$$
$$\alpha_{\tilde{L}}(\mathfrak{A}(D)) = \mathfrak{A}(LD) \qquad (L \text{ is the element of } \mathscr{P}_+^\uparrow \text{ corresponding to } \tilde{L} \text{ in } \tilde{\mathscr{P}}_+^\uparrow) \tag{6.65}$$

Furthermore $U(\tilde{L})$ is assumed to be gauge invariant.

$$U(g)U(\tilde{L}) = U(\tilde{L})U(g) \quad (g \in G, \tilde{L} \in \tilde{\mathscr{P}}_+^\uparrow) \tag{6.66}$$

**Definition 6.28**
Let $G$ be locally realizable as in Definition 6.27 with open double cones $D_1$ and $D_2$. $G$ and $\tilde{\mathscr{P}}_+^\uparrow$ are said to be *locally simultaneously realizable* if, for any open double cone $D_0$ satisfying $\bar{D}_0 \subset D_1$ and for any neighbourhood $\tilde{\mathscr{P}}_0$ of the unit element of $\tilde{\mathscr{P}}_+^\uparrow$ satisfying $\tilde{L}_0 D_0 \subset D_1$ for any $\tilde{L}_0 \in \tilde{\mathscr{P}}_0$, there exists a representation $V_L$ of $\tilde{\mathscr{P}}_+^\uparrow$ by unitary operators of $\mathscr{F}(D_2)$, satisfying

$$V_{\tilde{L}_0} F_0 V_{\tilde{L}_0}^* = \alpha_{\tilde{L}_0}(F_0) \tag{6.67}$$

for any $\tilde{L}_0 \in \tilde{\mathscr{P}}_0$ and any $F_0 \in \mathscr{F}(D_0)$, and commuting with all $V_g$ $(g \in G)$.

Correspondingly, we make the following definition for the Lie algebra $\mathfrak{g}$ of $G$ and the Lie algebra $\mathfrak{p}$ of $\tilde{\mathscr{P}}_+^\uparrow$.

**Definition 6.29**
For a pair of open double cones $D_1$ and $D_2$ satisfying $\bar{D}_1 \subset D_2$, a representation

$$u \in \mathfrak{g} \to J_u$$

of the Lie algebra $\mathfrak{g}$ by skew selfadjoint operators (generally unbounded) affiliated with $\mathscr{F}(D_2)$ is called a *local current algebra* of $\mathfrak{g}$ if it satisfies the following equations on a common dense $U(G)$ invariant domain consisting of their analytic vectors.

$$[J_u, J_v] = J_{[u,v]} \quad (u, v \in \mathfrak{g}), \tag{6.68}$$

$$[J_u, F] = \delta_u(F) \quad (F \in \mathscr{F}(D_1), u \in \mathfrak{g}), \tag{6.69}$$

$$\alpha_g(J_u) = J_{g(u)} \quad (g \in G, u \in \mathfrak{g}). \tag{6.70}$$

Here $\delta_u$ is the generating derivation of $\alpha_g$ $(g = e^{\lambda u} \in G)$ corresponding to $u$.

It follows from the representation theory of Lie algebras that if $G$ is locally realizable in the sense of Definition 6.27, then the local current algebra satisfying

$$V_{\exp \lambda u} = \exp \lambda J_u \quad (u \in \mathfrak{g}, \lambda \in \mathbf{R}) \tag{6.71}$$

is determined. It is called the *current algebra* associated to the local realization $V$ of $G$.

For the case that $G$ is a gauge group, the Casimir operators of the enveloping algebra of $\mathfrak{g}$ which distinguish different representations of $G$ play the role of generalized charges which distinguish different sectors. Their local version provided by the representation $J$ of $\mathfrak{g}$ is considered to represent the local charge. In the case of the Casimir operators, due to (6.62) and (6.70) they all commute with $U(g)$, and hence they are (unbounded) observables affiliated with $\mathfrak{A}$.

Next, we explain a sufficient condition which guarantees the validity of the aim formulated above.

**Definition 6.30**
If von Neumann algebras $M_1$ and $M_2$ satisfy the following relations, $M_2$ is said to *split-include* (or have a *split-inclusion* of) $M_1$.

(i) $M_2 \supset M_1$.
(ii) There exists a vector $\Omega$ which is separating and cyclic for any of $M_1$, $M_2$ and $M_1' \cap M_2$.
(iii) In between $M_1$ and $M_2$, there exists a type I factor $N$:

$$M_2 \supset N \supset M_1.$$

Here a type I factor is a von Neumann algebra which is $*$ isomomorphic to the von Neumann algebra $\mathscr{B}(\mathscr{H})$ formed by all bounded linear operators on a Hilbert space $\mathscr{H}$.

We will be using this condition for $\mathscr{F}(D_1) \subset \mathscr{F}(D_2)$. Putting off examination of the meaning of the conditions to a later discussion, first let us give the conclusion.

**Theorem 6.31**
If for any pair of open double cones $D_1$ and $D_2$ satisfying $\bar{D}_1 \subset D_2$, $\mathscr{F}(D_2)$ split-includes $\mathscr{F}(D_1)$, then $G$ and $\tilde{\mathscr{P}}_+^\uparrow$ are locally simultaneously realizable and the associated local current algebra can be constructed.

Doplicher [15] was the first who applied split-inclusion to local charges. The local realizability of $G$ has been investigated by Doplicher and Longo [16], and the local simultaneous realizability of $G$ and $\tilde{\mathscr{P}}_+^\uparrow$ by Buchholz, Doplicher, and Longo [17]. For proof we refer to the original papers. For the local charges defined by this method, it has also been proved [18] that if we expand the double cone to approach the total space, then they tend to the generating operators of the global gauge group $G$. The split $W^*$-inclusion of a local algebra system is conjectured by Borchers and has been proved by Buchholz [19] for free fields.

In the definition of the split $W^*$-inclusion, the separating and cyclic condition (ii) holds if the weak additivity of Definition 4.13 is assumed for a local system of observables $\{\mathfrak{A}(D)\}$ because the same property can then be derived for the operator algebra system of fields, and (ii) holds due to the Reeh–Schlieder theorem given by Theorem 4.14. In this case the normal commutation relations become useful in proving the property for $\mathscr{F}(D_1)' \cap \mathscr{F}(D_2)$. Since condition (i) simply follows from the monotone property of $\mathfrak{A}(D)$, the really new assumption is the assumption (iii) of the split $W^*$-inclusion, i.e. the existence of an intermediate type I factor. Let us discuss its physical meaning.

**Definition 6.32**
If for a normal state $\omega$ on $\mathfrak{A}(D)$ there exists a projection operator $E$ in $\mathfrak{A}$ with the following property, then $E$ is called a *local filter* of $\omega$.

Filter property: If $\varphi(E) \neq 0$ for any locally normal state $\varphi$ of $\mathfrak{A}$, then the reduced state

$$\varphi_E(A) \equiv \varphi(EAE)/\varphi(E), \tag{6.72}$$

when restricted to $\mathfrak{A}(D)$, is identical with $\omega$ independently of $\varphi$:

$$\varphi_E(A) = \omega(A) \quad (A \in \mathfrak{A}(D)). \tag{6.73}$$

**Theorem 6.33**
Suppose we are given an operator algebra system of local observables $\{\mathfrak{A}(D)\}$. In order to guarantee that $\mathfrak{A}(D_2)$ has split-inclusion of $\mathfrak{A}(D_1)$ for any pair of open double cones $D_1$ and $D_2$ satisfying $\bar{D}_1 \subset D_2$, the existence of the local filter $E$ in $D_2$ is necessary and sufficient.

The existence of the local filter for a state in a certain domain means that the state in that domain can be prepared by using some apparatus in a domain which is a little bigger than the original one, and this assumption is equivalent with the assumption of split-inclusion. The idea of local filters and proof of Theorem 6.33 are due to Buchholz, Doplicher, and Longo [17]. The existence of local filters has a physical meaning for observables. However, it is not known whether one can derive the split-inclusion property for operator algebra systems of fields from the same property of the operator algebra system of local observables.

Buchholz and Wichmann [20] have derived the properties of the split-inclusion from a different point of view as follows. Let $D_r$ be a double cone with a ball of radius $r$ at time $t = 0$ as its base. The energy operator is denoted as $P^0 = H$.

**Definition 6.34**
If a subset $N$ of a Hilbert space $\mathscr{H}$ satisfies the following condition, $N$ is said to be *nuclear*. There exists a sequence of linear functionals $l_n$, ($n \in \mathbf{N}$) defined on the linear hull of $N$ and a sequence of unit vectors $\Phi_n$ ($n \in \mathbf{N}$) such that

(i) $\lambda_n = \sup\{l_n(\Psi); \Psi \in N\}$ satisfies $\sum_n \lambda_n < \infty$,
(ii) for all $\Psi \in N$, $\sum_n l_n(\Psi)\Phi_n = \Psi$.

Here the *nuclearity index* $\nu$ of $N$ is defined as

$$\nu(N) = \inf \sum_n \lambda_n, \tag{6.74}$$

where the infimum is taken over all possible sequences $l_n$ and $\Phi_n$ satisfying conditions (i) and (ii).

Define

$$\mathscr{L}_r \equiv \{W\Omega; W \in \mathfrak{A}(D_r), W^* W = \mathbf{1}\} \tag{6.75}$$

and consider the following condition.

**Nuclearity condition**
For all $\beta > 0$, $e^{-\beta H}\mathscr{L}_r$ is nuclear and there exist positive numbers $c, n, r_0, \beta_0$ with $r \geq r_0$, $0 < \beta < \beta_0$ satisfying

$$\nu(e^{-\beta H}\mathscr{L}_r) \leq e^{cr^3\beta^{-n}}. \tag{6.76}$$

This condition is expected to hold for a local field theory for which the particle interpretation is possible and which shows good thermodynamical properties.

This condition implies the split-inclusion as follows.

**Theorem 6.35**
For a vacuum representation of an operator algebra system of local observables on a separable Hilbert space, assume that the vacuum vector $\Omega$ is a cyclic vector of any $\mathfrak{A}(D)$ and it satisfies the nuclearity condition. In this case for any $D_1$ there exists a $D_2 \supset D_1$ such that $\mathfrak{A}(D_2)$ has a split-inclusion of $\mathfrak{A}(D_1)$.

# APPENDIX A: HILBERT SPACE AND OPERATORS

## A.1 Hilbert space

The elements of a *Hilbert space* are called *vectors*. Between these vectors a linear operation is given. For two vectors $\Phi_1, \Phi_2$ and two complex numbers $c_1, c_2$, a *linear combination*

$$c_1\Phi_1 + c_2\Phi_2$$

is determined as a vector. We can also consider this as a combination of a product between vectors and complex numbers $\Psi_i = c_i\Phi_i$ $(i = 1, 2)$ and a sum of vectors $\Psi_1 + \Psi_2$. The fundamental properties of a linear space hold with respect to this operation. For example, we have properties like $1\Phi = \Phi$, $\Phi + (-1)\Phi = 0$.

A characteristic property of a Hilbert space is that we can define an *inner product* between two vectors $\Phi, \Psi$ given by a complex number $(\Phi, \Psi)$. Its basic properties are

1. **Linearity**: $(\Phi, c_1\Psi_1 + c_2\Psi_2) = c_1(\Phi, \Psi_1) + c_2(\Phi, \Psi_2)$. (A.1)
2. **Hermiticity**: $\overline{(\Phi, \Psi)} = (\Psi, \Phi)$.
3. **Positivity**: $(\Psi, \Psi) \geq 0$. If $(\Psi, \Psi) = 0$, then $\Psi = 0$.

Here $\overline{a + ib} = a - ib$ ($a, b$ are real numbers). From (1) and (2) the antilinearity in the first vector follows:

$$(c_1\Phi_1 + c_2\Phi_2, \Psi) = \bar{c_1}(\Phi_1, \Psi) + \bar{c_2}(\Phi_2, \Psi). \tag{A.2}$$

In the mathematical literature, usually the linearity like eq. (A.1) is assumed for the first component $\Phi$ of the inner product $(\Phi, \Psi)$ and therefore antilinearity like eq. (A.2) for the second component $\Psi$. Here we follow the convention in the physics literature.

From positivity we obtain the following *Cauchy–Schwarz inequality* (Lemma 2.9):

$$|(\Phi, \Psi)|^2 \leq (\Phi, \Phi)(\Psi, \Psi) \quad \text{(for any } \Phi, \Psi) \tag{A.3}$$

Using the basic properties of the inner product and the above inequality, we define the *norm*

$$\|\Psi\| = (\Psi, \Psi)^{1/2} \geq 0$$

and obtain its fundamental properties:

1. $\|\Psi\| \geq 0$. $\|\Psi\| = 0$ is equivalent to $\Psi = 0$,
2. $\|c\Psi\| = |c|\|\Psi\|$,
3. $\|\Psi_1 + \Psi_2\| \leq \|\Psi_1\| + \|\Psi_2\|$ (triangle inequality).

If a sequence (or directed family) of vectors $\Psi_n$ satisfies

$$\lim \|\Psi_n - \Psi\| = 0,$$

then we write $\lim \Psi_n = \Psi$ and the sequence is said to *converge strongly* to the vector $\Psi$ ("strongly" may be omitted). If

$$\lim(\Phi, \Psi_n - \Psi) = 0$$

holds for any vector $\Phi$, we write $w - \lim \Psi_n = \Psi$ and the sequence is said to *converge weakly* to $\Psi$. If we have strong convergence, then we also have weak convergence by (A.3). However, the converse does not necessarily hold.

If a sequence of vectors $\Psi_n$ satisfies

$$\|\Psi_m - \Psi_n\| \to 0 \qquad (m, n \to \infty)$$

then it is called a *Cauchy sequence*. A strongly convergent sequence is a Cauchy sequence. In a Hilbert space we require *completeness*, i.e. each Cauchy sequence must have a limit vector to which it converges strongly.

To summarize the above definition of a complex Hilbert space, it is a complex linear space with inner product which is complete with respect to the norm defined from the inner product.

## A.2 Pre-Hilbert space and completion

If we exclude the property of completeness in the definition of a Hilbert space given in the previous section, then we obtain the definition of a *pre-Hilbert space*. Let us explain its completion to a Hilbert space as its extension in a certain sense.

Consider the set of all Cauchy sequences $\{\Psi_n\}$ of vectors $\Psi_n (n = 1, 2, \ldots)$ in a given space $\mathscr{H}$. If $\mathscr{H}$ can be extended to a Hilbert space, then each such $\{\Psi_n\}$ must converge in the extended space. We intend to adopt Cauchy sequences themselves as vectors of the Hilbert space to be constructed, in place of their limit vectors.

Two Cauchy sequences $\Psi = \{\Psi_n\}$ and $\Phi = \{\Phi_n\}$ satisfying

$$\lim \|\Psi_n - \Phi_n\| = 0 \tag{A.4}$$

are defined to be equivalent. (We introduce this definition since their limit vectors (if they exist) must be equal.) We denote the set of all equivalence classes of Cauchy sequences by $\bar{\mathscr{H}}$ and define linear combination and inner product by

$$c\{\Phi_n\} + d\{\Psi_n\} = \{c\Phi_n + d\Psi_n\}, \tag{A.5}$$

$$(\{\Phi_n\}, \{\Psi_n\}) = \lim(\Phi_n, \Psi_n). \tag{A.6}$$

Then $\bar{\mathscr{H}}$ becomes a Hilbert space. It follows from the definition of a Cauchy sequence that the r.h.s. of (A.5) is again a Cauchy sequence, and the r.h.s. of (A.6) converges. We can also prove the completeness of $\bar{\mathscr{H}}$. If we identify each element $\Psi$ of $\mathscr{H}$ with the Cauchy sequence $\hat{\Psi} = \{\Psi_n\}$ for which $\Psi_n = \Psi$ for all $n$, then $\mathscr{H}$ is isomorphic to a linear subset $\hat{\mathscr{H}} = \{\hat{\Psi}\}$ of $\bar{\mathscr{H}}$ including the inner product, and $\hat{\mathscr{H}}$ is dense in $\bar{\mathscr{H}}$. Under this identification of $\mathscr{H}$ with $\hat{\mathscr{H}}$, $\bar{\mathscr{H}}$ is an extension of $\mathscr{H}$. The Hilbert space $\bar{\mathscr{H}}$ is called the completion of $\mathscr{H}$ where we usually identify $\mathscr{H}$ and $\hat{\mathscr{H}}$.

If the positivity of the inner product does not hold, but

3$'$. *Positive semidefiniteness*: $(\Psi, \Psi) \geq 0$ for all $\Psi \in \mathscr{H}$

holds (namely we may have $(\Psi, \Psi) = 0$ for $\Psi \neq 0$) instead, then we can obtain a Hilbert space by completion after gaining positivity in the following way.

From the positive semidefiniteness we can derive eq. (A.3). By defining a set $N$ of vectors $\Phi$ satisfying $(\Phi, \Phi) = 0$ as

$$N = \{\Phi \in \mathscr{H}; (\Phi, \Phi) = 0\},$$

then the elements $\Phi$ in $N$ are orthogonal to any vector $\Psi$ ($(\Phi, \Psi) = 0$) by (A.3) and form a linear subset. We define $\Psi_1$ and $\Psi_2$ to be equivalent if $\Psi_1 - \Psi_2 \in N$, and consider all equivalence classes of vectors in $\mathscr{H}$ (denoted by $\mathscr{H}/N$ and called the *quotient space*). By the linearity of $N$, $\mathscr{H}/N$ is also a linear space. If $\Phi_1 - \Phi_2 \in N$ and $\Psi_1 - \Psi_2 \in N$, then by the orthogonality of $N$ and $\mathscr{H}$ we get $(\Phi_1, \Psi_1) = (\Phi_2, \Psi_2)$. Hence, the inner product $(\Phi, \Psi)$ of $\mathscr{H}$ defines an inner product on $\mathscr{H}/N$ as it is. In particular, $(\Phi, \Phi) = 0$ occurs only if $\Phi \in N$, i.e. $\Phi$ is in the equivalence class of 0. Namely we obtain positivity.

## A.3 Orthonormal basis

As an example of a Hilbert space, we introduce the so-called $l_2$ *space*, which consists of sequences of complex numbers. We consider a set of complex numbers $\Psi_\alpha$

$$\Psi = \{\Psi_\alpha\}_{\alpha \in A}$$

labelled by elements $\alpha$ of a set $A$. We take the set of all such $\Phi$ satisfying the $l_2$ condition

$$\sum_\alpha |\Psi_\alpha|^2 < \infty \tag{A.7}$$

as a space $\mathscr{H}$ and define linear operation and inner product by

$$c\Phi + d\Psi = \{c\Phi_\alpha + d\Psi_\alpha\}_{\alpha \in A}, \tag{A.8}$$

$$(\Phi, \Psi) = \sum \bar{\Phi}_\alpha \Psi_\alpha. \tag{A.9}$$

Then we obtain a complex Hilbert space, which is called $l_2$ space.

In any Hilbert space, there exists a set of vectors

$$\{e_\alpha\}_{\alpha \in A} \qquad (e_\alpha \in \mathscr{H})$$

satisfying the following condition and which is called an *orthonormal basis*.

**Orthonormality:** $\|e_\alpha\| = 1$ and $(e_\alpha, e_\beta) = 0$ for $\alpha \neq \beta$.

**Completeness:** The set of all linear combinations of a finite number of basis vectors

$$\sum_{n=1}^{N} c_{\alpha_n} e_{\alpha_n} \quad (c_{\alpha_n} \text{is a complex number})$$

is dense in $\mathscr{H}$. Here $A$ is the set of the label $\alpha$ which distinguish the basis vectors $e_\alpha$.

In this case, any vector $\Psi$ in $\mathscr{H}$ has the expansion

$$\Psi = \sum \Psi_\alpha e_\alpha \qquad (\Psi_\alpha = (e_\alpha, \Psi) \text{ is a complex number}) \tag{A.10}$$

and the linear operation and inner product satisfy (A.5), (A.6). Thus we can write it in the form of an $l_2$ space. The cardinal number of the set $A$ of labels (number of elements) does not depend on the choice of the orthonormal basis and is called the *dimension* of the space $\mathscr{H}$. In particular, if the dimension is a natural number $n$, it becomes the ordinary $n$ dimensional (Euclidian) space. If $A$ has a one-to-one correspondence with the set of all natural numbers, $\mathscr{H}$ is of *countable infinite dimension*. In these two cases $\mathscr{H}$ is said to be *separable*.

## A.4 Unitary and antiunitary maps

For two Hilbert spaces $\mathscr{H}_1, \mathscr{H}_2$, a bijection $U$ (an isomorphism as a set) which maps $\mathscr{H}_1$ to $\mathscr{H}_2$, preserving the structure of the Hilbert spaces, i.e. satisfying the following two properties, is called a *unitary map*:

$$U(c_1\Psi_1 + c_2\Psi_2) = c_1 U\Psi_1 + c_2 U\Psi_2, \tag{A.11}$$

$$(U\Phi, U\Psi) = (\Phi, \Psi). \tag{A.12}$$

(The injectivity and the property (A.11) can be obtained from (A.12).) In this case the dimensions of $\mathscr{H}_1$ and $\mathscr{H}_2$ are the same. Conversely between Hilbert spaces of the same dimension there exists a unitary map. Thus, an equivalence class of Hilbert spaces (by unitary maps) can be completely classified by its dimension.

The inverse map $U^{-1}$ of a unitary map $U$ from $\mathscr{H}_1$ to $\mathscr{H}_2$ is defined by

$$U^{-1}(U\Psi) = \Psi \qquad (\Psi \in \mathscr{H}_1)$$

and is a unitary map from $\mathscr{H}_2$ to $\mathscr{H}_1$.

If a map $f$ assigning complex numbers $f(\Psi)$ to elements $\Psi$ of $\mathscr{H}$ satisfies linearity

$$f(c_1\Psi_1 + c_2\Psi_2) = c_1 f(\Psi_1) + c_2 f(\Psi_2), \tag{A.13}$$

then $f$ is called a *linear functional*. The continuity of $f$ with respect to strong convergence, the continuity with respect to weak convergence and the boundedness in the following sense are all equivalent.

$$|f(\Psi)| \le \lambda \|\Psi\|. \tag{A.14}$$

The functional

$$\Phi^*(\Psi) = (\Phi, \Psi) \tag{A.15}$$

determined by an element $\Phi$ of $\mathscr{H}$ is a bounded linear functional and conversely any bounded linear functional is of this form (*Riesz theorem*).

A linear combination of linear functionals can be naturally defined by the following equation:

$$(c_1 f_1 + c_2 f_2)(\Psi) \equiv c_1 f_1(\Psi) + c_2 f_2(\Psi). \tag{A.16}$$

The $\Phi^*$ defined by eq. (A.12) satisfies the following equation:

$$(c_1\Phi_1 + c_2\Phi_2)^* = \bar{c_1}\Phi_1^* + \bar{c_2}\Phi_2^* \tag{A.17}$$

Further, defining the inner product by

$$(\Phi_1^*, \Phi_2^*) = (\Phi_2, \Phi_1), \tag{A.18}$$

the set of all $\Phi^*$ becomes a Hilbert space. We denote it by $\mathscr{H}^*$ and call it the *conjugate space* of $\mathscr{H}$.

If a bijection $V$ from a Hilbert space $\mathscr{H}_1$ to a Hilbert space $\mathscr{H}_2$ satisfies the following two properties, it is called an *antiunitary map*:

$$V(c_1\Psi_1 + c_2\Psi_2) = \bar{c_1} V\Psi_1 + \bar{c_2} V\Psi_2, \tag{A.19}$$

$$(V\Phi, V\Psi) = (\Psi, \Phi). \tag{A.20}$$

The injectivity and (A.19) can be derived from (A.20). The map

$$V\Phi \equiv \Phi^*$$

from $\mathscr{H}$ to $\mathscr{H}^*$ is an antiunitary map.

If a map $U$ from $\mathscr{H}_1$ to $\mathscr{H}_2$ satisfies the linearity (A.11) it is called a *linear map*. Naturally, linear combination is defined as in (A.16) among linear maps. $U$ is called *bounded* if there exists a real number $\lambda$ satisfying

$$\|U\Psi\| \le \lambda \|\Psi\| \tag{A.21}$$

for any vector $\Psi$. The infimum of $\lambda$ satisfying (A.21) is denoted as $\|U\|$ and is called the *norm* of $U$. It satisfies the fundamental properties (see Section A.1) of a norm.

For a bounded linear map $U$, there exists a bounded linear map $U^*$ satisfying the following equation due to the Riesz theorem, and it is uniquely determined:

$$(\Phi, U^*\Psi) = (U\Phi, \Psi). \tag{A.22}$$

The operator $U^*$ is called the *Hermitian conjugate* of $U$. The unitarity of a map $U$ is equivalent to $U^* = U^{-1}$, i.e. to the properties

$$U^*U = \mathbf{1}_{(1)}, \quad UU^* = \mathbf{1}_{(2)}, \tag{A.23}$$

where $\mathbf{1}_{(i)}$ is the *identity operator* of $\mathscr{H}_i$.

$$\mathbf{1}_{(i)}\Psi = \Psi \qquad (\Psi \in \mathscr{H}_i)$$

Using the antilinearity (A.19) instead of the linearity (A.11) in the above definition, we obtain the definition of an *antilinear map*. In this case the *Hermitian conjugate* $V^*$ of $V$ is defined by

$$(\Phi, V^*\Psi) = (\Psi, V\Phi) \tag{A.24}$$

and is an antilinear map. $V^* = V^{-1}$ is the necessary and sufficient condition for the antilinear map to be an antiunitary map.

In the case $\mathscr{H}_1 = \mathscr{H}_2 = \mathscr{H}$ in the above definitions, we use the terminology *operator* instead of map. Then we have *unitary operators*, *bounded linear operators*, *antiunitary operators*, *bounded antilinear operators*, etc. on $\mathscr{H}$.

## A.5 Subspaces and projection operators

If a subset $\mathscr{K}$ of a Hilbert space $\mathscr{H}$ is closed under linear combination, namely if for any two elements $\Psi_1, \Psi_2$ in $\mathscr{K}$ and any complex numbers $c_1, c_2, c_1\Psi_1 + c_2\Psi_2$ is again in $\mathscr{K}$, then $\mathscr{K}$ is called a *linear subset*. If for any convergent sequence in $\mathscr{H}$

$$\lim \Psi_n = \Psi$$

with $\Psi_n \in \mathscr{K}$ for all $n$, $\Psi \in \mathscr{K}$ holds, then $\mathscr{K}$ is called a *subspace* of $\mathscr{H}$. Even if we use weakly convergent sequences instead of strongly convergent sequences in this definition, we obtain a definition equivalent to the above.

For a subset $S$ of $\mathscr{H}$, the set of all vectors $\Psi$ in $\mathscr{H}$ "orthogonal" to any $\Phi$ in $S$ in the sense that

$$(\Phi, \Psi) = 0$$

is denoted by $S^\perp$. This $S^\perp$ is always a subspace. A necessary and sufficient condition for $S$ itself to be a subspace is the following equation:

$$(S^\perp)^\perp = S.$$

For a subspace $\mathscr{K}$, $\mathscr{K}^\perp$ is called its *orthogonal complement*. Any vector $\Psi$ in $\mathscr{H}$ can be uniquely decomposed as

$$\Psi = \Psi_{||} + \Psi_\perp \qquad (\Psi_{||} \in \mathscr{K}, \quad \Psi_\perp \in \mathscr{K}^\perp). \tag{A.25}$$

The map

$$P(\mathscr{K})\Psi = \Psi_{||}$$

which assigns $\Psi_{||}$ to $\Psi$ is a bounded linear operator with the following properties:

$$P(\mathscr{K})^2 = P(\mathscr{K}), \qquad P(\mathscr{K})^* = P(\mathscr{K}) \tag{A.26}$$

Conversely, if a linear operator $P$ satisfies $P^2 = P = P^*$, the image of $P$

$$\mathscr{K} = P\mathscr{H} = \{P\Psi; \Psi \in \mathscr{H}\}$$

is a subspace and $P = P(\mathscr{K})$. Furthermore, the following equation is satisfied:

$$P(\mathscr{K}^\perp) = \mathbf{1} - P(\mathscr{K}). \tag{A.27}$$

A linear operator $P$ satisfying $P^2 = P = P^*$ is called an (orthogonal) *projection (operator)*.

## A.6 Direct sum and direct integral

For a family of Hilbert spaces $\mathscr{H}_\alpha$ labelled by elements $\alpha$ of a set $I$, their *direct sum*

$$\mathscr{H} = \oplus \mathscr{H}_\alpha$$

is defined as follows. $\mathscr{H}$ is the set of all collections of element $\Psi_\alpha$ from each $\mathscr{H}_\alpha$

$$\Psi = \{\Psi_\alpha\}_{\alpha \in I} \equiv \bigoplus_\alpha \Psi_\alpha$$

which satisfy the following $l_2$ condition:

$$\|\Psi\|^2 \equiv \sum_\alpha \|\Psi_\alpha\|^2 < \infty. \tag{A.28}$$

Linear combination and inner product are defined in terms of component vectors by

$$c(\bigoplus_\alpha \Phi_\alpha) + d(\bigoplus_\alpha \Psi_\alpha) = \bigoplus_\alpha (c\Phi_\alpha + d\Psi_\alpha), \tag{A.29}$$

$$(\bigoplus_\alpha \Phi_\alpha, \bigoplus_\alpha \Psi_\alpha) = \sum_\alpha (\Phi_\alpha, \Psi_\alpha). \tag{A.30}$$

Due to (A.28), (A.30) is absolutely convergent and $\mathscr{H}$ is a Hilbert space. For a bounded linear operator $A_\alpha$ on each $\mathscr{H}_\alpha$ satisfying

$$\|A_\alpha\| \leq \lambda$$

for all $\alpha \in I$, $\lambda$ being a positive number independent of $\alpha$, their direct sum

$$A = \bigoplus_\alpha A_\alpha, \quad A(\bigoplus_\alpha \Psi_\alpha) = \bigoplus_\alpha A_\alpha \Psi_\alpha \tag{A.31}$$

is a bounded linear operator on $\mathscr{H}$.

Next, let us introduce the *direct integral* as a continuous version of the direct sum. Consider a measure space $(\Omega, d\mu)$ and a collection of separable Hilbert spaces $\mathscr{H}_\omega$ labelled by points $\omega$ of $\Omega$. Suppose that the set $\{\omega : \dim \mathscr{H}_\omega = n\}$ is measurable for each $n$. For example for $\Omega = \mathbf{R}$ (the real numbers) or $\Omega = [0, 1]$ we take as a measure the Lebesgue measure $\mathrm{d}\mu(\omega) = \mathrm{d}\omega$. In order to define measurability, fix a countable number of families $e^n = \{e^n_\omega\}_{\omega \in \Omega}$, $(n = 1, 2, \ldots)$, consisting of a vector $e^n_\omega$ from each $\mathscr{H}_\omega$, such that for each $\omega$ the set of all linear combinations of a finite number of $e^n_\omega$ are dense in $\mathscr{H}_\omega$ and $(e^n_\omega, e^m_\omega)$ is $\mu$-measurable for any $n, m$. A family $\{\Psi_\omega\}$ of $\Psi_\omega \in \mathscr{H}_\omega$, such that $(e^n_\omega, \Psi_\omega)$ is $\mu$-measurable for any $n$ is called a *measurable family* and the direct integral

$$\mathscr{H} = \int \mathscr{H}_\omega \, \mathrm{d}\mu(\omega)$$

is defined to be the set of all measurable families (usually the symbol on the extreme right-hand side is used)

$$\Psi = \{\Psi_\omega\}_{\omega \in \Omega} \equiv \int \Psi_\omega \, \mathrm{d}\mu(\omega)$$

which satisfy the following $L_2$ condition:

$$\|\Psi\|^2 \equiv \int \|\Psi_\omega\|^2 \, \mathrm{d}\mu(\omega) < \infty. \tag{A.32}$$

More precisely speaking, a vector in $\mathscr{H}$ is defined to be an equivalence class where two measurable families are identified to be equivalent if their components $\Psi_\omega$ are equal almost everywhere with respect to $\mu$.

Linear combination and inner product are defined in terms of the component vectors by

$$c \int \Phi_\omega \, \mathrm{d}\mu(\omega) + d \int \Psi_\omega \, \mathrm{d}\mu(\omega) = \int (c\Phi_\omega + d\Psi_\omega) \, \mathrm{d}\mu(\omega), \tag{A.33}$$

and

$$\left( \int \Phi_\omega \, \mathrm{d}\mu(\omega), \int \Psi_\omega \, \mathrm{d}\mu(\omega) \right) = \int (\Phi_\omega, \Psi_\omega) \, \mathrm{d}\mu(\omega). \tag{A.34}$$

Equation (A.34) is absolutely integrable by (A.32) and $\mathscr{H}$ becomes a Hilbert space. If $A_\omega$ is a bounded linear operator on $\mathscr{H}_\omega$ and if $(e^m_\omega, A_\omega e^n_\omega)$ is a measurable function of $\omega$ for all $m, n$, then $\{\Psi_\omega\} \in \mathscr{H}$ implies $\{A_\omega \Psi_\omega\} \in \mathscr{H}$ and the direct integral

$$A = \int A_\omega \, \mathrm{d}\mu(\omega), \quad A \int \Psi_\omega \, \mathrm{d}\mu(\omega) = \int A_\omega \Psi_\omega \, \mathrm{d}\mu(\omega) \tag{A.35}$$

becomes a bounded linear operator on $\mathscr{H}$.

If each point in $\Omega$ has measure 1, then the direct integral becomes a direct sum.

### A.7 Spectral decomposition

A bounded linear operator $A$ satisfying $A^* = A$ is called *self-adjoint*. For bounded self-adjoint operators $A$ on a separable Hilbert space $\mathscr{H}$, there is a direct integral

$$\mathscr{L} = \int \mathscr{H}_\lambda \, \mathrm{d}\mu(\lambda)$$

over a measure space $(\mathbf{R}, \mathrm{d}\mu)$ on the real line $\mathbf{R}$ and a unitary map $U$ from $\mathscr{H}$ to $\mathscr{L}$, such that the action of $A$ on a vector $\Psi$ can be represented on its image

$$U\Psi = \int \Psi_\lambda \, \mathrm{d}\mu(\lambda)$$

in $\mathscr{L}$ as an operation of multiplying a real number $\lambda$:

$$UA\Psi = \int \lambda \Psi_\lambda \, \mathrm{d}\mu(\lambda). \tag{A.36}$$

This is called the *spectral representation* of $A$. Outside the interval $[-\|A\|, \|A\|]$ the measure $\mu$ is zero.

Let $B$ be a measurable set in $\mathbf{R}$ and $\chi_B$ be its defining function:

$$\chi_B(\lambda) = 1, \quad \text{for } \lambda \in B, \qquad \chi_B(\lambda) = 0, \quad \text{for } \lambda \notin B\,.$$

The set of all $\{\Psi_\lambda\} \in \mathscr{L}$ with vanishing components $\Psi_\lambda$ for $\lambda \notin B$ is a subspace of $\mathscr{L}$ and is denoted as

$$\mathscr{L}_B = \int_B \mathscr{H}_\lambda \, \mathrm{d}\mu(\lambda).$$

The projection operator onto $\mathscr{L}_B$ will be denoted as

$$P(\mathscr{L}_B) \int \Psi_\lambda \, \mathrm{d}\mu(\lambda) = \int \chi_B(\lambda) \Psi_\lambda \, \mathrm{d}\mu(\lambda)$$

The corresponding projection operator on $\mathscr{H}$

$$E_A(B) = P(U^{-1}\mathscr{L}_B)$$

is called a *spectral projection operator* of $A$. $E_A$ is a set function with projection operators as its values having the same properties as the measure and we can write

$$A = \int \lambda E_A(\mathrm{d}\lambda). \tag{A.37}$$

In particular, if the measure is zero outside a countable subset $\Lambda$ of $\mathbf{R}$, we set $E_\lambda \equiv E_A(\{\lambda\})$ for the set with only one element $\lambda$ for each $\lambda \in \Lambda$. Then $E_\lambda\mathscr{H}$ is an *eigenspace* of $A$ consisting of all *eigenvectors* of $A$ satisfying

$$A\Psi = \lambda\Psi$$

and $\lambda$ is an *eigenvalue* of $A$. Eigenspaces $E_\lambda\mathscr{H}$ with different eigenvalues $\lambda$ are mutually orthogonal, and $\mathscr{H}$ becomes their direct sum. The spectral projection operator $E_A(B)$ is simply the sum of all $E_\lambda$ over all eigenvalues $\lambda$ in $B$. (A.35) becomes

$$A = \Sigma\lambda E_\lambda. \tag{A.35$'$}$$

The union of all open subsets (of $\mathbf{R}$) with $\mu$-measure zero is the biggest open set with $\mu$-measure zero and its complement (being a closed set) is called the *support* of $\mu$ and the *spectrum* of $A$. $\mu$ is called the *spectral measure* of $A$. In the case of (A.35), the spectrum of $A$ is the closure of the set of eigenvalues.

In the case where a bounded linear operator $A$ commutes with $A^*$ ($AA^* = A^*A$) $A$ is called *normal*. For normal operators the spectral decomposition exists if we take the complex numbers $\mathbf{C}$ instead of $\mathbf{R}$ in the above discussion. Any unitary operator $U$ is normal because it satisfies $UU^* = U^*U$ ($= \mathbf{1}$). In this case the measure $\mu$ is zero outside the unit circle $\mathbf{T} = \{z \in \mathbf{C}; |z| = 1\}$. Thus we can write

$$U = \int_0^{2\pi} \mathrm{e}^{\mathrm{i}\theta}\,\mathrm{d}E(\theta)$$

Using the spectral decomposition we can define a *function* $f(A)$ of a self-adjoint or a normal operator $A$ as

$$f(A) = \int f(\lambda)E(\mathrm{d}\lambda), \quad Uf(A)\Psi = \int f(\lambda)\Psi_\lambda\,\mathrm{d}\mu(\lambda)$$

If $f(x)$ is a polynomial of $x$ this definition coincides with the algebraically defined polynomial $f(A)$ of $A$.

## A.8 Trace

A linear operator $A$ which satisfies

$$(\Psi, A\Psi) \geq 0$$

for any vector $\Psi$ is called a *positive operator*. Given a positive operator $A$ and an orthonormal basis $\{e_\alpha\}$, the *trace* of $A$ is defined by

$$\mathrm{Tr}\, A = \sum_\alpha (e_\alpha, Ae_\alpha), \tag{A.38}$$

$0 \leq \mathrm{Tr} A \leq \infty$, and its value does not depend on the orthonormal basis.

For any bounded linear operator $A$, $A^*A$ is a positive operator and its function

$$|A| \equiv (A^*A)^{1/2} \tag{A.39}$$

is also a positive operator. $|A|$ is called the *absolute value* of $A$. If

$$\|A\|_1 \equiv \mathrm{Tr}\,|A| < \infty$$

$A$ is said to be in the *trace class*, the set of all such $A$ is denoted by $\mathscr{T}(\mathscr{H})$ and is called the trace class.

Denoting the set of all bounded linear operators on $\mathscr{H}$ by $\mathscr{B}(\mathscr{H})$, $\mathscr{T}(\mathscr{H})$ is an ideal of $\mathscr{B}(\mathscr{H})$. Namely, $\mathscr{T}(\mathscr{H})$ is a linear subset and if $\rho \in \mathscr{T}(\mathscr{H})$ and $A \in \mathscr{B}(\mathscr{H})$, then $\rho A$ as well as $A\rho$ are in $\mathscr{T}(\mathscr{H})$. If $A$ is in the trace class, then even if $A$ is not a positive operator, the r.h.s. of (A.38) is absolutely convergent and the sum defines $\mathrm{Tr}\, A$, which is finite.

$\mathrm{Tr}\, A$ is a linear functional on $\mathscr{T}(\mathscr{H})$ and has the following properties:

*Positivity*: $\mathrm{Tr}(A^*A) \geq 0$

*Invariance*: $\mathrm{Tr}(UAU^*) = \mathrm{Tr} A$, (for any unitary operator $U$).

For a general bounded linear operator $A \in \mathscr{B}(\mathscr{H})$, the following equation holds:

$$\mathrm{Tr}(A^*A) = \mathrm{Tr}(AA^*). \tag{A.40}$$

If $A \in \mathscr{T}(\mathscr{H})$ and $B \in \mathscr{B}(\mathscr{H})$, then $AB \in \mathscr{T}(\mathscr{H})$ and the following equation holds:

$$\mathrm{Tr}(AB) = \mathrm{Tr}(BA)$$

If $A$ is a self-adjoint operator in the trace class, $A$ possesses a discrete spectral decomposition (A.35)′ and we have

$$\|A\|_1 = \Sigma|\lambda| \dim E_\lambda < \infty. \tag{A.41}$$

Here, $\dim E_\lambda$ denotes the dimension of the eigenspace $E_\lambda\mathscr{H}$.

### A.9 Unbounded operators

If the map $A$ which maps a linear subset $\mathscr{D}$ of a Hilbert space $\mathscr{H}$ into $\mathscr{H}$ possesses the linearity

$$A(c_1\Psi_1 + c_2\Psi_2) = c_1 A\Psi_1 + c_2 A\Psi_2 \tag{A.42}$$

with respect to any vectors $\Psi_1$, $\Psi_2$ in $\mathscr{D}$ and any complex numbers $c_1$, $c_2$, then $A$ is called a *linear operator* and $\mathscr{D}$ is called its *domain*. The case we are dealing with below is the case where $\mathscr{D}$ is dense in $\mathscr{H}$.

Let $\mathscr{D}$ be dense. For a given linear operator $A$ with domain $\mathscr{D}$, we define an operator $A^*$ with its domain $\mathscr{D}^*$ consisting of all vectors $\Phi$ for which there exists a vector $\Phi'$ satisfying

$$(\Phi, A\Psi) = (\Phi', \Psi) \tag{A.43}$$

for all vectors $\Psi$ in $\mathscr{D}$, by

$$A^*\Phi = \Phi'. \tag{A.44}$$

Then $A^*$ becomes a linear operator and is called the *adjoint* (*operator*) of $A$. (In order that $\Phi'$ is determined uniquely for each $\Phi$, we assumed the condition that $\mathscr{D}$ is dense.) The following concept of closable operators is important in the consideration of adjoint operators.

A linear operator $A$ is called *closable* if, for any sequence $\Psi_n$ of vectors in the domain $\mathscr{D}$ of $A$ converging to 0, the limit vector $\Phi$ of $A\Psi_n$, whenever it also converges, is zero. In this case, we denote the set of all $\Psi$ satisfying

$$\lim \Psi_n = \Psi, \quad \lim A\Psi_n = \Phi \tag{A.45}$$

for a sequence of vectors $\Psi_n$ in $\mathscr{D}$ by $\bar{\mathscr{D}}$ and define an operator $\bar{A}$ with the domain $\bar{\mathscr{D}}$ by

$$\bar{A}\Psi = \Phi. \tag{A.46}$$

$\bar{A}$ becomes a linear operator and is called the *closure* of $A$. If we put all $\Psi_n$ equal to $\Psi$ in the above for any $\Psi$ in $\mathscr{D}$, then $\Phi = A\Psi$. Hence $\bar{\mathscr{D}}$ contains $\mathscr{D}$ and $\bar{A}$ is identical with $A$ on $\mathscr{D}$. Namely $\bar{A}$ is an extension of $A$. An equivalent definition is obtained if we take weak convergence instead of strong convergence in the above definition.

In the direct product $\mathscr{H} \times \mathscr{H}$ as a linear topological space, the linear subset

$$G = \{(\Psi, A\Psi) \in \mathscr{H} \times \mathscr{H}; \Psi \in \mathscr{D}\}$$

is called the *graph* of $A$. In the opposite direction, a linear subset $\Gamma$ of $\mathscr{H} \times \mathscr{H}$ is the graph of a linear operator on $\mathscr{H}$ if there exists at most one $\Phi$ satisfying $(\Psi,\Phi) \in \Gamma$ for each $\Psi$. The condition that $A$ is closable is the condition that the closure $\bar{G}$ of the graph $G$ of $A$ in $\mathscr{H} \times \mathscr{H}$ becomes again a graph, and if this is the case, then $\bar{A}$ is the operator of which $\bar{G}$ is the graph.

The property that the domain $\mathscr{D}^*$ of the adjoint operator is dense is necessary and sufficient for $A$ to be closable. If $A$ is closable then the adjoint of its adjoint is its closure:

$$(\mathscr{D}^*)^* = \bar{\mathscr{D}}, \qquad (A^*)^* = \bar{A}. \tag{A.47}$$

$A$ is called a *closed operator* if $\bar{A} = A$. $A^*$ is in any case a closed operator. Also $\bar{A}$ is a closed operator ($\overline{(\bar{A})} = \bar{A}$). A bounded operator is closable and closable operators with domain $\mathscr{H}$ are bounded and closed.

A linear operator $A$ satisfying

$$(\Phi, A\Psi) = (A\Phi, \Psi) \tag{A.48}$$

for any vectors $\Psi$ and $\Phi$ in its domain is called a *Hermitian operator*. For a Hermitian operator with dense domain, $A^*$ coincides with $\bar{A}$ on the domain of $\bar{A}$. However it does not necessarily satisfy $A^* = \bar{A}$. If $A^* = A$, $A$ is called *self-adjoint*. In the case where $A^* = \bar{A}$, $A$ is called *essentially self-adjoint*. (In both cases the domain is to be dense, and (A.48) is naturally satisfied.)

A Hermitian operator cannot necessarily be extended to a self-adjoint operator. Even if it can be extended, self-adjoint extensions are not necessarily unique. A self-adjoint extension becomes unique only in the case where the Hermitian operator is essentially self-adjoint. A positive operator with dense domain ($(\Psi \mathbf{A} \Psi) \geq 0$ for any vector $\Psi$ in the domain) is a Hermitian operator and has a self-adjoint extension.

For self-adjoint operators, the spectral decomposition is exactly the same as for bounded self-adjoint operators.

### A.10 One-parameter group of unitary operators

A family of unitary operators $U(t)$ labelled by a real number $t$ is called a *one-parameter group* of unitary operators if the following two conditions are satisfied:

(i) $U(s)U(t) = U(t+s)$ ($s$, $t$ run over all real numbers),

(ii) $\lim_{s \to 0} U(s)\Psi = \Psi$ (for all vectors $\Psi$).

Since $U(s)$ is unitary and has an inverse, if we take $t = 0$ in (i) then we obtain $U(0) = 1$. If we set $t = -s$ in (i), then we get $U(-s) = U(s)^{-1} = U(s)^*$. Applying (ii) to (i) we obtain continuity

$$\lim_{t' \to t} U(t')\Psi = U(t)\Psi \tag{A.49}$$

for arbitrary $t$.

For a one-parameter group of unitary operators, a self-adjoint linear operator $H$ (which is called the *generator*) satisfying

$$U(s) = \exp(isH) \tag{A.50}$$

exists and is unique. Using the spectral representation of $H$ ((A.37) for $H$ instead of $A$) we obtain

$$U(s) = \int e^{is\lambda} E(d\lambda).$$

The condition that the vector $\Psi$ is in the domain of $H$ is given by

$$\int \lambda^2 \, d\mu_\Psi(\lambda) < \infty, \qquad \mu_\Psi(B) \equiv (\Psi, E(B)\Psi)$$

Such a $\Psi$ can also be characterized as a vector for which the following derivative exists (in strong topology):

$$\frac{d}{dt} U(t)\Psi \equiv \lim_{s\to 0} s^{-1}(U(t+s) - U(s))\Psi = iHU(t)\Psi = iU(t)H\Psi.$$

## A.11 Tensor product

Suppose we are given Hilbert spaces $\mathscr{H}$, $\mathscr{K}$ and their orthonormal bases $\{e_m\}$, $\{f_n\}$. A Hilbert space with an orthonormal basis consisting of symbols $e_m \otimes f_n$ (where we consider any combinations of $m, n$) is denoted by $\mathscr{H} \otimes \mathscr{K}$ and is called the *tensor product* of $\mathscr{H}$ and $\mathscr{K}$. A vector in this space is of the form

$$\Psi = \sum_{m,n} \Psi_{m,n} e_m \otimes f_n, \qquad \sum_{m,n} |\Psi_{m,n}|^2 < \infty$$

($\Psi_{m,n}$ are complex numbers). Its linear combination and inner product are given by

$$c \sum_{m,n} \Phi_{m,n} e_m \otimes f_n + d \sum_{m,n} \Psi_{m,n} e_m \otimes f_n = \sum_{m,n} (c\Phi_{m,n} + d\Psi_{m,n}) e_m \otimes f_n,$$

$$\left( \sum_{m,n} \Phi_{m,n} e_m \otimes f_n, \sum_{m,n} \Psi_{m,n} e_m \otimes f_n \right) = \sum_{m,n} \overline{\Phi_{m,n}} \Psi_{m,n}.$$

For the vectors

$$\xi = \sum_m c_m e_m \in \mathscr{H}, \qquad \eta = \sum_n d_n f_n \in \mathscr{K}$$

in $\mathscr{H}$ and $\mathscr{K}$, we define the vector $\xi \otimes \eta$ in $\mathscr{H} \otimes \mathscr{K}$ by

$$\xi \otimes \eta \equiv \sum_{m,n} c_m d_n e_m \otimes f_n. \tag{A.51}$$

$\xi \otimes \eta$ is linear in $\xi$ and $\eta$:

$$(c\xi_1 + d\xi_2) \otimes \eta = c(\xi_1 \otimes \eta) + d(\xi_2 \otimes \eta), \tag{A.52}$$

$$\xi \otimes (c\eta_1 + d\eta_2) = c(\xi \otimes \eta_1) + d(\xi \otimes \eta_2). \tag{A.53}$$

The tensor product has also the following characteristic properties.

$$(c\xi) \otimes \eta = \xi \otimes (c\eta) = c(\xi \otimes \eta), \tag{A.54}$$

$$(\xi_1 \otimes \eta_1, \xi_2 \otimes \eta_2) = (\xi_1, \xi_2)(\eta_1, \eta_2). \tag{A.55}$$

For any orthonormal bases $\{e'_\alpha\}$ and $\{f'_\beta\}$ of $\mathscr{H}$ and $\mathscr{K}$, $\{e'_\alpha \otimes f'_\beta\}$ constructed according to the above definition (A.51) becomes an orthonormal basis of $\mathscr{H} \otimes \mathscr{K}$. Therefore, we see that $\mathscr{H} \otimes \mathscr{K}$ does not depend on the choice of the orthonormal bases of $\mathscr{H}$ and $\mathscr{K}$ used in the above definition.

For linear operators $A$ and $B$ on $\mathscr{H}$ and $\mathscr{K}$, respectively, a linear operator $A \otimes B$ on $\mathscr{H} \otimes \mathscr{K}$ is defined by

$$(A \otimes B) \sum_{m,n} \Psi_{m,n} e_m \otimes f_n = \sum_{m,n} \Psi_{m,n}(Ae_m \otimes Bf_n). \tag{A.56}$$

The following formulae hold:

$$(A \otimes B)(\xi \otimes \eta) = (A\xi) \otimes (B\eta), \tag{A.57}$$

$$(\xi \otimes \eta, (A \otimes B)(\xi' \otimes \eta')) = (\xi, A\xi')(\eta, B\eta'). \tag{A.58}$$

# APPENDIX B: OPERATOR ALGEBRAS

Below we consider a $*$ algebra consisting of bounded linear operators on a Hilbert space. Here the $*$ represents Hermitian conjugation of an operator, and by a $*$ algebra we mean an algebra (with coefficient field of complex numbers) in which $A$ and $A^*$ are always simultaneously included.

As for the topology of the operators, there are many choices which are useful depending on the situation. A $*$ algebra which is closed in one of these topologies turns out to be either a $C^*$ algebra or a von Neumann algebra, both of which are important objects to be studied. Below we give a simple summary of introductory and basic materials.

## B.1 $C^*$ algebras

First we give an abstract definition.

**Definition B.1**
A set $\mathfrak{A}$ is called a $C^*$ *algebra* if it has the following properties.
(a) $\mathfrak{A}$ is an algebra with complex numbers as the coefficient field.
(b) On $\mathfrak{A}$ a bijection $A \in \mathfrak{A} \to A^* \in \mathfrak{A}$ is defined and satisfies the following properties, where $A_j \in \mathfrak{A}$, $c_j \in \mathbf{C}$ and $\bar{c}$ is the complex conjugate of $c$:

$$(c_1 A_1 + c_2 A_2)^* = \bar{c}_1 A_1^* + \bar{c}_2 A_2^*, \qquad (A_1 A_2)^* = A_2^* A_1^*$$
$$(A^*)^* = A$$

(c) A positive number $\|A\|$ (called the *norm* of $A$) satisfying the basic properties of a norm (see Section A.1) with respect to the linear operation of $\mathfrak{A}$ is given for each element $A$ of $\mathfrak{A}$, and $\mathfrak{A}$ is complete with respect to this norm.
(d) The norm has the following property (the property of a $C^*$ *norm*):

$$\|A^* A\| = \|A\|^2. \tag{B.1}$$

If the above properties are satisfied, the norm of A possesses the following properties.

$$\|\mathbf{1}\| = 1, \quad \|A^*\| = \|A\|, \qquad \|A_1 A_2\| \leq \|A_1\| \cdot \|A_2\|$$

In simple words, a $C^*$ algebra is a $*$ Banach algebra, the norm of which satisfies the property (B.1) of a $C^*$ norm.

The set of all bounded linear operators on a Hilbert space $\mathscr{H}$ is denoted by $\mathscr{B}(\mathscr{H})$. If the map $\pi$ from a $C^*$ algebra $\mathfrak{A}$ into $\mathscr{B}(\mathscr{H})$ preserves the algebraic calculus of $\mathfrak{A}$ in the following sense, $\pi$ is called a *representation* of $\mathfrak{A}$ (on $\mathscr{H}$).

$$\pi(c_1A_1 + c_2A_2) = c_1\pi(A_1) + c_2\pi(A_2)$$

$$\pi(A_1A_2) = \pi(A_1)\pi(A_2)$$

$$\pi(A^*) = \pi(A)^*$$

A representation is automatically continuous:

$$\|\pi(A)\| \leq \|A\|. \tag{B.2}$$

If a representation $\pi$ is injective (namely, $\pi(A) = 0$ only if $A = 0$), then $\pi$ is called *faithful*. A faithful representation preserves the norm:

$$\|\pi(A)\| = \|A\|. \tag{B.3}$$

The following theorem shows that essentially a $C^*$ algebra belongs to those $*$ algebras which are formed by bounded linear operators on a Hilbert space, closed in the norm topology, and the above definition gives its abstract version.

**Theorem B.2**
The norm

$$\|A\| = \sup\{\|A\Psi\|/\|\Psi\|; \Psi \in \mathscr{H}, \Psi \neq 0\} \tag{B.4}$$

of bounded linear operators on a Hilbert space $\mathscr{H}$ satisfies (B.1) and a $*$-subalgebra of $\mathscr{B}(\mathscr{H})$ closed in the norm topology is a $C^*$ algebra. Conversely, for any $C^*$ algebra $\mathfrak{A}$, a faithful representation $\pi$ on an appropriate Hilbert space $\mathscr{H}$ exists.

For a faithful representation $\pi$, $\pi(\mathfrak{A})$ is a $*$ subalgebra of $\mathscr{B}(\mathscr{H})$ which is closed in the norm topology and $\mathfrak{A}$ and $\pi(\mathfrak{A})$ are isomorphic with respect to all algebraic calculus. Moreover, since they are also isometric due to (B.3) it is possible to identify $\mathfrak{A}$ and $\pi(\mathfrak{A})$. However, as we shall explain later, among the representations of a $C^*$ algebra there are many inequivalent representations in general and it is characteristic for theory of $C^*$ algebras that whole varieties of representations become subjects of investigation. Therefore, the abstract definition independent of a specific representation given above is meaningful.

As has been proved in Section 2.3, a representation of a given $C^*$ algebra can be constructed from states $\varphi$ by the GNS construction. The direct sum representation of all representations $\pi_\varphi$ obtained in this way

$$\pi_u \equiv \bigoplus_{\varphi} \pi_\varphi \tag{B.5}$$

over all states $\varphi$ is called the *universal representation* and it represents one example of a faithful representation mentioned in Theorem 4.2.

Below let us consider non-degenerate representations (Definition 2.16). (Namely, for a representation $\pi$ on $\mathscr{H}$, if we say $\pi = 0$, then $\mathscr{H} = 0$ is meant.) As explained after Definition 2.12 in Section 2.2, for two representations $\pi_1$ and $\pi_2$, if there is no intertwining map

$$T\pi_1(A) = \pi_2(A)T \quad (A \in \mathfrak{A}) \tag{B.6}$$

from $\mathscr{H}_1$ to $\mathscr{H}_2$ except for $T = 0$, $\pi_1$ and $\pi_2$ are said to be *disjoint*, and if the unitary map $T$ satisfying (B.6) exists, these representations are said to be *unitarily equivalent* (or simply *equivalent*). Situated between these two situations there is another concept of quasi-equivalence which is weaker than equivalence. If the representations $\hat{\pi}_1$ and $\hat{\pi}_2$ given on the tensor product (see Section A.11) of the representation spaces $\mathscr{H}_1$ and $\mathscr{H}_2$ as

$$\hat{\pi}_1(A) = \pi_1(A) \otimes \mathbf{1}, \quad \hat{\pi}_2(A) = \mathbf{1} \otimes \pi_2(A)$$

are unitarily equivalent, then $\pi_1$ and $\pi_2$ are said to be *quasi-equivalent*. (The usual definition using von Neumann algebras will be presented later.) For example, on a finite-dimensional Hilbert space $\mathscr{H}_1 = \mathbf{C}^n$ there is a natural representation $\pi_1$ of the $C^*$ algebra $\mathfrak{A}$ of all $n \times n$ matrices. However, their $k$-fold direct sum

$$\mathscr{H}_2 = \mathscr{H}_1 \oplus \cdots \oplus \mathscr{H}_1 \approx \mathscr{H}_1 \otimes \mathbf{C}^k$$
$$\pi_2(A) = \pi_1(A) \oplus \cdots \oplus \pi_1(A) \approx \pi_1(A) \otimes \mathbf{1}$$

is not unitarily equivalent to the irreducible representation $\pi_1$, but quasi-equivalent. As in this example, representations which are equivalent only after adjustment of the difference in multiplicity are quasi-equivalent.

For any pair of representations $\pi_1$, $\pi_2$ of a $C^*$ algebra, there exist three disjoint representations $\pi_{12}$, $\pi_1'$, $\pi_2'$ such that $\pi_1$ is quasi-equivalent to $\pi_{12} \oplus \pi_1'$, $\pi_2$ is quasi-equivalent to $\pi_{12} \oplus \pi_2'$ and quasi-equivalence classes of these three representations are unique. If $\pi_{12} = 0$ then $\pi_1$ and $\pi_2$ are disjoint. On the other hand, if $\pi_1' = \pi_2' = 0$ then $\pi_1$ and $\pi_2$ are quasi-equivalent.

Among the irreducible representations, quasi-equivalence and unitary equivalence are the same and, if two representations are not equivalent, they are disjoint. Furthermore, as a fundamental class of representations which include irreducible representations, a representation is defined to be *primary* if any subrepresentation is quasi-equivalent to the original one. From the point of view of quasi-equivalence classes it is the representation of the minimal unit and, in the terminology of von Neumann algebras given in the next section, it is called a *factor representation*.

## B.2 Von Neumann algebras

Let us consider the following topologies which are weaker than the norm topology of $\mathscr{B}(\mathscr{H})$.

(1) **The strong operator topology**: For each set of a natural number $n$, $n$ vectors $\xi_1, \ldots, \xi_n$ in $\mathscr{H}$ and $n$ positive real numbers $\varepsilon_1, \ldots, \varepsilon_n$ arbitrarily chosen, consider the set $N(A)$ determined as

$$N(A) = \{A' \in \mathscr{B}(\mathscr{H});\ \|A'\xi_j - A\xi_j\| < \varepsilon_j\}. \tag{B.7}$$

They generate a neighbourhood system of $A \in \mathscr{B}(\mathscr{H})$ of this topology. For a directed family of operators $A_\alpha \in \mathscr{B}(\mathscr{H})$, it converges to $A \in \mathscr{B}(\mathscr{H})$ in this topology if

$$\lim \|A_\alpha \xi - A\xi\| = 0$$

for any $\xi \in \mathscr{H}$.

(2) **The weak operator topology**: For a set of a natural number $n$, $2n$ vectors $\xi_1, \ldots, \xi_n, \eta_1, \ldots, \eta_n$ in $\mathscr{H}$ and $n$ arbitrarily chosen positive real numbers $\varepsilon_1, \ldots, \varepsilon_n$, consider the set $N(A)$ determined by

$$N(A) = \{A' \in \mathscr{B}(\mathscr{H});\quad |(\xi_j, A'\eta_j) - (\xi_j, A\eta_j)| < \varepsilon_j\}. \tag{B.8}$$

They generate a neighbourhood system of $A \in \mathscr{B}(\mathscr{H})$ in this topology. For a directed family of operators $A_\alpha \in \mathscr{B}(\mathscr{H})$, it converges to $A \in \mathscr{B}(\mathscr{H})$ in this topology if

$$\lim(\xi, A_\alpha \eta) = (\xi, A\eta)$$

for any $\xi, \eta \in \mathscr{H}$.

(3) **The $\sigma$ weak topology**: For a set of two sequences of vectors $\xi_j$, $\eta_j$ with finite sums of their norm squared ($j = 1, 2 \ldots$; $\Sigma\|\xi_j\|^2 < \infty$, $\Sigma\|\eta_j\|^2 < \infty$) and an arbitrarily chosen real number $\varepsilon$, consider the set $N(A)$ determined by

$$N(A) = \left\{A' \in \mathscr{B}(\mathscr{H});\ \left|\sum(\xi_j, (A' - A)\eta_j)\right| < \varepsilon\right\}. \tag{B.9}$$

They generate a neighbourhood system of $A \in \mathscr{B}(\mathscr{H})$ in this topology. For a directed family of operators $A_\alpha \in \mathscr{B}(\mathscr{H})$, it converges to $A \in \mathscr{B}(\mathscr{H})$ in this topology if

$$\lim \operatorname{Tr}(\rho A_\alpha) = \operatorname{Tr}(\rho A)$$

for any density matrix $\rho$ (see Theorem 2.7).

Furthermore, one can also consider the $\sigma$ strong topology, the $*$ strong topology and the $*\sigma$ strong topology. It turns out that a $*$-subalgebra of $\mathscr{B}(\mathscr{H})$ closed in one of these topologies is closed in all above mentioned topologies.

**Definition B.3**

A $*$-subalgebra $M$ of $\mathscr{B}(\mathscr{H})$, closed in the weak operator topology (the same for the strong topology and the $\sigma$ weak topology) and containing the identity operator $\mathbf{1}$, is called a *von Neumann algebra*.

If a $*$-subalgebra $M$ of $\mathscr{B}(\mathscr{H})$ is closed in the weak operator topology (the same applies for the strong topology and others), then it possesses a unit element $e$ as an algebra and in the direct sum decomposition

$$\mathscr{H} = \mathscr{H}_e + \mathscr{H}_0, \quad \mathscr{H}_e = e\mathscr{H}, \quad \mathscr{H}_0 = (\mathbf{1} - e)\mathscr{H}$$

any $A \in M$ vanishes on $\mathscr{H}_0$ and the restriction of $M$ to $\mathscr{H}_e$ is a von Neumann algebra. A $C^*$ algebra which is isomorphic to a von Neumann algebra is called a *$W^*$ algebra*, and the $M$ in this example is a $W^*$ algebra. Since we usually consider the case $\mathscr{H}_0 = 0$ (the non-degenerate case), the condition $\mathbf{1} \in M$ is included in the definition of a von Neumann algebra from the beginning.

For any subset $S$ of $\mathscr{B}(\mathscr{H})$, define $S'$ to be the set of all elements in $\mathscr{B}(\mathscr{H})$ which commute with every element of $S$:

$$S' = \{A \in \mathscr{B}(\mathscr{H});\ B \in S \Rightarrow [B, A] = 0\} \tag{B.10}$$

This $S'$ is called the *commutant* of $S$. Performing this procedure twice, denoted as $S'' = (S')'$, $S''$ is called the *double commutant*. The following theorem characterizes a von Neumann algebra as a self-adjoint double commutant.

**Theorem B.4**
If $S$ is a $*$-invariant subset of $\mathscr{B}(\mathscr{H})$ (if $A \in S$, then also $A^* \in S$), then the commutant $S'$ of $S$ is a von Neumann algebra. The double commutant $S''$ of $S$ is the smallest von Neumann algebra containing the original $*$-invariant set $S$ and is the set of all strong limits of the polynomials of elements of $S$ and the identity operator $\mathbf{1}$ (the von Neumann algebra generated by $S$). In particular, if $M$ is a von Neumann algebra, then $M'' = M$.

For a von Neumann algebra $M$, $M \cap M'$ is the set of all elements of $M$ which commute with any element in $M$, called the *centre* of $M$. In the case where the centre is trivial, namely if

$$M \cap M' = \mathbf{C}\mathbf{1}$$

$M$ is called a *factor*.

The quasi-equivalence of two representations $\pi_1, \pi_2$ of a $C^*$ algebra described in Section B.1 is equivalent to the existence of an isomorphism (bijection which preserves the algebraic operations and the $*$ operation) of von Neumann algebras generated by the representations

$$M_1 = \pi_1(\mathfrak{A})'', \quad M_2 = \pi_2(\mathfrak{A})''$$

which maps the representing operator $\pi_1(A)$ to $\pi_2(A)$ for any $A \in \mathfrak{A}$, (in other words, the map $\pi_1(A) \to \pi_2(A)$ can be extended to an isomorphism of weak closures), and usually quasi-equivalence is defined by this condition. In order that a representation $\pi$ of a $C^*$ algebra is primary, it is necessary and sufficient that $\pi(\mathfrak{A})''$ is a factor. Therefore it is also called a *factor representation*.

An isomorphism between two von Neumann algebras $M_1$ and $M_2$ is automatically continuous in the strong topology, the weak topology and the weak $\sigma$ topology.

If $M_1$ and $M_2$ satisfy

$$UM_1U^* = M_2$$

for some unitary map $U$ between the Hilbert spaces $\mathscr{H}_1$ and $\mathscr{H}_2$ on which $M_1$ and $M_2$ are respectively acting, $M_1$ and $M_2$ are said to be *spatially isomorphic*. In that case $M_1$ and $M_2$ are clearly algebraically isomorphic. However the converse statement does not hold. The difference of the two is the multiplicity or the size of the commutant. The following result clarifies this point.

For a von Neumann algebra $M_1$ on a Hilbert space $\mathscr{H}_1$, let a subspace $\mathscr{H}_2$ of $\mathscr{H}_1$ be invariant under $M_1$ (for any vector $\xi$ in $\mathscr{H}_2$ and any operator $A$ in $M_1$, $A\xi \in \mathscr{H}_2$). Such an $\mathscr{H}_2$ is in a one-to-one correspondence to a projection operator $E$ of the commutant $M_1'$ of $M_1$ through the relation $E\mathscr{H}_1 = \mathscr{H}_2$. (If $E^2 = E^* = E \in M_1'$, $E\mathscr{H}_1$ is invariant under $M_1$ and, if $\mathscr{H}_2$ is $M_1$ invariant, the projection operator $E = E^2 = E^*$ to $\mathscr{H}_2$ is in $M_1'$.) In this case $M_2 = EM_1$ as a von Neumann algebra on $\mathscr{H}_2$ is called the *reduction* of $M_1$ to $\mathscr{H}_2$ and denoted by $(M_2)_E$. (Its commutant is $EM_2'E$ on $\mathscr{H}_2$ and is denoted by $(M_2')_E$.) When $M_2$ on $\mathscr{H}_2$ is given first, $M_1$ is called an *amplification* of $M_2$ if $\mathscr{H}_2$ is embedded (we may call this a unitary map) as a subspace of $\mathscr{H}_1$ satisfying the above relations and the map $A \in M_1 \to AE \in M_2$ is injective. The following theorem shows that an isomorphism of von Neumann algebras is always a product of an amplification, a unitary map and a reduction.

**Theorem B.5**
A necessary and sufficient condition that von Neumann algebras $M_1$ and $M_2$ are isomorphic is that an amplification of $M_1$ and an amplification of $M_2$ are unitarily equivalent.

Given a von Neumann algebra $M$ on $\mathscr{H}$ and a vector $\xi$ in $\mathscr{H}$. If $M\xi$ is dense in $\mathscr{H}$ then $\xi$ is said to be *cyclic* or a *cyclic vector*. If the equation $A\xi = B\xi$ for $A, B \in M$ implies $A = B$, then $\xi$ is said to be *separating*. If $\xi$ is cyclic for $M$, then it is separating for $M'$ and, if it is separating for $M$, then it is cyclic for $M'$. If $M$ possesses a cyclic vector and a separating vector, then it possesses a cyclic and separating vector and $M$ or the pair $M, \Omega$ is said to be *standard*. An isomorphism between standard von Neumann algebras is equivalent to unitary equivalence.

Given a positive linear map $\phi$ between two von Neumann algebras (the positivity of $\phi$ means that if $A \geq 0$ then $\phi(A) \geq 0$). If

$$\lim \phi(A_\alpha) = \phi(\lim A_\alpha) \tag{B.11}$$

holds for any bounded monotone increasing directed family $A_\alpha$, then $\phi$ is said to be *normal*. In the case of representations, $\phi$ is called a *normal representation* and in the case of states ( i.e. mappings to **C**) it is called a *normal state*.

The normal states $\varphi$ of a von Neumann algebra $M$ on $\mathscr{H}$ are exactly those states represented by a trace class operator $\rho$ on $\mathscr{H}$ as

$$\varphi(A) = \mathrm{Tr}(\rho A).$$

The closure in the norm topology defined by the norm

$$\|\varphi_1 - \varphi_2\| = \sup\{|\varphi_1(A) - \varphi_2(A)|; \quad A \in M, \|A\| \leq 1\}$$

of the convex hull of all states $(\Psi, A\Psi)$ represented by unit vectors $\Psi$ in $\mathscr{H}$, is identical with the set of all normal states. The linear hull of all normal states is identical with the set of all linear functionals on $M$ which are continuous in the $\sigma$ weak topology, and it forms a Banach space with respect to the above norm. This space is denoted by $M_*$ and is called the *predual* of $M$. If we consider an $A$ of $M$ as a linear functional of $\varphi \in M_*$ given by $\hat{A}(\varphi) \equiv \varphi(A)$, then $M$ is exactly the dual space of $M_*$ (the set of all continuous functionals on $M_*$).

$$M = (M_*)^*. \tag{B.12}$$

A $W^*$ algebra, which is an abstract von Neumann algebra, is characterized as a $C^*$ algebra which is the dual space of some Banach space (*Sakai's theorem*). An $M_*$ satisfying (B.12) is determined uniquely as an isomorphism class of a Banach space.

As a Banach space, the linear hull of all states of a $C^*$ algebra $\mathfrak{A}$ is identical with the dual space $\mathfrak{A}^*$ of $\mathfrak{A}$ and its dual space $\mathfrak{A}^{**}$ is identical with the weak closure of the universal representation of $\mathfrak{A}$ (see Section B.1) and is a von Neumann algebra.

$$\mathfrak{A}^{**} = \pi_u(\mathfrak{A})''$$

Murray and von Neumann gave a classification of factors in terms of the following dimension theory. In the set $P(M)$ of all projection operators of a factor $M$, we introduce an equivalence relation by saying that if there exists a $u \in M$ satisfying $p = u^*u$ and $q = uu^*$, then two projections $p$ and $q$ are *equivalent*. ($u$ gives a unitary map from $p\mathscr{H}$ to $q\mathscr{H}$ and is called a *partial isometry*.) For this equivalence class $[p]$ we introduce an order relation by saying that if $p \leq q$, then $[p] \leq [q]$. Then we obtain a totally ordered set for a factor $M$. To represent faithfully this structure by a positive real number we consider any positive valued function $d(p)$ (which is not identically zero) on the equivalence classes of $P(M)$ (i.e. if $p$ is equivalent to $q$ then $d(p) = d(q)$), satisfying

$$p \perp q \rightarrow d(p+q) = d(p) + d(q)$$

Up to a constant coefficient it is uniquely determined. By taking appropriate coefficients, this function takes one of the following ranges:

1. Type $I_n : d(p) \in \{0, 1, \ldots, n\}$ $(n = 1, \ldots, \infty)$
2. Type $II_1 : d(p) \in [0, 1]$ (finite interval)
3. Type $II_\infty : d(p) \in [0, \infty] = R_+ \cup \infty$
4. Type $III : d(p) \in \{0, \infty\}$ (for $p \neq 0, d(p) = \infty$).

In each case, factors $M$ are called type $I_n$ (or simply type $I$), type $II_1$ or type $II_\infty$ (these two cases together called type $II$), or type $III$. A type $I_n$ factor is isomorphic to $\mathscr{B}(\mathscr{H})$ of an n-dimensional Hilbert space $\mathscr{H}$ and for the projection operator $p$ of $\mathscr{B}(\mathscr{H})$ the function $d(p)$ is the dimension of the subspace $p\mathscr{H}$. For the other cases, $d(p)$ is also called the *dimension function*. This classification into type $I_n$, type $II_1$, type $II_\infty$ and type $III$ is generalized to von Neumann algebras which are not factors.

Defining the conjugate linear operator $S$ for a cyclic and separating vector $\xi$ of a standard von Neumann algebra $M$ by

$$SA\xi = A^*\xi \quad (A \in M)$$

$S$ is in general unbounded but closable. By the polar decomposition of the closure $\bar{S}$ of $S$

$$\bar{S} = J\Delta^{\frac{1}{2}} ,$$

we define a positive self-adjoint operator $\Delta = S^*\bar{S}$ and an involutive antiunitary operator $J$ (involutive means $J^2 = 1$) and call them the *modular operator* and *modular conjugation operator*, respectively. They have the following properties.

$$\Delta^{it} M \Delta^{-it} = M, \quad JMJ = M'.$$

The elements of the one parameter group of automorphisms $\sigma_t^\xi$ of $M$ defined by

$$A \in M \to \sigma_t^\xi(A) = \Delta^{it} A \Delta^{-it}$$

are called *modular automorphisms*. This is the core of the Tomita–Takesaki theory and plays a very important role in the development of the theory of operator algebras and its applications.

# APPENDIX C: FREE FIELDS

## C.1 Charged scalar field

A system of two types of particles which are described by mass $m > 0$, spin 0 irreducible representations without mutual interaction is described by a tensor product $F_+(\mathscr{H}) \otimes F_+(\mathscr{H})$ of two Fock spaces as given in Section 3.5. To distinguish the creation and annihilation operators of each particle we use the symbols

$$(b^*, h) = (a^*, h) \otimes \mathbf{1}, \quad (h, b) = (h, a) \otimes \mathbf{1}$$
$$(c^*, h) = \mathbf{1} \otimes (a^*, h), \quad (h, c) = \mathbf{1} \otimes (h, a)$$

For a well-behaved complex-valued function $h(x)$ on the four-dimensional space $M$ we define

$$A(h) = (b^*, \hat{h}) + (\hat{\bar{h}}, c). \tag{C.1}$$

On the domain $D_0^+$ we obtain

$$A^*(h) \equiv A(\bar{h})^* = (c^*, \hat{h}) + (\hat{\bar{h}}, b). \tag{C.2}$$

As in the case of Section 3.5, we can show the following properties:

$$(\Box + m^2)A(x) = 0, \tag{C.3}$$

$$[A(x), A(y)] = [A^*(x), A^*(y)] = 0, \tag{C.4}$$

$$[A^*(x), A(y)] = \mathrm{i}\Delta(x - y) \tag{C.5}$$

$A(x)$ is called a *free charged scalar field* and it is a Wightman field. Due to (C.4) and (C.5), the local commutativity holds also for this field like in (3.73). However, between the real free scalar fields constructed from the individual creation and annihilation operators and this free charged scalar field there are no relations such as commutativity at a spacelike distance. With the terminology of Chapter 4, two free neutral scalar fields and a free charged scalar field define different systems of algebras of local observables although they are constructed from the same creation and annihilation operators.

The adjective "charged" originates from the following situation. In the Fock space, an observable which represents the number of particles can be defined as

$$\text{if } \Psi \in F_+^n(\mathscr{H}) \text{ then } N\Psi = n\Psi \tag{C.6}$$

Thus, on $F_+(\mathscr{H}) \otimes F_+(\mathscr{H})$ we define

$$Q = N \otimes \mathbf{1} - \mathbf{1} \otimes N. \tag{C.7}$$

Then this becomes an observable representing the total charge when we assign charge 1 and $-1$ for particles created by $(b^*, h)$ and $(c^*, h)$, respectively. The Wightman field $A(h)$ increases the charge by 1 and hence it is called a charged field in the sense that it carries the charge 1. Correspondingly, $A^*(h)$ decreases the charge by 1.

## C.2 One particle system with positive mass and arbitrary spin

The irreducible representation of $SU(2)$ with spin $j$ can be defined on a $(2j+1)$-dimensional space

$$\mathscr{K}_j \equiv S_{2j}^+ (\mathbf{C}^2)^{\otimes 2j} \approx \mathbf{C}^{(2j+1)} \tag{C.8}$$

as

$$D_j(A) = A^{\otimes 2j} \quad (A \in SU(2)). \tag{C.9}$$

Here $S_{2j}^+$ is the complete symmetrization operator used in (3.53). A vector of the representation space, written in terms of its components, is of the form

$$u = \{u_{\rho_1 \cdots \rho_{2j}}\}$$

and is called a *spinor*. Each index takes the values 1 and 2, and the spinor is completely symmetric in its indices.

We can extend this representation to that of $SL(2, C)$ without changing the dimension $2j+1$ of the representation space. To this end, there are two possible extensions, $D_{[j,0]}$ and $D_{[0,j]}$:

$$D_{[j,0]}(A) = A^{\otimes 2j}, \tag{C.10}$$

$$D_{[0,j]}(A) = \hat{A}^{\otimes 2j} \quad (\hat{A} = \sigma_2 \bar{A} \sigma_2 = \varepsilon \bar{A} \varepsilon^{-1}). \tag{C.11}$$

Here $\bar{A}$ is the complex conjugate matrix of $A$ with all matrix elements being replaced by their complex conjugates and $\varepsilon = \mathrm{i}\sigma_2$ ($\varepsilon_{12} = 1$, $\varepsilon_{kl} = -\varepsilon_{lk}$). If $A \in SU(2)$ then $\hat{A} = A$.

Correspondingly, the representation of $\tilde{\mathscr{P}}_+^\uparrow$ with mass $m > 0$ and spin $j$ can be described in two ways as follows. The Hilbert space for the representation consists of functions $u(p)$ on the orbit $m_+$, taking values in the Hilbert space $\mathscr{K}_j$, and their inner product is given by

$$(u, v) = \int_{m+} (u(p), n_\alpha(p) v(p))\, \mathrm{d}\mu(\mathbf{P}), \tag{C.12}$$

where the two cases are distinguished by the index $\alpha$ taking values $\alpha = 1, 2$. In terms of the notation of (3.25) used for $\tilde{p}$, the matrix $\eta_\alpha(p)$ is defined by

$$\eta_1(p) = (m/\tilde{p})^{\otimes 2j}, \quad \eta_2(p) = (\tilde{p}/m)^{\otimes 2j} \tag{C.13}$$

and the representing operator is given by

$$[U((a, A))u](p) = e^{i(p,a)} D_\alpha(A) u(\pi(A)^{-1} p), \tag{C.14}$$

where

$$D_1(A) = D_{[j,0]}(A), \quad D_2(A) = D_{[0,j]}(A). \tag{C.15}$$

These two representations are equivalent by the following unitary transformation.

$$u(p) \in L_2(m_+, \mathrm{d}\mu, \mathscr{K}_2) \to \eta_2(p)u(p) \in L_2(m_+, \mathrm{d}\mu, \mathscr{K}_1). \tag{C.16}$$

One method to understand the above definition is the spinor notation. We denote an index, which represents a component of a vector transforming under $A$, by lower indices (suffixes). On the other hand, the components of a vector transforming under $\bar{A}$ are represented by lower dotted indices like $\dot{\rho}$. A vector of the representation space of $D_{[j,0]}$ such as $v_{\rho_1 \ldots \rho_{2j}}$ is a symmetric spinor, symmetric in its $2j$ lower indices. In the case of $D_{[0,j]}$ we have a spinor $v_{\dot{\rho}_1 \ldots \dot{\rho}_{2j}}$ symmetric in its $2j$ dotted lower indices. The indices take the two values $1, 2$. The components of quantities arising from contraction with the antisymmetric matrix $\varepsilon$ are represented by upper indices. Due to the formula ${}^t A \varepsilon A = (\det A)\varepsilon$ (${}^t A$ is the transposed matrix of $A$), we obtain an invariant expression for $A \in SU(2C)$ when the upper and lower indices are contracted. (Contraction means summation over indices after setting pairs of upper and lower indices equal; the pairing being always between upper and lower undotted indices and between upper and lower dotted indices.) The matrix elements of $\varepsilon$ itself are represented by two upper indices, and those of $\varepsilon^{-1} = -\varepsilon$ by two lower indices. By (3.27) the matrix elements of $\tilde{p}$ can be represented as $(\tilde{p})_{\rho\dot{\sigma}}$. From the above, the invariance of the inner product (C.12) and hence the unitarity of $U$ become apparent. By the formula $\det \tilde{p} = (p, p)$, the matrix $(\tilde{p}/m)$ has determinant 1 on the orbit $m_+$ and thus belongs to $SL(2, C)$. Furthermore, since it is self-adjoint and satisfies $\mathrm{Tr}\, \tilde{p} = 2p^0 > 0$ as well as $\det \tilde{p} > 0$, it is a positive matrix. Therefore, the inner product (C.12) is positive definite.

## C.3 Examples of free fields with positive mass and integer spin

For the case of an integer spin $j$, we define the symmetric spinor field $A(x)^{\rho_1 \ldots \rho_{2j}}$ with $2j + 1$ components as an operator-valued distribution on the Fock space of a particle with the unitary irreducible representation of $\mathscr{P}_+^\uparrow$ of mass $m$ and spin $j$ given in the previous section as follows.

For complex-valued functions

$$h(x) = \{h(x)_{\rho_1\cdots\rho_{2j}}\} \quad \text{(completely symmetric in indices)}$$

with $2j+1$ components, we define

$$A(h) = \sum_\rho \int A(x)^{\rho_1\cdots\rho_{2j}} h(x)_{\rho_1\cdots\rho_{2j}}\, \mathrm{d}^4x = (a_1^*, \hat{h}_+) + (\hat{h}_-, a_2) \tag{C.17}$$

on a dense domain $D_0^+$ of $F_+(\mathscr{H})$, where the Fourier transform $\tilde{h}$ of $h$ according to (3.69), restricted to the orbit $m_+$, is denoted by $\hat{h}_+$ and is considered as a vector of the representation space of spin $j$ (the case of $\alpha = 1$ in (C.12)), as defined in the previous section. $\hat{h}_-$ is considered as a vector of the representation space (this time $\alpha = 2$) by the same procedure as $\hat{h}_t$ after applying (here, applying means contracting with) $S = \varepsilon^{\otimes 2j}$ to the complex conjugate $\bar{h}(x)$ of $h(x)$, thus changing undotted indices to dotted indices and then raising the lower indices to upper indices. The two representations ($\alpha = 1, 2$) are to be identified by the unitary map (C.16). However, depending on which representation space vectors the functions are considered to be, we distinguish the creation and annihilation operators $a_\alpha^*$, $a_\alpha$ by the index $\alpha = 1, 2$.

Denoting the contraction of $A(x)^{\rho_1\cdots\rho_{2j}}$ with $S$ (due to $S = S^{-1}$ for integer $j$) by $A_{\rho_1\ldots\rho_{2j}}$, we compute the transformation properties under $\Gamma(a,\Lambda) \equiv \Gamma(U(a,A))$ as follows:

$$\Gamma(a,\Lambda)A(h)\Gamma(a,\Lambda)^* = A((a,A)h), \tag{C.18}$$

$$\Gamma(a,\Lambda)A(x)\Gamma(a,\Lambda)^* = (A^{-1})^{\otimes 2j} A(\Lambda x + a). \tag{C.19}$$

Here, $\Lambda = \pi(A)$ (we can also take $\Lambda = \pi(-A)$), and

$$[(a,A)h](x) = A^{\otimes 2j} h(\Lambda^{-1}(x-a)). \tag{C.20}$$

The action of $A^{\otimes 2j}$ is the action of a matrix $A$ on each of the $2j$ indices as follows:

$$(A^{\otimes 2j}h)_{\rho_1\cdots\rho_{2j}} = \sum_\sigma A_{\rho_1}^{\sigma_1} \cdots A_{\rho_{2j}}^{\sigma_{2j}} h_{\sigma_1\cdots\sigma_{2j}}$$

(C.19) is the transformation formula for $A(x)_{\rho_1\ldots\rho_{2j}}$, where the action of $(A^{-1})^{\otimes 2j}$ is similar as in the above equation. In order to obtain (C.19) from (C.18) we use ${}^t A \varepsilon A = \varepsilon$.

We compute the commutation relations of the fields on $D_0^+$ as follows:

$$[A(x)_\rho, A(y)_\sigma] = \mathrm{i}\Delta(x-y)S_{\rho\sigma}, \tag{C.21}$$

$$[A(x)_\rho, A(y)^*_{\dot\sigma}] = \mathrm{i}(\eta_2^x)_{\rho\dot\sigma}\Delta(x-y), \tag{C.22}$$

$$\eta_2^x = (-\mathrm{i}(\partial/\partial x^0 - \sigma_1\partial/\partial x^1 - \sigma_2\partial/\partial x^2 - \sigma_3\partial/\partial x^3)/m)^{\otimes 2j}$$
$$= \left(\sum_\mu(-\mathrm{i}\partial/\partial x_\mu)\sigma_\mu/m\right)^{\otimes 2j}. \tag{C.23}$$

Here we abbreviate the notation by writing the indices $\rho_1 \ldots \rho_{2j}$ as $\rho$, and similarly for $\dot\sigma$. $A^*(x)$ is related with $A(x)$ by the following formula:

$$A(x)^*_{\dot\rho} = (S\eta_1^x A(x))_{\dot\rho} = \sum_{\dot\sigma\kappa} S_{\dot\rho\dot\sigma}(\eta_1^x)^{\dot\sigma\kappa}A(x)_\kappa, \tag{C.24}$$

$$\eta_1^x = \left(\sum_\mu(-\mathrm{i}\partial/\partial x^\mu)\sigma_\mu/m\right)^{\otimes 2j}. \tag{C.25}$$

In the above equations, $\eta_2^x$ is $\eta_2(p)$ written in terms of differential operators, $\eta_1^x$ stands for $\eta_1(p) = S'\eta_2(p)S^{-1}$ (a relation derived by using $\varepsilon' A\varepsilon^{-1} = A^{-1}$ for $\det A = 1$), written as a differential operator using $\varepsilon'\sigma_\mu\varepsilon^{-1} = \Sigma_\nu g_{\mu\nu}\sigma_\nu$.

Each component of $A(x)$ satisfies the Klein-Gordon equation.

$$(\Box/m^2)A(x) = -\eta_1^x\eta_2^x A(x) = -\eta_2^x\eta_1^x A(x) = -A(x). \tag{C.26}$$

If we consider the tensor product of two Fock spaces of a particle with mass $m$ and spin $j$ (namely, a particle and an antiparticle), we obtain a *charge symmetric free spinor field* by considering creation and annihilation operators with indices $\alpha = 1, 2$ acting on different Fock spaces. For this case, among the above formulae (C.21) becomes zero, (C.24) does not hold, but (C.19) and (C.22) hold as they stand.

In all cases the fields discussed above are examples of Wightman fields. For example, if $j = 1$ and eq. (C.24) holds, we obtain a self-adjoint antisymmetric second rank tensor field.

## C.4 Examples of free fields with positive mass and half odd integer spin

The construction we use is completely the same as for the charge symmetric spinor field. However, as for the Fock space of a particle and an antiparticle we use the fermion Fock space $F_-(\mathscr{H})$. Changing the symbol for the field from $A$ to $\Psi$ we define

$$\Psi(h) = \sum_\rho \int \Psi(x)^\rho h(x)_\rho \mathrm{d}^4x = (a_1^*, \hat h_+) + (\hat h_-, a_2), \tag{C.27}$$

$$\rho = (\rho_1, \ldots, \rho_{2j}),$$
$$\Psi(x)_\rho = \sum_\sigma (S^{-1})_{\rho\sigma}\Psi(x)^\sigma. \tag{C.28}$$

For half-odd integer $j$, $S = (\varepsilon)^{\otimes 2j}$ satisfies $S^{-1} = -S$. In addition, we have $(-p)\tilde{} = -\tilde{p}$, and the anticommutation relation (3.62) holds. Compared with the case of integer spin $j$, there appear only some changes of the sign. Similar to the case of the charge symmetric spinor field of the previous section, we obtain the Wightman fields for a Fermion by considering the tensor product of two different Fock spaces for $\alpha = 1$ and $\alpha = 2$ as follows:

$$\Gamma(a, A)\Psi(x)\Gamma(a, A)^* = (A^{-1})^{\otimes 2j} A(\Lambda x + a), \tag{C.29}$$

$$[\Psi(x)_\rho, \Psi(y)_\sigma]_+ = [\Psi(x)^*_{\dot{\rho}}, \Psi(y)^*_{\dot{\sigma}}]_+ = 0, \tag{C.30}$$

$$[\Psi(x)_\rho, \Psi(y)^*_{\dot{\sigma}}]_+ = \mathrm{i}(\eta_2^x)_{\rho\dot{\sigma}} \Delta(x - y). \tag{C.31}$$

Each component of $\Psi(x)$ satisfies the Klein–Gordon equation.

Let us consider the case of $j = 1/2$ as an example. $\Psi(x)_\rho$ has two components for $\rho = 1, 2$. Together with this, we consider

$$\dot{\Psi}(x)^{\dot{\sigma}} = \sum_\rho (\eta_1^x)^{\dot{\sigma}\rho} \Psi(x)_\rho. \tag{C.32}$$

$(\Psi(x)_\rho, \dot{\Psi}(x)^{\dot{\rho}})$ has four components. Noting that

$$\eta_1^x = \sum_\mu (-\mathrm{i}\partial/\partial x^\mu)\sigma_\mu/m,$$

$$\eta_2^x = \sum_\mu (-\mathrm{i}\partial/\partial x_\mu)\sigma_\mu/m,$$

$$\eta_1^x \eta_2^x = \eta_2^x \eta_1^x = \Box/m^2,$$

we obtain the following equation due to $(\Box + m^2)\Psi = 0$:

$$\begin{pmatrix} 0 & \eta_2^x \\ \eta_1^x & 0 \end{pmatrix} \begin{pmatrix} \Psi \\ \dot{\Psi} \end{pmatrix} = \begin{pmatrix} \Psi \\ \dot{\Psi} \end{pmatrix}, \tag{C.33}$$

which is the *Dirac equation.*

The result is similar for general $j$, if we use the above $\eta_1^x$ for each index ($\eta_1^x$ for a general $j$ is the product of these $2j$ $\eta_1$s) to raise each lower undotted index to an upper dotted index and obtain the whole set of components $\Psi(x)^{\dot{\sigma}_1 \dots \dot{\sigma}_k}_{\rho_{k+1} \dots \rho_{2j}}$, which satisfy the first order Dirac-type differential equations.

# GENERAL LITERATURE AND REFERENCES

The main subject of this book is the formulation of quantum field theory by local observables. As a textbook on this field we recommend

1. R. Haag: *Local Quantum Physics* (Springer, 1992; revised 1996)

This book is a comprehensive summary by a founder of this field. It includes the development until a very recent time and a fairly complete list of references. A book complementary to the above is the following one which explains mathematical aspects of local operator algebra systems.

2. H. Baumgärtel and M. Wollenberg: *Causal Nets of Operator Algebras* (Academic Press, 1992).

The reference below treats fairly restricted mathematical aspects.

3. S.S. Horuzhy: *Introduction to Algebraic Quantum Field Theory* (Kluwer, 1990).

The following textbooks treat the axiomatic field theory closely related to the content of this book.

4. R.F. Streater and A.S. Wightman: *PTC, Spin and Statistics and All That* (Benjamin, 1964).
5. R. Jost: *The General Theory of Quantized Fields* (American Mathematical Society, 1965).
6. N.N. Bogoliubov, A.A. Logunov, and I.T. Todorov: *Mathematical Methods in Quantum Field Theory* [Original 1969, Japanese translation 1972 (Tokyotosho), English translation 1975 (Dover Publications)].
7. N.N. Bogoliubov, A.A. Logunov, A.I. Oksak and I.T. Todorov: *General Principles of Quantum Field Theory* [Original 1987] (Kluwer, 1990).
8. H. Ezawa and A. Arai: *Quantum Field Theory and Statistical Mechanics* [in Japanese] (Nihonhyoronsha, 1988).

Among the above textbooks, both books by Bogoliubov *et al.* (including the one translated by H. Ezawa) mention the formulation by local observables.

Concerning the theory of operator algebras which occupies an important part of the mathematical foundation of this book, there are the following textbooks.

9. O. Bratteli and D.W. Robinson: *Operator Algebras and Quantum Statistical Mechanics*, 2 volumes (Springer-Verlag, volume I 1979, volume II 1981; volume II revised 1997).
10. J. Dixmier: *Von Neumann Algebras*, revised version (North Holland, 1981).
11. J. Dixmier: *$C^*$ Algebras*, revised version (North Holland, 1982).

12. R.V. Kadison and J.R. Ringrose: *Fundamentals of the Theory of Operator Algebras* (Academic Press, volume I 1983, volume II 1986, volume III 1991, volume IV 1992).

13. M.A. Naimark: *Normed Rings* (P. Noordhoff, 1972).

14. G.K. Pedersen: *C* Algebras and their Automorphism Groups* (Academic Press 1979).

15. S. Sakai: *C* Algebras and W* Algebras* (Springer, 1971).

16. S. Sakai: *Operator Algebras in Dynamical Systems* (Cambridge University Press, 1991).

17. S. Stratila: *Modular Theory in Operator Algebras* (Edituria Academiei, Abacus Press, 1981).

18. S. Stratila and L. Zsido: *Lectures on von Neumann Algebras* (Edituria Academiei, Abacus Press, 1979).

19. M. Takesaki: *Theory of Operator Algebras* (Springer, volume I, 1979).

20. M. Takesaki: *Structure of Operator Algebras* (in Japanese, Iwanami Publ. 1983).

There is also the following related literature which does not fit the above classification.

21. G.E. Emch: *Algebraic Methods in Statistical Mechanics and Quantum Field Theory* (Wiley).

Many parts of this book up to section V are based on a lecture given by the author at the Swiss Federal Institute of Technology (ETH) in Zürich, Switzerland, which was distributed by the ETH as

22. H. Araki: *Einführung in die Axiomatische Quantenfeldtheorie*, I and II (1962).

The mathematical terminology in this book follows the *Encyclopedic Dictionary of Mathematics* (*EDM*), edited by the Mathematical Society of Japan (Iwanami Publ., 3rd edition 1985; English translation: The MIT Press, Cambridge, 2nd edition 1986).

The general theory of the probabilistic description in section I was developed in section I of the above-mentioned lecture notes in Zürich, and its summary is given in the appendix of the following paper.

23. H. Araki: *Prog. Theor. Phys.* **64** (1980) 719.

The content in Section 1.6 and Section 1.7 is a reformulation (in the general framework of Chapter 1) of the theory developed in the $C^*$-algebra framework in the following paper.

24. H. Haag and D. Kastler: *J. Math. Phys.* **5** (1964) 848–861.

The $C^*$ algebra formulation of quantum mechanics introduced in Section 2 originates in

25. I.E. Segal: *Ann. Math.* **48** (1947) 930.

The research on the lattice of projections originates in

26. G. Birkhoff and J. von Neumann: *Ann. Math.* **37** (1936) 823.

The general considerations about the symmetry introduced here have been originated by

27. E.P. Wigner: *Group Theory and Quantum Mechanics* [Original 1931, Japanese translation by M. Morita and R. Morita (Yoshioka Publ., 1971)].

There is also the following review on this subject:

28. A.S. Wightman: In *Dispersion Relations and Elementary Particles*, ed. by C. DeWitt and R. Omnes (Wiley, 1960) p. 161.

The following papers are fundamental about group theoretical investigations of relativistic invariance

29. E.P. Wigner: *Ann. Math.* **40** (1939) 149.

30. V. Bargmann: *Ann. Math.* **48** (1947) 568.

31. V. Bargmann: *Ann. Math.* **59** (1954) 1.

There are the following textbooks:

32. Y. Ohnuki: *Unitary Representations of the Poincaré Group and Relativistic Wave Equations* (World Sci. Publ., Teaneck, NJ. 1988).

33. I.M. Gel'fand, R.A. Minlos and Z.Ya. Shapiro, Representations of the rotation and Lorentz groups and their applications (Pergaman Press, 1963).

The content after Chapter 4 is the theory of local observables and relevant references are the book [1] by Haag and the references cited therein.

Concerning the content in Chapter 6, there is a detailed description in the book [2] by Baumgärtel and Wollenberg.

**References for Chapter 1**

1. For example, see K. Yosida, *Functional Analysis* (Springer Verlag, 4th edition, 1974), Chapter XII, Sect. 1, p. 362.

2. For example, see K. Yosida, ibid. Chapter 0, Sect. 2, p. 6.

**References for Chapter 2**

1. The original paper is

A.M. Gleason, Measures on the closed subspaces of a Hilbert space, *J. Math. Mech.* **6** (1957) 885–893.

An introductory textbook is

Shuichiro Maeda, *Lattice Theory and Quantum Logic* (Maki Shoten, 1980) (in Japanese).

Survey on later generalizations can be found in

Shuichiro Maeda, Probability measures on projections in von Neumann algebras, *Rev. Math. Phys.* **1** (1989) 235–290.

G. Kalmbach, *Measures and Hilbert Lattices* (World Scientific, 1986).

2. For example, see General Ref. [14], p. 54, Theorem 3.6.4.

3. For the existence proof, see e.g. ibid. p. 11, Theorem 1.4.2.

4. See General Ref. [24].

Also see

J.M.G. Fell, The dual spaces of C*-algebras, *Trans. Amer. Math. Soc.* **94** (1960) 365–403.

5. The original proof is in General Ref. [27]. Also see General Ref. [28].

A. Barut and A.S. Wightman, Relativistic invariance and quantum mechanics, *Nuovo Cim.* **14** (1959) 81–94.

V. Bargmann, Note on Wigner's theorem on symmetry operations, *J. Math. Phys.* **5** (1964) 862–868.

A detailed list of references is given in

U. Uhlhorn, Representation of symmetry transformations in quantum mechanics, *Arkiv Fysik* **23** (1963) 307–340.

Also see the following article where Kadison's formulation is quoted and explained.

B. Simon, Quantum dynamics: From automorphism to Hamiltonian, in

E.H. Lieb, B. Simon, and A.S. Wightman (eds), *Studies in Mathematical Physics* (Princeton Univ. Press, 1976), pp. 327–349.

6. H.A. Dye, On the geometry of projections in certain operator algebras, *Ann. Math.* **61** (1955) 73–89.

7. R.V. Kadison, Isometries of operator algebras, *Ann. Math.* **54** (1951) 325–338.

8. R.V. Kadison, Transformations of states in operator theory and dynamics, *Topology*, Vol. 3, Suppl. 2 (1965) pp. 177–198.

## References for Chapter 3

1. The original reference is

E.C. Zeeman, Causality implies the Lorentz group, *J. Math. Phys.* **5** (1964) 490–493.

Also see

H.J. Borchers and G.C. Hegerfeldt, The structure of space time transformations I, *Commun. Math. Phys.* **28** (1972) 259–262.

2. See General Ref. [29],[30],[31],[32],[33] and Chapter 7, Sect. 7.2 of General Ref. [7]. Also see

A.S. Wightman, L' invariance dans la mécanique relativiste, in C. de Witt and R. Omnes (eds), *Dispersion Relations and Elementary Particles* (Hermann, Paris, 1961), pp. 159–226.

3. This name is in General Ref. [4], p. 92. It is a generalization of Stone's theorem to more than one parameters and can be inferred from General Ref. [10], Chapter 2.

F. Riesz and B.Sz. Nagy, *Functional Analysis* (Unger, 1955), Chapter 10.

4. For the Fock space, see, for example, General Ref. [9], II, Sect. 5.2.1 and Chapter 7, Sect. 7.3 of General Ref. [7].

For free fields, see, for example, Chapter II of General Ref. [5], Chapter 3, Sect. 4 of General Ref. [6] and Chapter 8, Sect. 8.4 of General Ref. [7].

5. C. Eckart, Application of group theory to the quantum dynamics of monatomic systems, *Rev. Mod. Phys.* **2** (1930) 305–380.

**References for Chapter 4**

1. H. Araki, On the algebra of all local observables, *Progr. Theoret. Phys.* **32** (1964) 844–854.

2. For example, see Chapter XI, Sect. 16 p. 357 of Ref. 1 for Chapter 1.

3. Original papers are

R. Jost and H. Lehmann, Integral Darstellung kausaler Kommutatoren, *Nuovo Cim.* **5** (1957) 1598–1610.

F.J. Dyson, Integral representations of causal commutators. *Phys. Rev.* **110** (1958) 1460–1464.

For a detailed explanation, see Chapter 4 of General Ref. [7] and A. S. Wightman, Analytic functions of several complex variables, Sect. VII in

C. De Witt, and R. Omnes (eds), *Dispersion Relations and Elementary Particles* (Hermann, Paris, 1961), pp. 227–315.

Generalizations are in

J. Bros, Al Messiah, and R. Stora, A problem of analytic completion related to the Jost–Lehmann–Dyson formula, *J. Math. Phys.* **2** (1961) 639–651.

J. Bros, C. Itzykson, and F. Pham, Representations integrales de fonctions analytiques et formule de Jost–Lehmann–Dyson, *Ann. Inst. Henri Poincaré* **5** (1966) 1–35.

4. H. Araki, K. Hepp, and D. Ruelle, On the asymptotic behaviour of Wightman functions in space-like directions, *Helv. Phys. Acta* **35** (1962) 164–174.

5. H. Reeh and S. Schlieder, Bemerkungen zur Unitäräquivalenz von Lorentzinvarianten Feldern, *Nuovo Cim.* **22** (1961) 1051–1068.

6. H.J. Borchers, Über die Vollständigkeit lorentzinvarianter Felder in einer zeitartigen Röhre, *Nuovo Cim.* **19** (1961) 787–793.

7. H. Araki, A generalization of Borchers theorem, *Helv. Phys. Acta* **36** (1963) 132–139.

8. The original paper is

A.S. Wightman, Quantum field theory in terms of vacuum expectation values, *Phys. Rev.* **101** (1956) 860–866.

For textbooks, see, for example, General Ref. [4], [5], [6], and [7].

9. See, for example, Chapter 4, Sect. 4.4 of General Ref. [4], Chapter V, Sect. 3 of General Ref. [5], Chapter 5, Sect. 3.3 of General Ref. [6] and Chapter 6, Sect. 9.3. D of General Ref. [7].

10. D. Buchholz, On quantum fields that generate local algebras, *J. Math. Phys.* **31** (1990) 1839–1846.

**References for Chapter 5**

1. R. Haag, Quantum field theories with composite particles and asymptotic conditions, *Phys. Rev.* **112** (1958) 669–673.

D. Ruelle, On asymptotic condition in quantum field theory, *Helv. Phys. Acta* **35** (1962) 147–163.

2. L.J. Landau, Asymptotic locality and the structure of local internal symmetries, *Commun. Math. Phys.* **17** (1970) 156–176.

3. H. Araki and R. Haag, Collision cross sections in terms of local observables, *Commun. Math. Phys.* **4** (1967) 77–91.

4. H. Lehmann, K. Symanzik, and W. Zimmermann, Zur Formulierung quantisierter Feldtheorien, *Nuovo Cim.* **1** (1955) 120–134.

H. Lehmann, K. Symanzik, and W. Zimmermann, On the formulation of quantized field theories II, *Nuovo Cim.* **6** (1957) 319–333.

5. K. Hepp, On the connection between the LSZ and Wightman quantum field theory, *Commun. Math. Phys.* **1** (1965) 95–111.

Also see the following Lecture note:

K. Hepp, The connection between Wightman and LSZ quantum field theory, in M. Chrétien and S. Deser (eds), *Axiomatic Field Theory* (Gordon and Breach 1966) 135–246.

6. W. Zimmermann, Yang-Feldman Formalismus und einzeitige Wellenfunktionen, *Nuovo Cim.* **11** (1954) 577–589.

7. For example, see the following textbooks:

S. Bochner and W.T. Martin, *Several Complex Variables* (Princeton Univ. Press, 1948).

R.C. Gunning and H. Rossi, *Analytic Functions of Several Complex Variables* (Prentice Hall, 1965)

L. Hörmander, *An Introduction to Complex Analysis in Several Variables* (Van Nostrand, 1966)

L. Nachbin, *Holomorphic Functions, Domains of Holomorphy and Local Properties* (North Holland, 1970)

Also see A.S. Wightman, Analytic functions of several complex variables, Sect. VII in

C. De Witt and R. Omnes (eds), *Dispersion Relations and Elementary Particles* (Herman, 1961), pp. 227–315.

8. V. Glaser, H. Lehmann, and W. Zimmermann, Field operators and retarded functions, *Nuovo Cim.* **6** (1957) 1122–1128.

9. O. Steinmann, Über den Zusammenhang zwischen den Wightmanfunktionen und den retardierten Kommutatoren, *Helv. Phys. Acta.* **33** (1960) 257–298.

10. H. Epstein, Generalization of the "Edge of the Wedge" theorem, *J. Math. Phys.* **1** (1960) 524–531.

11. H. Araki, Generalized retarded functions and analytic function in momentum space in quantum field theory, *J. Math. Phys.* **2** (1961) 163–177.

H. Araki, Wightman functions, retarded functions and their analytic continuations, *Progr. Theoret. Phys.* Suppl. No. 18 (1961) 83–125.

12. Ref. 9 above and

O. Steinmann, Zur Definition der retardierten und zeitgeordneten Produkte, *Helv. Phys. Acta* **36** (1963), 90–112.

13. D. Ruelle, Connection between Wightman functions and Green functions in p-space, *Nuovo Cim.* **19** (1961) 356–376.

14. H. Epstein, CTP invariance of the S-matrix in a theory of local observables, *J. Math. Phys.* **8** (1967) 750–767.

15. R. Jost, Eine Bemerkung zum CTP-Theorem, *Helv. Phys. Acta.* **30** (1957) 409–416.

16. J. Bros, H. Epstein, and V. Glaser, Some rigorous analyticity properties of the four-point function in momentum space, *Nuovo Cim.* **31** (1964) 1265–1302.

17. J. Bros, H. Epstein, and V. Glaser, A proof of the crossing property for two-particle amplitudes in general quantum field theory, *Commun. Math. Phys.* **1** (1965) 240–264.

18. H. Epstein, V. Glaser, and A. Martin, Polynomial behaviour of scattering amplitudes at fixed momentum transfer in theories with local observables, *Commun. Math. Phys.* **13** (1969) 257–316.

19. For further references for analyticity properties, see the following review articles:

H. Epstein, Some analytic properties of scattering amplitudes in quantum field theory, in

M. Chrétien and S. Deser (eds), *Axiomatic Field Theory* (Gordon and Breach, New York, 1966) pp. 1–134.

J. Bros, On some analyticity properties implied by the two particle structure of Green's functions in general quantum field theory, in

R.P. Gilbert and R.G. Newton (eds), *Analytic Methods in Mathematical Physics* (Gordon and Breach, 1970), pp. 85–133.

J. Bros, Analytic structure of Green's functions in quantum field theory, in

K. Osterwalder (ed.), *Mathematical Problems in Theoretical Physics* (Lecture Notes in Phys. vol. 116, Springer, 1980) pp. 166–199.

J. Bros, Integral relations in complex space and the global analytic and meromorphic structure of Green's functions in quantum field theory: Some general ideas and recent results, in

D. Iagolnitzer (ed.), *Complex Analysis, Microlocal Calculus and Relativistic Quantum Theory* (Lecture Notes in Phys. Vol. 126, Springer, 1980) pp. 254–262.

J. Bros, Derivation of asymptotic crossing domains for multiparticle processes in axiomatic quantum field theory: A general approach and a complete proof for $2 \to 3$ particle processes, *Phys. Rep.* **134** (1986) 325–390.

as well as the following original articles:

J. Bros, V. Glaser, and H. Epstein, Local analyticiy properties of the $n$ particle scattering amplitude, *Helv. Phys. Acta* **45** (1972) 149–181.

M. Lassalle, Analyticity properties implied by the many-particle structure of the $n$-point functions associated with a graph, *Commun. Math. Phys.* **36** (1974) 243–226.

J. Bros and M. Lassalle, Analyticity properties and many-particle structure in general quantum field theory II. One-particle irreducible $n$-point functions, *Commun. Math. Phys.* **43** (1975) 279–309.

J. Bros and M. Lassalle, Analyticity properties and many-particle structure in general quantum field theory III. Two-particle irreducibility in a single channel, *Commun. Math. Phys.* **54** (1977) 33–62.

H. Epstein, V. Glaser, and D. Iagolnitzer, Some analyticity properties arising from asymptotic completeness in quantum field theory, *Commun. Math. Phys.* **80** (1981) 99–125.

**References for Chapter 6**

1. The idea of incoherence and superselection rule is introduced in the following article by taking the univalence superselection rule as a concrete example:

J.C. Wick, A.S. Wightman, and E.P. Wigner, The intrinsic parity of elementary particles, *Phys. Rev.* **88** (1952) 101–105.

2. The original work consists of two sets of papers. The first group analysing the given algebra of fields and gauge group are:

(a) S. Doplicher, R. Haag, and J.E. Roberts, Fields, observables and gauge transformations I, *Commun. Math. Phys.* **13** (1969) 1–23; II, *Commun. Math. Phys.* **15** (1969) 173–200.
(b) K. Drühl, R. Haag, and J.E. Roberts, On parastatistics, *Commun. Math. Phys.* **18** (1970) 204–226.
S. Doplicher and J.E. Roberts, Fields, statistics and non-Abelian gauge groups, *Commun. Math. Phys.* **28** (1972) 331–348.

The second group, starting from observable algebras and constructing fields and gauge group via sector theory are:

(c) S. Doplicher, R. Haag, and J.E. Roberts, Local observables and particle statistics I, *Commun. Math. Phys.* **23** (1971) 199–230; II, *Commun. Math. Phys.* **35** (1974) 49–85.

This DHR analysis is completed in Ref. 13 below by the powerful mathematical results in Ref. 11 below.

Reviews can be found in the following and articles quoted therein.

S. Doplicher, Abstract compact group duals, operator algebras and quantum field theory, in

The Mathematical Society of Japan (ed.), *Proceedings of the International Congress of Mathematicians 1990* (Springer, 1991) vol. 2, pp. 1319–1333.

3. H.J. Borchers, A remark on a theorem of B. Misra, *Commun. Math. Phys.* **4** (1967) 315–323.

4. D. Buchholz and K. Fredenhagen, Locality and the structure of particle states, *Commun. Math. Phys.* **84** (1982) 1–54.

5. Lemma 3.8 of Ref. 2 (c) I above.

6. Appendix of Ref. 2 (c) II above.

For infinite statistics, see

K. Fredenhagen, Superselection sectors with infinite statistical dimensions, in

H. Araki, Y. Kawahigashi, and H. Kosaki (eds), *Subfactors* (World Scientific, 1994) pp. 242–258.

7. R. Longo, Index of subfactors and statistics of quantum fields I, *Commun. Math. Phys.* **126** (1989) 217–247.

8. K. Fredenhagen, K.H. Rehren, and B. Schroer, Superselection sectors with braid statistics and exchange algebras I. General theory, *Commun. Math. Phys.* **125** (1989) 201–226; II. Geometric aspects and conformal covariance, *Rev. Math. Phys.* **4** Special issue (1992), 111–157.

K. Fredenhagen, Structure of superselection sectors in low-dimensional quantum field theory, in L.-L. Chau and W. Nahm (eds), *Differential Geometric Methods in Theoretical Physics* (Plenum, NATO ASI Series B, vol. 245, 1990), pp. 95–104.

J. Fröhlich, F. Gabgiani, and P.-A. Marchetti, Superselection structure and statistics in three-dimensional local quantum theory, in G. Lusanna (ed.), *Knots, Topology and Quantum Field Theories* (World Scientific, 1989), 335–415.

J. Fröhlich and P.-A. Marchetti, Spin-statistics theorem and scattering in planar quantum field theory with braid statistics, *Nucl. Phys.* **B356** (1991) 533–573.

J. Fröhlich and F. Gabbiani, Braid statistics in local quantum theory, *Rev. Math. Phys.* **2** (1991) 251–353.

J. Fröhlich, F. Gabbiani, and P.-A. Marchetti, Braid statistics in three-dimensional local quantum field theory, in *Proceedings of the Banff Summer School in Theoretical Physics, "Physics, Geometry and Topology" 1989.*

J. Fröhlich, F. Gabbiani and P.-A. Marchetti, Braid statistics in three-dimensional local quantum theory, in D. Kastler (ed.), *The Algebraic Theory of Superselection Sectors* (World Scientific Publ. Co., 1990), 259–332.

J. Fröhlich and P.-A. Marchetti, Quantum field theory of anions, *Lett. Math. Phys.* **16** (1988) 347–358; Quantum field theories of vortices and anyons, *Commun. Math. Phys.* **121** (1989) 177–223.

F. Gabbiani and J. Fröhlich, Operator algebras and conformal field theory, *Commun. Math. Phys.* **155** (1993) 569–640.

K.-H. Rehren and B. Schroer, Einstein causality and Artin braids, *Nucl. Phys.* **B312** (1989) 715–750.

K.-H. Rehren, Braid group statistics and their superselection rules, in

D. Kastler (ed.), *The Algebraic Theory of Superselection Sectors* (World Scientific, 1990), 333–355.

K.-H. Rehren, Quantum symmetry associated with braid group statistics, in *Lecture Notes in Phys.* (Springer), vol. 370 (1990), pp. 318–339.

K.H. Rehren, Field operators for anyons and plektons, *Commun. Math. Phys.* **145** (1992) 123–148.

B. Schroer, Scattering properties of anyons and plektons, *Nucl. Phys.* **B369** (1992) 478–498.

9. For example, see General Ref. [4].

10. See Ref. 2 (c) above.

11. S. Doplicher and J. E. Roberts, A new duality theory for compact groups, *Invent. Math.* **98** (1989) 157–218.

12. See Ref. 2 (a) and (b) above.

13. S. Doplicher and J.E. Roberts, Why there is a field algebra with compact gauge group describing the superselection structure in particle physics, *Commun. Math. Phys.* **131** (1990) 51–107.

14. H. Araki, On the connection of spin and commutation relations between different fields, *J. Math. Phys.* **2** (1961) 267–270.

15. S. Doplicher, Local aspects of superselection rules, *Commun. Math. Phys.* **85** (1982) 73–86.

16. S. Doplicher and R. Longo, Local aspects of superselection rules II., *Commun. Math. Phys.* **88** (1983) 399–409.

17. D. Buchholz, S. Doplicher, and R. Longo, On Noether's theorem in quantum field theory, *Ann. Phys.* **170** (1986) 1–17.

18. C.D. Antoni, S. Doplicher, K. Fredenhagen, and R. Longo, Convergence of local charges and continuity properties of W*-inclusions, *Commun. Math. Phys.* **110** (1987) 325–348.

19. D. Buchholz, Product states of local algebras, *Commun. Math. Phys.* **36** (1974) 287–304.

20. D. Buchholz and E.H. Wichmann, Causal independence and the energy-level density of states in local quantum field theory, *Commun. Math. Phys.* **106** (1986) 321–344.

**Literature for appendices**

For Appendix A, see, for example,

N.I. Akhiezer and I.M. Glazman, *Theory of Linear Operators in Hilbert Space* (English translation: Ungar, 1961, 1963)

F. Riesz and B.Sz. Nagy, *Functional Analysis* (Ungar, 1955)

M.H. Stone, *Linear Transformations in Hilbert Space and their Applications to Analysis* (Amer. Math. Soc. Colloqu. Publ. vol. 15, 1932)

For Appendix B, see General Ref. [9–20].

For Appendix C, see Ref. 4 for Chapter 3.

# INDEX

The manufacturer's authorised representative in the EU for product safety is Oxford University Press España S.A. of el Parque Empresarial San Fernando de Henares, Avenida de Castilla, 2 – 28830 Madrid (www.oup.es/en or product.safety@oup.com). OUP España S.A. also acts as importer into Spain of products made by the manufacturer.

www.ingramcontent.com/pod-product-compliance
Ingram Content Group UK Ltd.
Pitfield, Milton Keynes, MK11 3LW, UK
UKHW051020210726
13857UKWH00006B/622

* 9 7 8 0 1 9 9 5 6 6 4 0 2 *